DO NOT SPECIAL LOAN

		DATE	

Organometallic compounds in the environment

Organometallic compounds in the environment

Principles and Reactions

A WILEY-INTERSCIENCE PUBLICATION

Editor P.J. Craig, BSc, PhD, CChem, FRSC

School of Chemistry,
Leicester Polytechnic,
Leicester, UK.

JOHN WILEY & SONS
New York

Longman Group Limited
Longman House, Burnt Mill, Harlow
Essex CM20 2JE, England
Associated companies throughout the world

Published in the United States of America
by Wiley-Interscience, a Division of John Wiley & Sons, Inc., New York
© Longman Group Limited 1986

First published 1986

ISBN: 0-471-84727-5

Printed and Bound in Great Britain at The Bath Press, Avon

The book is dedicated to my parents Wilfrid and Leslie Craig of Manchester, England

Contents

P J Craig, School of Chemistry, Leicester Polytechnic, PO Box 143,
Leicester, LE1 9BH, UK and F E Brinckman, Surface Chemistry and
Bioprocesses Group, National Bureau of Standards, Gaithersburg,
MD 20899, USA.

P J Craig, School of Chemistry, Leicester Polytechnic, PO Box 143,
Leicester LE1 9BH, UK

S J Blunden and A Chapman, International Tin Research Institute,
Fraser Road, Perivale, Greenford, Middlesex, UB6 7AQ, UK

C N Hewitt, Department of Environmental Science, University of Lancaster, Lancaster LA1 4YQ, UK and R M Harrison, Department of Chemistry, University of Essex, Wivenhoe Park, Colchester, CO4 3SQ, UK

Chapter 5 Organoarsenic compounds in the environment   198

M O Andreae, Department of Oceanography, The Florida State University, Tallahassee, Florida 32306, USA

A W P Jarvie, Department of Chemistry, University of Aston, Gosta
Green, Birmingham B4 7ET, UK

Chapter 7 Organic Group VI elements in the environment 254

Y K Chau, National Water Research Institute and P T S Wong, Great
Lakes Fisheries Research Branch; Canada Centre for Inland Waters,
Burlington, Ontario, Canada L7R 4A6

Chapter 8 Methyl transfer reactions of environmental significance involving naturally occurring and synthetic reagents 279

S Rapsomanikis, Department of Chemistry, University of Essex, Wivenhoe Park, Colchester, CO4 3SQ, UK and J H Weber, Department of Chemistry, Parsons Hall, University of New Hampshire, Durham NH 03824, USA

Preface

Organometallic compounds have found increasing commercial use over the past twenty years and in much of this use there is a direct interaction with the natural environment. Prime examples of this involvement of organometallics come from their use as pesticides (organomercury and organotin compounds), gasoline additives (methyl- and ethylleads), polymers (organosilicons), additives and catalysts, and the annual consumption of all organometallics for use is in the region of several hundred thousand tonnes. This has initiated intensive research in recent years on their toxicity, pathways and transformations in the environment and on their ultimate fate and disposal. This work was prompted particularly by the serious environmental or toxicity problems caused by the occurrence of organomercury compounds in the environment in Iraq, Japan, Sweden and North America in the period, 1950–1975. Although much research material now exists in the literature, it is either in scattered or large-scale form and tends to be somewhat inaccessible to the reader who wishes to form an understanding of the topic as a whole.

The subject of organometallic compounds in the environment was first treated as a single topic at a symposium of the American Chemical Society held at Anaheim in March 1978. The individual presentations of this meeting were published in 1978 in what is still the only source of data on this field available in a single volume. It is the desire to incorporate much of the new research since 1977 in this field that has prompted the preparation of this volume. Since 1977 for example, much information about the existence of new and unsuspected organoarsenic forms in the oceanic environment has come to light, and during the past year the discovery of methyl derivatives of germanium and antimony as natural species in the oceans has been made. Although there are some detailed reviews on the properties of organometallic compounds in the environment they occur in widely disparate publications or as parts of multi-volume series. The present volume constitutes the only recent single volume source of information in this area.

In defining the boundaries of what constitutes an organometallic and whether or not it is environmentally important, a generally pragmatic approach has been taken. The word 'organometallic' means a compound possessing a direct metal to carbon bond. Complex organic ligands bound to metals by oxygen, nitrogen or sulphur ligands are not organometallic unless they also possess a metal–carbon bond. Metals and metalloids are included by criteria of normal practice, extent of usage and coverage elsewhere. The normally accepted metals are all included as are metalloidal elements including silicon, selenium and arsenic. These all have important environmental roles. Phosphorus and sulphur compounds are covered where they cast light on this area but they are normally considered as non-metals and are treated selectively in this work. Use, chemical reactions in the environment, environmental transport pathways and toxicity are all aspects of the environmental role of substances and are covered in the present work. The specific area of organometallic interaction with living organisms follows from their environmental role and is the subject of a work by Thayer.[1] In view of *environmental* problems arising from drug use usually being small, this topic is covered in less detail than the main themes although use of organometallic compounds as drugs is noted and environmental implications are discussed. The use of unstable precious metal organometallic compounds as industrial catalysts is covered in numerous works. In general, there may be some loss of *metals* to the products or environment in these processes, but not in the form of the organometallic catalyst which will normally decompose under aqueous aerobic conditions. The use of catalysts in this category does not have general environmental implications in the context of the present work. It should be seen as part of the overall behaviour of metals in the environment.

The essential chemical properties of organometallic compounds have been researched in detail during the past thirty years and the interested reader is referred to some important sources describing purely chemical properties of individual organometallic compounds. It is interesting to note that in the nine-volume standard reference source of organometallic chemistry,[2] planned several years ago, the coverage of environmental aspects required 40 out of nearly 10 000 pages. The pace of research in this area can be judged by the current expansion of that article into the present work. Some of the most useful sources on the chemistry of organometallic compounds are listed on page xxi. There are a number of works in existence which cover the environmental and biogeochemical behaviour of metals and some of these are also listed. The references here provide a comprehensive guide to the chemical, environmental, industrial and biogeochemical properties of organometallic compounds and with the present work can be seen as core sources for the subject.[3–9]

It is intended that this work be read either as a whole, or as a series of self-contained chapters which stand for a particular topic in their own right. For example, the chapter on organotin compounds constitutes a complete discussion on this area, but it can also be seen as just one aspect of a single theme, with key reactions and properties common to most metals. The purpose of Chapter 1 is to describe the underlying properties of organometallic molecules that are relevant to their environmental behaviour. The reader may choose to consider these properties first and then go on to look at the role of individual elements in the environment. Equally he may consider elements separately and then draw the topic area together by following Chapter 1. After considering Chapter 1 and the behaviour of some of the bulk organometallic compounds (particularly those of mercury, tin, lead and arsenic) the reader will see the environmental importance of vitamin B_{12} derivatives particularly methyl cobalamin, CH_3CoB_{12}. The role of CH_3CoB_{12} towards metals is discussed in Chapter 8 and a brief resumé of the biological properties of CH_3CoB_{12} is given in Chapter 10. In a similar way, while silicon compounds are dealt with in Chapter 6, further environmental behaviour of silicones as high polymer elastomers is covered in Chapter 9.

In order to allow for sequential or separate progress through the book, a small amount of redundancy is built into certain points of the work in order to allow these sections to become fully meaningful on their own. This element of repetition is intentional and it is hoped not over-burdensome to the reader who follows the work sequentially.

A list of the units used is given immediately following the preface and links to other commonly used units are also given.

Editor's Acknowledgements

I am happy to acknowledge the expert help of the following colleagues at Leicester Polytechnic who have typed and arranged this and numerous other manuscripts in recent years. For their help and assistance I am glad to thank Debbie Coulbeck, Secretary, School of Chemistry; Dorothy Bryan, Amanda Ardington and Marlene Liburd of the Science Faculty Office; Beryl Hood, Irene Miller, Greta Day, Joan Johnson, Sheila Evans and Mary Daniel of the Polytechnic Typing Section. I am also pleased to acknowledge grants in recent years, chiefly from the Science and Engineering, and the Natural Environment Research Councils, UK; the Royal Society, London; the Nuffield Foundation, London; the Royal Society of Chemistry, London; the British Council, London; the North Atlantic Treaty Organization; the University of Minnesota; and Leicester Polytechnic. I am pleased to thank Imperial Chemical Industries Ltd, particularly Dr David Taylor of the Brixham Laboratory, for much help and assistance in fieldwork and sample collection. I am also grateful for the help given to me by a number of able and enthusiastic research students, including Drs Stephen Morton, Paul Bartlett, Peter Moreton and Spyridon Rapsomanikis. I particularly thank Dr Graham

Lawson of Leicester Polytechnic for help and useful discussion during the preparation of this work. Finally I thank Yvonne Lusack of the Leicester Polytechnic library for much appreciated help on information retrieval during the past few years.

The following colleagues have read and commented on parts of this work: M O Andreae (Florida State University), M Fox (Leicester Polytechnic), R M Harrison (University of Essex), A Jarvie (University of Aston), G Lawson (Leicester Polytechnic), J M Pratt (University of Surrey), P Smith (International Tin Research Institute, London) and D Taylor (ICI, Plc, Brixham, UK). For their help and comments I am grateful. Any mistakes remaining in this work are, of course, my own responsibility.

I am also grateful for the advice, help and encouragement I have received from the staff of Longman. Finally my thanks to my wife Carole for help in the preparation of the manuscript.

P J Craig
August 1985

Acknowledgements

We are grateful to the following for permission to reproduce copyright material:

The American Chemical Society for table 6.2 from table 6.1 (ref 49); the author, Dr M O Andreas and the American Chemical Society for tables 10.1, 10.2 (refs 1,8); Elsevier Applied Science Publishers Ltd for fig 2.6 from fig 1 (ref 182); International Marine Coatings for fig 3.2; the author, Dr J A Johnson and Cambridge University Press for data in table 1.8 from table 7.2 (ref 45); Dr W R A de Jonghe for fig 4.6 from fig 7 (ref 146); Dr W W Kellogg and the American Association for the Advancement of Science for fig 7.2 from fig 2 Copyright 1972 by AAAS (ref 70); Dr J A Kerr and the Chemical Rubber Company for data in table 1.6 (ref 64); Dr A Kudo and Pergamon Press Ltd for table 2.7 from table 3 Copyright 1978 Pergamon Press Ltd (ref 183); Macmillan Journals Ltd for fig 2.1 from figs 1,2 Copyright 1969 Macmillan Journals Ltd (ref 16) & figs 4.2–4.4 from fig 1 a,b,c Copyright 1978 Macmillan Journals Ltd (ref 16); Open University Press for fig 1.4 adapted from fig 22 & data in tables 1.2, 1.3 from table 22.2 (ref 39); the author, Professor R G Pearson for table 1.10 (ref 77); Pergamon Press Ltd for fig 2.5 from fig 2 Copyright 1983 Pergamon Press Ltd (ref 142); Plenum Publishing Corpn for table 7.5 (ref 80); the author, Dr D Rabenstein, the American Chemical Society and the National Research Council for data in table 1.11 from tables I,II (ref 78); Dr G Roderer and Academic Press Inc for fig 4.5 from fig 1 (ref 136); Dr A Rodriguez-Vasquez and Pergamon Press Ltd for fig 2.4 adapted from fig 1 & table 2.5 adapted from table 1 Copyright 1978 Pergamon Press Ltd (ref 135); Van Nostrand Reinhold Co Inc for table 5.2 from fig 1 (ref 12); John Wiley & Sons Inc for data in table 2.1 from table 2.7; the author, Dr R W Wollast and Plenum Publishing Corpn, the author, Dr P Benes and Elsevier Biomedical Press BV for fig 2.3 from fig 7.4 (ref 139); the author, Professor J M Wood and the American Association for the Advancement of Science for table 1.1 adapted from table 1 Copyright 1977 by AAAS (ref 1); the author, Professor J M Wood and Walter de Gruyter & Co for fig 1.2 (ref 4).

Units and abbreviations

The full English language version of chemical names is used unless length makes them too clumsy. Conventional abbreviations are used for the various instrumental techniques mentioned. The book was produced in England, and English rather than American usage is adopted (e.g. sulphur not sulfur). The following concentration units are used

$\mu g\ g^{-1}$ = micrograms per gram = parts per million = $mg\ kg^{-1}$
$ng\ g^{-1}$ = nanograms per gram = parts per billion = $\mu g\ kg^{-1}$
$\mu g\ m^{-3}$ or $ng\ m^{-3}$ = weight per cubic metre (for atmospheric measurements)
$\mu g\ dm^{-3}$ or $ng\ dm^{-3}$ = weight per decimetre cube (litre) for measurements in water

1 tonne = 1000 kg = 0.984 ton(UK)
1 nm = 10 Å = 10^{-9} m

Analytical techniques are referred to by their normal abbreviations, e.g. atomic absorption (AA), gas chromatography (GC), mass spectroscopy (MS)

Standard sources of reference

1. Thayer J S 1984 *Organometallic Compounds and Living Organisms.* Academic Press, New York

2. Abel E W, Stone F G A, Wilkinson G (eds) 1982 *Comprehensive Organometallic Chemistry.* Pergamon, Oxford

3. Coates G E, Green M L H, Wade K 1967 *Organometallic Compounds.* Methuen, London, vols 1 and 2

4. Coates G E, Aylett B J, Green M L H, Mingos D M P, Wade K 1979 *Organometallic Compounds*. Chapman and Hall, London, Several volumes from 1979 onwards

5. Aylett B J, Lappert M F, Pauson P L (eds) 1984 *Dictionary of Organometallic Compounds*. Chapman and Hall, London; New York

6. Salomons W, Forstner U 1984 *Metals in the Hydrocycle*. Springer Verlag, Berlin; Heidelberg; New York; Tokyo

7. Nriagu J O 1984 *Changing Metal Cycles and Human Health*. Springer Verlag, Berlin; Heidelberg; New York; Tokyo

8. Twigg M V (ed) 1983 *Mechanisms of Inorganic and Organometallic Reactions*. Plenum, New York

9. Kochi J K 1978 *Organometallic Mechanisms and Catalysis*. Academic Press, New York; San Francisco; London

Chapter 1

Occurrence and pathways of organometallic compounds in the environment — general considerations

1.1 Occurrence of organometallic species in the environment

Organometallic species (i.e. compounds, complexes or ions) may be found in the natural environment either because they are *formed* there or because they are *introduced* there. In general the behaviour of the latter group is better understood, and their environmental impact has been assessed by studies of their direct toxicities, their stabilities and routes to decay and by toxicity studies of their decay products. Organometallic compounds entering the environment may be deliberately introduced as products whose properties relate to the environment (e.g. biocides) or they may enter peripherally to a separate, main function (e.g. petrol additives, polymer stabilizers). Compounds of arsenic, mercury, tin and lead have important uses as organometallic compounds. Their role and behaviour in the environment are covered in the appropriate chapters of this work (Chs 2–5). The behaviour of other organometallic species in the natural environment is also covered (Chs 6, 7, 9 and 10). However, not all organometallics found in the environment are introduced there — some in fact are formed after entry as inorganic species and are components of global biogeochemical cycles.

1.2 Formation of organometallic species in the environment

There is by now no doubt that some organometallic compounds may be formed in the natural environment from inorganic precursors. Essentially this is a process of environmental methylation (sometimes called biomethylation).

There is little evidence for the formation of other than methyl derivatives. Methylation of mercury and arsenic (Chs 2 and 5) in the environment is well known although mechanistic details differ considerably between the two cases. Methyltin derivatives have also been found in oceans and rivers, and methyllead formation in the environment has also been discussed (Chs 3 and 4). Abiotic, model work in laboratories tends to show that environmental methylation for tin and lead may occur but the evidence is not conclusive. The situation for lead is clouded by the widespread use of methyl- and ethylleads as petrol additives. Finding methyllead compounds in the environment is not usually evidence for their having been formed there. Other methyl metal species occur in the natural, particularly oceanic, environment for which there is a presumption that they are formed there (Ch. 10). Analytical or chemical evidence for environmental methyl metal formation occurs for antimony, selenium, tellurium and germanium particularly.

The general routes to environmental methyl metal formation are known. These and other questions have been reviewed by Wood[1-4] and Craig.[5-7] For elements in their normal highest oxidation states without lone electron pairs (e.g. mercury(II)) the methyl group will be transferred to the metal as a carbanion (CH_3^-). Methyl carbanions will usually arise from the naturally occurring biological methylating agent methyl cobalamin (CH_3CoB_{12})[8-10] but they may also arise from other organometallic species present (e.g. $(CH_3)_3Sn^+$, $(CH_3)_3Pb^+$). The role of CH_3CoB_{12} in the environment and methylation is discussed later (Chs 2, 8 and 10) but it may be noted here as the only known carbanion transfer reagent in biochemistry. Other natural methylating agents transfer methyl groups as radicals ($CH_3^{\bullet}$) or as carbonium ions (CH_3^+). CH_3CoB_{12} although complex, can essentially be considered as a methylcobalt(III) derivative in an octahedral ligand environment (Fig. 1.1).

The methyl to cobalt linkage is stable but allows methyl transfer to certain metal species. For mercury ions in aqueous solution this mechanism may be exemplified as follows (equation 1.1, type 1 mechanism, direct substitution):

Type I mechanism

$$CH_3CoB_{12} + Hg^{2+} \xrightarrow{\;H_2O\;} CH_3Hg^+ + H_2OCoB_{12}^+ \qquad [1.1]$$

In this case there is no formal change of oxidation state for mercury. For elements in low oxidation states capable of further oxidation (i.e. those possessing lone electron pairs) methylation may take place by radical or carbonium ion attack on the lone pair, resulting in a formal oxidation state change, as the methyl group is more electron attracting than the metal and produces an electron drift away from the metal orbitals. Numerous sources for radical or carbonium ion attack are known, including CH_3CoB_{12}, which under certain conditions may lose methyl groups as $CH_3^{\bullet}$ or CH_3^+. Other sources include the natural methylating agents S-adenosylmethionine, which is involved in arsenic methylation, and N-methyltetrahydrofolate. In addition there are large numbers of biochemically derived molecules (e.g. sulphonium species,

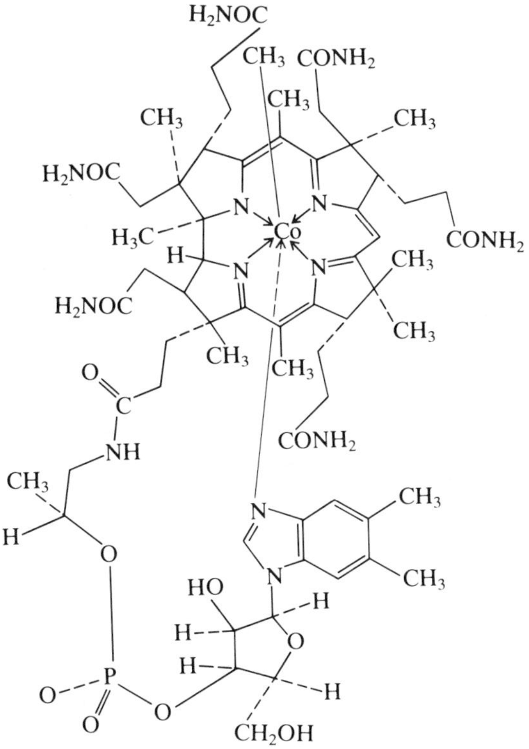

[Axial group is benzimidazole, B_z]

Fig. 1.1 Structure of methylcobalamin coenzyme (CH_3CoB_{12}) – charges not shown

Methylcobinamide differs from CH_3CoB_{12} by hydrolysis of the phosphate group resulting in an alcohol function, and coordination of H_2O *trans* to CH_3. The formal charge of cobalt in CH_3CoB_{12} is III, but it will be omitted on structures and formulae of this molecule and its derivatives for simplicity. Replacement of CH_3 by CN or OH gives vitamin B_{12}. Replacement by 5′-deoxyadenosyl group gives another coenzyme, $AdenCoB_{12}$.

5′-deoxyadenosyl =

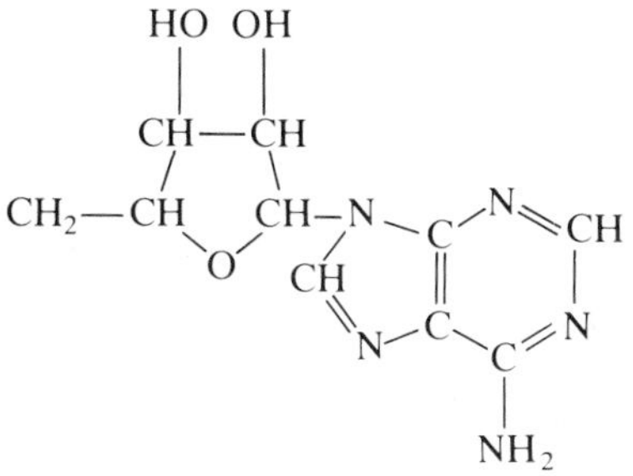

Structure diagrams provided by S Rapsomanikis and J H Weber

iodomethane, betaine) which may formally oxidize and methylate metal species. In addition numerous photochemically derived radicals exist in the atmosphere (sect. 1.3.4). Methyl transfer possibilities for methyl cobalamin are shown in Fig. 1.2.

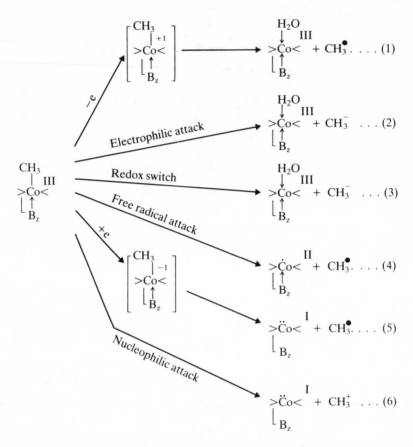

Fig. 1.2   Reaction mechanisms for the cleavage of the cobalt–carbon bond in CH_3CoB_{12}. Reproduced with permission from Wood J M and Fanchiang Y T 1979. From Vitamin B_{12}, pub Walter de Gruyter, Berlin, New York (Charges on CoB_{12} products not shown). (Ref. 4)

Mechanisms for free radical methyl group transfer to metals have been discussed particularly by Wood.[4] The radical source was assumed to be CH_3CoB_{12} but, as noted, others may exist. For CH_3CoB_{12}, homolytic cleavage of the methyl–cobalt bond produces a methyl radical which transfers to the lower oxidation state of a metal capable of oxidation by one unit. A methylation mechanism for tin including initial oxidation to tin(III) by CH_3CoB_{12} has been suggested (equation 1.2, type II mechanism, radical addition).[11]

Type II mechanism

$$CH_3CoB_{12} + Sn(II) \rightarrow CH_3Sn(III)^\bullet + CoB_{12}^\bullet$$

$$CH_3Sn(III)^\bullet + O_2 \rightarrow CH_3Sn(IV)^+ + O_2^- \text{ (aerobic)} \qquad [1.2]$$

$$\text{or } CH_3Sn(III)^\bullet + H_2OCoB_{12}^+ \rightarrow CH_3Sn(IV)^+ + H_2O + CoB_{12}^\bullet \text{(anaerobic)}$$

This route depends upon facile oxidation of a lower to a stable higher oxidation state for the metal, and is related by Wood to the standard reduction potentials ($E^\ominus$) for the metallic elements. Elements which react in their highest oxidation states with carbanions (CH_3^-), and give methyl metals in the same oxidation state have been shown to be those with $E^\ominus$ values $> + 0.8$ volts,[1] i.e. these are elements where the lowest oxidation state is not easily oxidized, which would take place if the radical mechanism occurred. More easily oxidized elements, i.e. those having low $E^\ominus$ values, may undergo the formal oxidation process involved in the radical type II mechanisms.[1] Table 1.1 shows the metals of interest in this context and predicts the methylation mechanism.

Table 1.1   Redox potential ($E^\ominus$) and methylation mechanism

Redox couple	$E^\ominus$ (volts)	Mechanism*
Pb(IV)/Pb(II)	+1.46	Type I
Tl(III)/Tl(I)	+1.26	Type I
Se(VI)/Se(IV) (acid)	+1.15	Type I
Pd(II)/Pd(O)	+0.987	Type I
Hg(II)/Hg(O)	+0.854	Type I
Sb(V)/Sb(III)	+0.678	Types II–IV
As(V)/As(III) (acid)	+0.662	Types II–IV
S(VI)/S(IV) (acid)	+0.20	Types II–IV
Sn(IV)/Sn(II)	+0.154	Types II–IV
Pb(II)/Pb(O)	−0.13	Types II–IV
Ge(IV)/Ge(II)	−0.13	Types II–IV
Sn(II)/Sn(O)	−0.14	Types II–IV
Cr(III)/Cr(II)	−0.41	Types II–IV
As(V)/As(III) (base)	−0.67	Types II–IV
Si(IV)/Si(O) (acid)	−0.84	Types II–IV

(Adapted and extended from Ridley W P, Dizikes L J, Wood J M 1977, *Science* **197**: 329–32) Copyright 1977 by American Association for the Advancement of Science. Quoted with permission.

* See text for full discussion.
Type I – Direct substitution of M^+ by CH_3^-. No change in metal oxidation state.
Types II–IV – Oxidation of low oxidation state metal to a state one unit higher (type II) or two units higher (type III or IV) *or* direct carbanion substitution of the higher oxidation stage only with no change in oxidation state.

A third general methylation route, allied to the type II mechanism above, also exists. The oxidative addition process common in abiotic organometallic chemistry[12] may convert easily oxidized inorganic species in the environment to stable methyl metal species having a formal oxidation number two units greater (equation 1.3, type III mechanism, oxidative addition). In equation 1.3 a model tin(II) complex, exemplifying environmental tin−oxygen co-ordination, is used.

Type III mechanism

$$CH_3I + Sn(II)(acac)_2 \rightarrow CH_3Sn(IV)(acac)_2I \qquad [1.3]$$

(acac = bidentate acetylacetonate ligand, $CH_3COC\bar{H}COCH_3$)

Here the naturally occurring compound iodomethane undergoes an oxidative addition to tin(II) converting it to a methyltin(IV) derivative.[13] It might be expected that low redox potential metals susceptible to the type II mechanism discussed previously would also undergo the kind of oxidative addition shown in equation 1.3. It has been pointed out that there are a number of naturally occurring molecules in addition to iodomethane which are capable of formally oxidizing and methylating metals in this way (see also Ch. 2).

Closely allied to the oxidative addition, where both components of the oxidizing molecule add to the metal, is the case of direct transfer of a methyl carbonium ion to a low oxidation state of a metal to give a methyl metal derivative where the formal oxidation state has also increased by two units. Methyl sources include *S*-adenosylmethionine, which with *N*-methyltetra-hydrofolate, exists as a biological methyl transfer agent operating by transfer of carbonium ions (CH_3^+). Such a carbonium ion transfer is known to occur in the case of arsenic methylation (equation 1.4, type IV mechanism, carbonium ion transfer)[14]

Type IV mechanism

$$CH_3^+ + As(III)(OH)_3 \rightarrow CH_3As(OH)_3^+ \qquad [1.4]$$

$$CH_3As(OH)_3^+ \rightarrow CH_3As(V)O(OH)_2 + H^+$$

In the example given above the methyl transfer process is enzymatic.

It can be seen that, effectively, mechanisms II−IV are related in that they result in oxidation of the metal. Although a variety of mechanisms leading to methyl metal species from inorganic precursors may operate, both in the laboratory and in the natural environment, it should also be noted that forma-tion of a methyl metal species does not imply that it is stable in the environ-ment. For example the methylation of chromium(II) by the CH_3CoB_{12} − dependent type II mechanism leads to a methylchromium(III) species (equa-tion 1.5)[15]

$$CH_3CoB_{12} + Cr(II)(H_2O)_6^{2+} \rightarrow CH_3Cr(III)(H_2O)_5^{2+} + CoB_{12}^{\bullet} + H_2O \quad [1.5]$$

However, under aqueous conditions the methyl−chromium bond is unstable and is rapidly cleaved to methane and $Cr(H_2O)_6^{3+}$. For organometallic compounds the question of stability in the environment is of great importance. This is discussed in section 1.3.

There has been much debate over whether or not methylation processes are enzymatic, but for *practical* purposes the question is semantic. Methylation of metals in the environment does occur, and it requires the presence of methyl donor molecules. In the aquatic or sediment environment a number of these exist and most are biologically synthesized. Non-enzymatic methylation is methylation by the products (metabolites) of biological activity and this must be presumed to be the normal environmental process. Enzymatic methylation would require the presence of actively living organisms and would postulate the intervention of the metal within the metabolic pathways of a living cell. Wood[16] has suggested a number of enzymatic processes for mercury(II) involving methionine synthetase, acetate synthetase and methane synthetase (see Ch. 2). Although crucially different in terms of biochemistry, both enzymatic and non-enzymatic methylation would produce methyl metal products whose environmental importance will depend on their stability, toxicity and pathways rather than on their mode of production. No environmental experiments have been carried out which clearly demonstrate whether or not the proximate methyl source for mercury methylation (CH_3CoB_{12}) is a free metabolite or whether it is enzymically complexed, e.g. endocellular. For this reason it is generally preferable (and safer) to use the term environmental methylation for the process rather than the presumptive term 'biomethylation'. In any case abiotic methylation has been demonstrated from various chemical species including those discussed above and also humic and fulvic materials which exist as products of biological activity (Ch. 2).

Methylation by carbonium ion routes is more likely to be genuinely enzymatic owing to the well-known enzymatic synthesis of *S*-adenosylmethionine (containing a CH_3—$\overset{+}{S}$ linkage) and consequent transfer of a $CH_3{}^+$ group. The involvement of *S*-adenosylmethionine in methyl transfer in biochemistry is well known. Hence the term 'biomethylation' for arsenic seems not inappropriate.[17] In this context it should also be noted that the process of demethylation of methylmercury compounds, which occurs also in the environment, may be either chemical or enzymatic (see Ch. 2).

It is sometimes assumed that methylation in the environment is a process by which organisms detoxify themselves, i.e. remove unwanted heavy metals from their cells. Where there is little evidence that these reactions occur within living cells methylation is best regarded as a chance abiotic reaction between an available metal and a biologically formed methylating agent outside the cell, i.e. a normal chemical reaction albeit in a complex environment. In any case most methyl metals are more toxic than their inorganic precursors so methylation as a heavy metal removal process seems unlikely. For some metalloids (e.g. arsenic) the methyl derivatives are less toxic than inorganic forms, and detoxification here might be the explanation for the methylation process (see Ch. 5).

Coordination grouping	Species	Occurrence
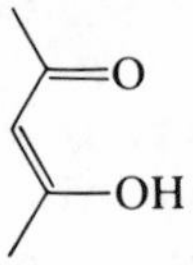 [structure: C=O and C-OH]	Flavenoids, lignins, quinones, carbohydrates Fulvic, humic acid	Plants, fungi, marine animals
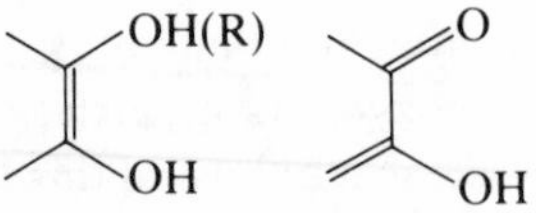 [structure: C-OH(R), C-OH and C=O, C-OH]	Flavenoids, anthocyanins, carbohydrates (sugars). Fulvic, humic acid	Plants, flowers, fruits, sediments, plants moulds
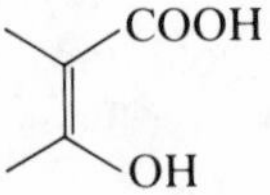 [structure: C-COOH, C-OH]	Fulvic, humic acid	,,
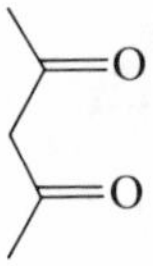 [structure: C=O, CH$_2$, C=O]	Fulvic, humic acid	,,
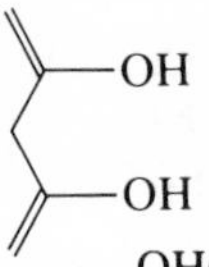 [structure: C-OH, CH$_2$, C-OH]	Fulvic, humic acid	,,
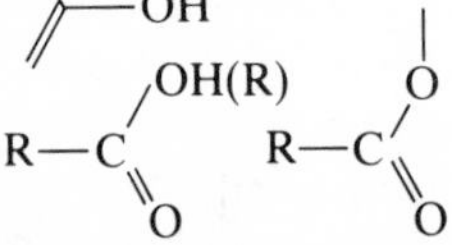 [structures: R-C with OH(R) and O; R-C with O and O]	Lipidic fatty acids, waxy esters, carotenoids Fulvic, humic acid	Plants, animals
$RCH{=}O$, RCH_2OH, C_6H_5OH	Terpenoids	Plants, trees
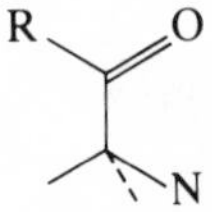 [structure: R, O, N]	Amino acids (if R=OH), alkaloids. Nearly all amino acids are α coordinated but β-alanine exists	Plants, animals
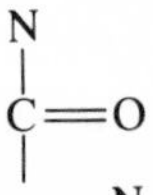 [structure: N, C=O, N]	Alkaloids, e.g. piperine, capsaicin	Plants
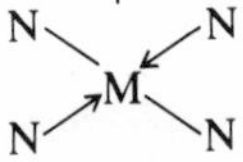 [structure: N, N, M, N, N]	Porphyrin systems with fused heterocyclic rings. See Fig 1.1 as general example	Plants, animals
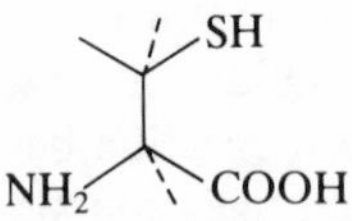 [structure: SH, NH$_2$, COOH]	Cysteine	Protein
$CH_3SCH_2CH_2C(NH_2)HCOOH$	Methionine	,,
$[-SCH_2C(NH_2)HCOOH]_2$	Cystine	,,

In coordination, 5-membered rings are favoured for steric and entropy reasons

Fig. 1.3 Coordinating fragments of natural ligand molecules

1.3 Stability of organometallic compounds under environmental conditions

1.3.1 Overview

The stability of organometallic compounds under laboratory/abiotic conditions to heat, water, oxygen, light and biological systems is discussed in the following sections. The discussion is also extended to encompass conditions found in the natural environment. Two general observations can be made at this point.

First, as stability is so dependent upon effective coordination and ligand binding at the metal centre (sect. 1.3.2), then organometallic cationic species (e.g. CH_3Hg^+, $(CH_3)_3Pb^+$) might be expected to be more stable in the aqueous and sedimentary natural environment than they are, for example, in aqueous solution in the laboratory. Many natural coordinating ligands, some with aromatic character, exist in natural waters (as suspended particulate matter) and in sediments and these bind strongly to metals and organometallic species present. These ligands have sulphur, nitrogen or oxygen donor ligands and are often multidentate. Such binding will enhance the stability of the coordinated organometallic species. Examples of natural coordinating ligands are shown in Fig. 1.3.

In the atmosphere the reverse holds true. Stability to light of equal intensity and character to that available in the atmosphere is often exhibited by organometallic species when measurements are made under laboratory conditions. However, in the real atmosphere radicals are often present (which lead to decomposition particularly in polluted locations) as also are numerous particulate species whose surfaces offer the opportunity for heterocatalytic decomposition. Hence a given organometallic compound will be very much less stable in the real atmosphere than physico-chemical measurements in the laboratory would suggest.

1.3.2 Thermodynamic and kinetic routes to decomposition

It was at one time thought that transition metal organometallic compounds particularly, were an unstable class of materials. Thirty years ago attempts to prepare them often led to pyrophoric phenomena or to the production of intractable organic oils. This instability was ascribed to *thermodynamic* factors, chiefly an inherent weakness of the metal−carbon bond. This was thought to be responsible for the observed negative free energies of decomposition ($\Delta G_D^\ominus$) of organometallic compounds and the instability of the transition metal derivatives. In general, the word 'stability' used in this (environmental) context refers to stability at room temperature, to water, or to air. More recent quantitative measurements of bond energies, however, have shown that metal−carbon bond energies do in fact lie within the normal range of chemical bond energies. The problem of instability lay more within the area of *kinetics*, i.e. there existed low energy pathways leading to decomposition at room temperatures. The existence of low activation energies allowing easy decomposition has been amply solved by synthetic organometallic chemists in recent years. Greater sophistication in molecular design now allows the incorporation of kinetic stability (i.e. high activation energy) in molecules that are thermodynamically unstable (i.e. $\Delta G_D^{\ominus}$ is negative). This may be achieved by use of strongly bonding ligand molecules also attached to the metal–carbon moiety (i.e. complexation), a procedure that is often duplicated by the natural ligands available in the environment. An energy level diagram showing the factors for thermal decomposition is shown in Fig. 1.4. It has to be considered that just as organometallic compounds may be

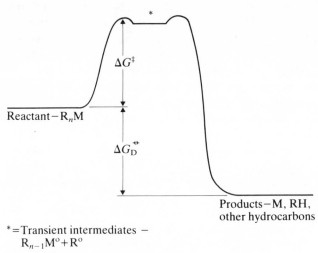

$* =$ Transient intermediates $-$
    $R_{n-1}M^{\circ} + R^{\circ}$

$\Delta G_D^{\ominus}$ is negative, favouring decomposition

If $\Delta G^{\ddagger}$ is large there is no low energy pathway for decomposition and the compound is kinetically stable under normal conditions

Fig. 1.4 Energy diagram for thermal or light-induced decomposition of organometallic molecules. Reproduced with permission, Open University Press. (ref 39)

stabilized by complexation in the laboratory, that analogous processes also arise in the natural environment. Of particular interest would be the stabilization of monoalkyllead compounds which have not been fully characterized in the laboratory,[18] but which must be intermediates if methylation reactions in the environment are to occur for lead (for a full discussion see Ch. 4). Some natural ligand and ligand groupings capable of enhancing kinetic stability of organometallic species are shown in Fig. 1.3.

Conversely, the natural environment provides routes for decomposition considerably enhanced over those available for compounds stored in the laboratory, where inert atmospheres, darkness and refigeration are all available. In the next sections the role of thermal, oxidative, hydrolytic and biological factors on the stability of organometallic compounds in the natural environment is considered.

Above all it should be borne in mind that organometallic compounds are thermodynamically unstable with respect to their constituent elements (M, C and H_2), and also to their actual decomposition products – which may differ (e.g. for thermal decomposition; M, RH and H_2). However, in many cases they are kinetically stable at room temperature to air and water and hence may be assured of more than transient existence under environmental conditions.

1.3.3   Thermal stability

Table 1.2 lists standard molar enthalpies ($\Delta H_F^{\ominus}$) of formation for some metal methyl compounds of environmental interest. Although free energies of formation ($\Delta G_F^{\ominus}$, a more fundamental parameter) are not always known, $\Delta H_F^{\ominus}$ often provides a good indication of thermal stability when the entropy changes ($\Delta S^{\ominus}$) are of similar orders for each case. It can be seen that there is a fairly even spread between exo- and endothermic compounds, and although Table 1.2 is only an approximate guide (decomposition in the absence of oxygen is not to metals, carbon and hydrogen but usually to a mixture of hydrogen, hydrocarbons and metal) it might be expected that, on this basis, methyls of silicon, germanium, tin, boron, aluminium, gallium and

Table 1.2   Standard molar enthalpies of formation ($\Delta H_F^{\ominus}$) for $(CH_3)_nM$ compounds of environmental interest (in kJ mol^{-1})

Group III n = 3		Group IV n = 4		Group V n = 3		Group VI n = 2		Group IIB n = 2	
B	− 122	C	− 167	N	− 24	O	+ 195	Zn	+ 55
Al	− 88	Si	− 238	P	− 96	S	−	Cd	+ 110
Ga	− 39	Ge	− 71	As	+ 15	Se	− 57†	Hg	+ 93
In	+ 172	Sn	− 19	Sb	+ 31	Te	−		
Tl	−	Pb	+ 137	Bi	+ 192	Po	−		

$\Delta H_F^{\ominus}$ is for formation from the elements at 298°K and 1 atmosphere pressure.

Compounds with $\Delta H_F^{\ominus}$ endothermic (i.e. + ve) would decompose exothermally (i.e. heat would be evolved, $\Delta H_D^{\ominus}$ would be − ve).

NB   All of the above can be isolated and stored at room temperature – some require inert atmospheres.

†   Data for $(C_2H_5)_2Se$. Data from Ref 39 with permission (Open University Press).

phosphorus might be more stable than those of mercury, lead, cadmium and bismuth. However, it should also be considered that, as gaseous products are evolved, this will input a large positive entropy value ($\Delta S^{\ominus}$) into the overall free energy equation for the decomposition ($\Delta G_D^{\ominus} = \Delta H_D^{\ominus} - T\Delta S_D^{\ominus}$). This will render exothermic compounds of slight stability even less stable than appears at first sight, as the overall $\Delta G_D^{\ominus}$ for decomposition is now rendered negative. However, taking tetramethyllead (($CH_3)_4Pb$) for example, despite its thermodynamic instability to thermal decomposition to metal, hydrocarbons and hydrogen, it is in practice quite stable for kinetic reasons at normal temperatures – there is no route with a low enough activation energy ($\Delta G^{\ddagger}$) which would allow decomposition to occur.

In a similar way bond dissociation energies (D) can be used as a guide to stability. These are the enthalpy changes when one bond is homolytically cleaved, CH_3—M giving $M^{\bullet}$ and $CH_3{}^{\bullet}$ radicals. Mean values ($\bar{D}$) are $\Sigma D/n$ for each methyl metal bond in a $(CH_3)_nM$ molecule. Bond enthalpies refer to the conversion of the molecules into separate atoms but for individual metal–carbon bonds trends are similar to bond dissociation energies. Table 1.3 gives energy values for the process shown in equation 1.6.

$$R_NM(g) \rightarrow M(g) + nR^{\circ}(g) \quad \bar{D} = \Sigma D/n \qquad\qquad [1.6]$$

The values of $\bar{D}$ given in Table 1.3 are the mean values of the energies involved in breaking the several methyl metal bonds in the molecules.

Table 1.3   Mean bond dissociation energies ($\bar{D}$) for $(CH_3)_nM$ compounds of environmental interest (in kJ mol^{-1})

Group III n = 3		Group IV n = 4		Group V n = 3		Group VI n = 2		Group IIB n = 2	
B	364	C	347	N	314	O	358	Zn	176
Al	276	Si	320	P	276	S	289	Cd	138
Ga	247	Ge	247	As	230	Se	247	Hg	123
In	163	Sn	218	Sb	218	Te	–		
Tl	–	Pb	155	Bi	141	Po	–		

D = enthalpy change when *one* mole of the bond in question is homolytically cleaved, reactants and products being in the ideal gas state at 1 atm and 298.15 K, i.e. $\bar{D}$ is for $R_nM(g) \rightarrow M(g)$ + $nR^{\bullet}(g)$ $\bar{D} = \Sigma D/n$.

D is also known as bond disruption enthalpy. Data from Ref 39 with permission (Open University Press).

Tables 1.2 and 1.3 do not always reflect the actual mode of decomposition and therefore give only an approximate guide to thermodynamic stability. Tetramethyllead normally decomposes on heating to lead (O) and C_2H_6, not to its constituent elements. Taking into account the standard enthalpy of ethane $(-85 \text{ kJ mol}^{-1})$ the enthalpy change for the reaction is -307 kJ mol^{-1} compared to -137 kJ mol^{-1} for decomposition into the elements, i.e. it should be much less thermally stable than Table 1.2 suggests.

However, as the *first* dissociation step (D_1) leading to $(CH_3)_3Pb^\bullet$ and $CH_3^\bullet$ requires more energy than the mean energy $\bar{D}$ averaged over the four bonds (205 v. 155 kJ mol^{-1}), in fact tetramethyllead is a rather stable compound thermally, beginning to decompose at 265 °C. Tetraethyllead decomposes at over 110 °C. Tetramethyltin is stable to about 200 °C and dimethylmercury undergoes no noticeable decomposition at room temperature. The mercury, tin and lead methyl metal halogen derivatives (except monomethyllead) are all thermally stable at normal temperatures, as are the methyl- and dimethylarsenic(V) acidic derivatives $((CH_3)_2AsOOH$ and $CH_3AsO(OH)_2)$. Even the cadmium dialkyls are stable thermally at environmental temperatures, although they are subject to oxidation and hydrolysis (sect. 1.3.6). Alkyl cadmium halides are also thermally stable at normal temperatures. Those organometallic compounds which are not met in the environment are often unstable for reasons other than thermal instability at natural temperatures.

Although not inherently weak, changes in metal–carbon bond strengths (bond dissociation energies) down groups of the Periodic Table do point out *trends* for ease of decomposition of metal alkyls. In general the mean metal–carbon bond strengths $(\bar{D})$ for main group metals decrease down the groups, the opposite trend to that occurring for transition elements (Tables 1.3 and 1.4).

The rather loose connection between bond dissociation energies and environmental stability is shown by the generally greater values observed for the thermally and environmentally unstable Group IVB transition metal alkyls (Table 1.4), compared to the actually stable mercury derivatives which have lower values for $\bar{D}$.

From this survey of bond energies it may be concluded that thermal stability in the environment depends on factors additional to the intrinsic strength of the metal to carbon bonds. The actual routes to such decomposition lie through the production of *hydrocarbyl radicals*, in the possibility of decay by elimination of a chemical species from a *β or α carbon atom*, and in whether or not the metal is *coordinatively saturated*.[19]

Ease of free radical formation is linked directly to metal–carbon bond strengths. Radicals when produced are reactive and will lead directly to stable end-products (alkanes, alkenes or hydrogen). It is the thermodynamic stability of these products which makes $\Delta G_D^\ominus$ for radical decomposition negative, and their kinetic stability which prevents recombination to produce the original organometallic. It should be noted that actual stability against radical production for organometallics is usually greater than that implied by the mean bond energies shown in Tables 1.3 and 1.4. There the quoted figure $(\bar{D})$ is the average or mean value for *each* metal–carbon bond calculated from thermochemical experiments on the decomposition of all the metal–carbon bonds

Table 1.4   Mean bond dissociation energies ($\bar{D}$)* for mercury and some transition metal compounds (in $kJ\,mol^{-1}$)

Compound	D	Compound	$\bar{D}$†	Compound	$\bar{D}$‡
CH_3HgCl	269	$(CH_3)_2Hg$	123	$(CH_3)_4Ti$	260
CH_3HgBr	260	$(C_2H_5)_2Hg$	101	$(CH_3)_4Zr$	311
CH_3HgI	248	$(nC_3H_7)_2Hg$	103	$(CH_3)_4Hf$	332
C_2H_5HgCl	252	$(iC_3H_7)_2Hg$	88	$(CH_3)_5Ta$	260
C_2H_5HgBr	244	$(CH_2{=}CH)_2Hg$	141	$(CH_3)_6W$	160
C_2H_5HgI	227	$(C_6H_5)_2Hg$	137	$CH_3Mn(CO)_5$	118§
C_6H_5HgCl	277	$Hg(CN)_2$	302	$CH_3Re(CO)_5$	206§
C_6H_5HgBr	269			$(C_6H_5CH_2)_4Ti$	203
C_6H_5HgI	260			$(C_6H_5CH_2)_4Zr$	252
				$(CH_3)_5Re$	75–90
				$((CH_3)_3CCH_2)_4Ti$	188
				$((CH_3)_3CCH_2)_4Zr$	227
				$((CH_3)_3CCH_2)_4Hf$	224

* For each of the carbon–metal bonds, in $kJ\,mol^{-1}$. For definition see Table 1.3.

† Note D_1 values for RHg on Table 1.5. $\bar{D}$ here $= \dfrac{D_1 + D_2}{2}$

‡ Transition metal values (From Connor J A 1977 *Topics Current Chem* **71**: 71–111)

§ Methyl metal bond. M—C bond strengths are similar to those for M—CO

in the molecule (i.e. $\Sigma(D/n)$. Actually the energy required to break the *first* bond in a radical process is greater than that to break subsequent metal–carbon bonds. This makes multi-methyl compounds more stable in the environment than might at first appear from Tables 1.3 and 1.4. Conversely the remaining metal–carbon bonds are weak against radical production (usually very weak) and hence further dissociation of the molecule should be rapid. If the decay of dimethylmercury (($CH_3)_2Hg$) as shown in equations 1.7 and 1.8 is taken, it can be seen that bond energy D_1 is considerably greater than the mean metal–carbon bond energy in the molecule ($\bar{D}$) but that D_2 is very much less (Table 1.5). It would therefore be expected that thermal or light-induced decomposition of dimethylmercury in the atmosphere would occur to produce methane, ethane and metallic mercury with monomethylmercury species being short-lived unless they were stabilized by other factors, e.g. complexation.

$$CH_3HgCH_3 \xrightarrow[\text{slow}]{\Delta,\ uv,\ hv} CH_3{}^\bullet + CH_3Hg^\bullet \qquad D_1 \qquad\qquad [1.7]$$

Table 1.5 Bond energies in alkylmetal compounds (in kJ mol^{-1})

Compound	Mean bond enthalpy, $\bar{D}$†	D_1	D_2
$(CH_3)_2Hg$	123	218	29
$(C_2H_5)_2Hg$	101	179	25
$(iC_3H_7)_2Hg$	88	113	63
$(CH_2=CH)_2Hg$	141	202	80
$Hg(CN)_2$	302	517	94
	180	252	97$^\square$
$(CH_3)_4Sn$	218	270	
$(CH_3)_4Pb$	155	205	
$(CH_3)_2Zn$	172	197	147‡
$(CH_3)_2Cd$	141	189	92‡

† $\bar{D} = \dfrac{D_1 + D_2}{2}$ etc

‡ But note environmental instability despite $\bar{D}$ and D_1 values compared to $(CH_3)_2Hg$

$\square$ Separate estimates

$$CH_3Hg^\bullet \xrightarrow{\text{fast}} CH_3^\bullet + Hg \qquad D_2 \qquad\qquad [1.8]$$

A similar decay pathway operates for silicon, germanium, tin and lead methyls on homolysis.

Table 1.5 illustrates this trend. Compounds which produce stable radicals tend to decompose more easily, e.g. i-C_3H_7 compared to n-C_3H_7. It should follow that environmental processes which lead to loss of the first organic group would rapidly lead to decomposition of the whole molecule. From Tables 1.4 and 1.5, the influence of electronegative groups such as halide anions in increasing the thermodynamic stability of the metal–carbon bond should be noted (e.g. D_1 $(CH_3)_2Hg = 218$ kJ mol^{-1}, D for CH_3HgCl is 269 kJ mol^{-1}). This is a general phenomenon in organometallic chemistry.

Direct homolytic dissociation of metal–carbon bonds as shown in equations 1.7 and 1.8, followed by organic free radical reactions, is not usually the main route to decomposition of most organometalics. Isotopic labelling experiments for the decomposition of CD_3CH_2Cu have suggested the following autocatalytic route which is more typical (equations 1.9–1.11)[20]

$$CD_3CH_2Cu + Cu^* \rightarrow CD_3CH_2CuCu^* \qquad\qquad [1.9]$$

$$CD_3CH_2CuCu^* \rightarrow CD_2=CH_2 + DCu + Cu^* \qquad\qquad [1.10]$$

$$DCu + CD_3CH_2Cu \rightarrow CD_3CH_2D + 2Cu \qquad\qquad [1.11]$$

The CuD intermediate is believed to arise as a consequence of a β-elimination mechanism well known in organometallic chemistry. This is shown in equation 1.10. D-labelling shows that the hydrogen in the β carbon is transferred to the metal in a concerted migration elimination step, whose general pathway is shown in equation 1.12.

$$
\begin{array}{ccccc}
\text{M} - \text{C}_\alpha - & & \text{M} ----- \text{C}_\alpha & & \text{M}^* + \text{C}_\alpha \\
& \rightleftharpoons & & \rightleftharpoons & \\
\text{H} - \text{C}_\beta - & & \text{H} ----- \text{C} & & \text{H} \qquad \text{C}_\beta
\end{array} \qquad [1.12]
$$

* Further reactions may give alkanes and other products.

This path generally offers a lower energy route compared to simple metal–carbon bond cleavage owing to favourable energies for the formation of the M—H and the new C=C bonds. It can be seen from this that metal–alkyl groups other than methyl, i.e. those with β carbon–hydrogen bonds should be less stable than methyl metals. This accounts for the greater environmental stability of methyl metal compounds compared to others, and for the faster decay rates of ethyl compared to methyl derivatives in the environment.[21] By the same token it might be expected that phenyl, benzyl, $(CH_3)_3CCH_2$ or $(CH_3)_3SiCH_2$ metal derivatives will also have enhanced stability as β-elimination is not possible. In an analogous way β-elimination of halide or alkoxy groups from organometallics with a β halogen– or alkoxy–carbon bond also takes place rapidly, so β halo organometallics may not be expected to have great stability in the environment. In fact phenyl complexes are not as stable as might be expected as ortho (β) hydrogen atoms may be abstracted and the stability of the benzyl radical promotes instability.

Following the loss of olefin as outlined in equation 1.12, generation of alkanes may occur by hydrogen transfer and also by reductive coupling (equation 1.13), without the presence of alkyl radicals being necessary.

$$ RM \rightleftharpoons R(\text{olefin}) + M—H $$

$$ RM + MH \rightarrow 2M + RH \qquad [1.13] $$

$$ 2RM \rightarrow R—R + 2M $$

α-Elimination processes, which of course may apply to methyl metal complexes, have been identified (equation 1.14).[22] Metal–carbene complexes are the products of the elimination.

$$ R_2CHM \rightarrow R_2C = M + R^1H \qquad [1.14] $$
$$ \quad | $$
$$ \quad R^1 $$

Whether or not such metal–carbene complexes play any part in the decomposition of methyl metal compounds in the environment is unknown, but the hypothetical $CH_2{=}Mn$ bond has been calculated to have a bond strength of 210 kJ mol^{-1} compared to 118 kJ mol^{-1} for the $CH_3{-}Mn$ linkage, so $CH_2{=}M$ bonds in view of this driving force could play a role following thermolysis or light exposure of methyl metals. For example $(CH_3)_5As$ may decompose this way at 100–120 °C.[23]

It can be seen then, that although direct homolysis of metal–carbon bonds after thermolysis or exposure to light may produce radicals (equations 1.7 and 1.8), in fact non-radical pathways are more important. The metal–carbon bond is usually broken by elimination, reductive disproportionation or combination processes. The latter processes are shown in equations 1.15 and 1.16 and one is exemplified in equation 1.17.

$$2RM \Bigg\langle \begin{array}{l} \longrightarrow RH + R(-H) + 2M \qquad\qquad [1.15] \\[2ex] \longrightarrow RR + 2M \qquad\qquad\qquad\qquad [1.16] \end{array}$$

$$2\,CH_3CH_2Ag \longrightarrow C_4H_{10} + 2\,Ag \qquad\qquad [1.17]$$

Metal–carbon bond cleavage by *reductive elimination*[19] is well known particularly for the later transition metals, but environmental applications are not known apart from a route proposed for selenium methylation (Ch. 7). A laboratory example is given in equation 1.19.

$$LM^{n+}\Big\langle \begin{array}{l} {}^X \\[-0.3ex] {}_Y \end{array} \longrightarrow LM^{(n-2)+} + X - Y \qquad\qquad [1.18]$$

$$(C_6H_5)_5Sb \rightarrow (C_6H_5)_3Sb + C_6H_5{-}C_6H_5 \qquad\qquad [1.19]$$

Sterically crowded complexes of metals having a higher oxidation state with availability of a lower, stable oxidation state two units less may undergo this reaction, e.g. equation 1.20.

$$CH_3Ir(III)Cl_2(CO)\,[P(C_6H_5)_3]_2 \rightleftharpoons Ir(I)Cl(CO)\,[P(C_6H_5)_3]_2 + CH_3Cl \;[1.20]$$

The role of β-elimination in organometallic instability depends to an extent on the metal centre being *coordinatively unsaturated*, i.e. that the metal has an empty low energy orbital into which the hydrogen atom may bond. This process is inhibited if the metal has all of its orbitals occupied in bonding to strongly attached ligands, i.e. full coordination to ligands hinders β-elimination. Natural, strongly binding, ligands in the environment will therefore enhance the stability of most organometallics present. For alkyl cobalamins by contrast, recent work suggests that β elimination here does not require an additional coordination site on cobalt.

Full coordination also stabilizes methyl metals from decomposition in thermolysis or light exposure, where the initial step is promotion of a metal–carbon bonding electron to a higher empty metal orbital, thereby weakening the former. If no metal orbitals are empty this process will not occur. $(CH_3)_4Ti$ for example decomposes even at ~ -78 °C having empty metal orbitals available but ligand adducts, e.g. $(CH_3)_4TiL_2$, are coordinatively saturated and more stable,[24] decomposing at 20 °C for L = bipyridyl. Again this suggests a role in the environment for the many oxide, nitro and sulphur ligands present, in the stabilization (by occupation of coordination sites) of organometallic cations, viz. CH_3Hg^+, $(CH_3)_3Sn^+$ and $(CH_3)_3Pb^+$. $(CH_3)_3Sn^+$ may expand its coordination to 5 giving greater stability. Accordingly, simple alkyl derivatives of *transition* metals are hardly accessible environmentally owing to numerous empty metal orbitals facilitating ligand, radical and β-elimination. They will be little found environmentally and if introduced into the environment they will decompose rapidly. It should be pointed out that certain coordination reactions with oxygen and sulphur ligands may, however, lead to *labilization*, not *stabilization*, through dismutation processes (see Chs 2, 3 and 4). Examples are shown in equations 1.21 and 1.22.

$$2CH_3Hg^+ + S^{2-} \rightarrow (CH_3Hg)_2S \xrightarrow{\Delta,h\nu} (CH_3)_2Hg + HgS \qquad [1.21]$$

$$6(CH_3)_3Sn^+ + 3S^{2-} \rightarrow 3\,((CH_3)_3Sn)_2S \xrightarrow{\Delta,h\nu} 3(CH_3)_4Sn + c((CH_3)_2SNS)_3$$
$$[1.22]$$

The driving force for these labilizations is the formation of stable products, e.g. HgS or $c((CH_3)_2Sn)_3$. In practice these processes have been shown to reduce the stability of the above organometallic species in the natural environment (see particularly Ch. 2).

Empty orbitals on the metal also facilitate decomposition by accommodating electrons promoted (e.g. by light) from metal carbon σ orbitals. This weakens the σ bond and is a first step to rupture. Strong π-bonding ligands will not only occupy such orbitals but will also increase the energy gap between them and the metal carbon σ orbitals thereby making promotion more difficult. This role for π-bonding ligands has been discussed in numerous reviews[25] and is particularly relevant to transition metals having partially occupied d-orbitals. However numerous saturated complexes are unstable; saturation by no means always confers stability.

Empty orbitals can also facilitate decomposition by accommodating hydrogen atoms on the metal from α route decompositions. As noted, $(CH_3)_4Ti$ with empty orbitals decomposes above -78 °C, but $(CH_3)_4Pb$ decomposes above 265 °C and $(CH_3)_4Si$ is stable above 500 °C.[26]

It can be seen that numerous decomposition routes exist for the rupture of the metal–carbon bond by thermal or radiolytic routes, and that certain electronic configurations exist which allow facile low energy decompositions to take place. Despite metal–carbon bond strengths being of similar orders, it might be expected that main group metal alkyls are more stable in the environment owing to fewer low energy decomposition pathways existing for them than for transition metal alkyls. This is borne out by experience. A full list of

Table 1.6   Bond dissociation energies* of diatomic molecules (in kJ mol^{-1})

Main group elements†

III	M—C	M—O	IV	M—C	M—O	V	M—C	M—O	VI	M—C	M—O
B	449	806	C	605	1080	N	773	—	O	1080	500
Al	—	449	Si	437	790	P	517	596	S	701	521
Ga	—	299	Ge	462	660	As	—	479	Se	584	424
In	—	323	Sn	—	525	Sb	—	391	Te	—	382
Tl	—	—	Pb	—	416	Bi	—	302	Po	—	—

Transition elements and others†

TiC	538	HH	437	SS	424
VC	559	HC	340	CBr	281
RuC	638	HO	428		
RhC	589	HCl	433		
CeC	458	HN	315		
IrC	626	CCl	391		
PtC	613	CF	554		
CU	466	CI	210		

From Kerr J A 1983 Strengths of chemical bonds, *Handbook of Chemistry and Physics*, Chemical Rubber Company 65 edn, F171–181 Reprinted with permission. Copyright CRC Press Inc, Boca Raton, Fla, USA (Ref 64)

* $D(M—X) = M^• + X^•$, i.e. bond dissociation enthalpy; loosely, bond strength.

† In kJ mol^{-1} measured spectroscopically (mass spectrum) from transient diatomic molecules as above, at 25 °C.

metal–carbon bond strengths based on spectroscopic measurements on dissociation of the diatomic molecules is given in Table 1.6. This list provides little direct guide as to stabilities in the environment but it does give guidance to relative bond strengths and likely stability patterns, and it also indicates the relative tendencies towards conversion to metal oxides (see sect. 1.3.5).

1.3.4   Stability to light

The primary radiolytic decomposition process for organometallic compounds is electronic absorption leading to organic radical formation. The absorption may lead to d–d electronic transitions in the case of transition elements or to charge transfer to or from metal orbitals. The former often causes the dissociation of metal–ligand bonds, and hence coordinative unsaturation, and the latter may facilitate either nucleophilic or electrophilic attack at the metal. For organometallic compounds photo properties are more dependent on the wavelength of excitation radiation than for organic compounds. Light stability is more relevant for volatile, i.e. $R_nM°$ species, as it is they that enter the atmosphere, not the organometallic cationic derivatives which are complexed in sediments, water, etc. The most important are dimethylmercury, tetramethyllead and tetramethyltin.

Photolysis of dimethylmercury at 254 nm in the gas phase produces $CH_3Hg^•$ and $CH_3^•$ radicals, further reactions producing ethane and methane by hydrogen abstractions. At normal temperatures ethane is formed; methane occurs at higher temperatures. Methylmercuriciodide in organic solvents at 313 nm forms $CH_3^•$ by breaking of the mercury–carbon bond. In the gas phase, however, the mercury–halogen bond breaks. Diphenylmercury in organic solvents is photolysed to $C_6H_5^•$ and also decomposes thermally.[27]

Degradation of methyltin halides in water at about 200 nm was observed to produce inorganic tin via sequential degradation,[28] and irradiation of alkyllead compounds at 254 nm also leads to breakdown.[29] These wavelengths are available in the homosphere (see below) and hence these materials would be expected to decay if they volatilize to the atmosphere. Atmospheric fates are discussed in detail in the appropriate chapters.

Processes in the real atmosphere are obviously more complex and, in general, lead to reduced stability from that suggested by laboratory experiments. There is the additional presence of oxygen, other free radicals and surfaces on which enhanced decomposition will take place. Where this has been measured the lifetime of organometallics in the atmosphere may be in terms of hours or days, not years. The lifetime of dimethylmercury in the atmosphere, for example, has been estimated as several days.[30]

From Fig. 1.5 laboratory processes using wavelengths < 340 nm might be thought less relevant under normal conditions for the lower atmosphere as radiation of these wavelengths hardly penetrates to the earth's surface.[31] However, up to 85 km height the atmosphere is homogeneous (homosphere)[32] and volatile materials released to it will, if stable, eventually circulate to that height and be subject to interaction with radiation penetrating the levels below 85 km. A wavelength of 120 nm is equivalent to 998 kJ mol^{-1}; 240 nm is equivalent to 499 kJ mol^{-1} and 340 nm equals 352 kJ mol^{-1}. These are generally sufficient to homolyse metal–carbon bonds (sect. 1.3.3). The main absorbing medium at wavelengths below 340 nm is ozone, whose importance to biology is obvious in view of the toxicity of short wavelength radiation. It might, therefore, be inferred that dimethylmercury, tetramethyllead and tetramethyltin will photolyse in the atmosphere to methyl radicals and $(CH_3)_nM°$ and that further reactions to produce methane, ethane and other hydrocarbons will occur. However this is not the main decomposition process.

In addition to direct photolysis, reactions with other species produced by atmospheric photochemistry will also occur (e.g. OH$^•$, O^{3}P and O$_3$). These occur faster than alternative heterogeneous processes available after adsorption on particles. Tetramethyllead has been most thoroughly investigated[33] from this point of view and details are given in Chapter 4 but hydroxyl radical

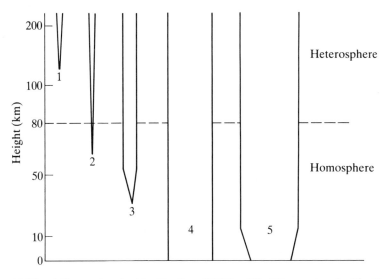

1 UVλ<120 nm, absorbed by N₂, O₂   2 UVλ=120–180 nm, absorbed by O₂
3 UVλ=180–340 nm, absorbed by O₃   4 UV-visλ = 340–700 nm
5 IRλ=>700 nm, slight absorbtion by O₂, H₂O (near IR)

Fig. 1.5 Absorption of light in the atmosphere

> **Notes** Stable molecules released at ground level will eventually rise to the higher atmosphere. Above 25 km radiation of wavelengths 200–220 nm is beginning to penetrate. This is able to photolyse chemical bonds. However, other atmospheric chemical or physico-chemical reactions will reduce the energy required for decomposition (see text).

attack seems to be the most important first step reaction here,[33] viz. equations 1.23–1.25. These chemical decomposition routes are much faster in fact than the photolytic decomposition processes described above when the real atmosphere is considered.

$$(CH_3)_4Pb + OH^\bullet \rightarrow (CH_3)_3PbCH_2^\bullet + H_2O \qquad [1.23]$$

$$(CH_3)_4Pb + O_3 \rightarrow (CH_3)_3PbOOH + CH_2O \qquad [1.24]$$

$$(CH_3)_4Pb + O(^3P) \rightarrow (CH_3)_3PbCH_2^\bullet + OH^\bullet \qquad [1.25]$$

In view of these being reactions of a general nature, i.e. abstraction of hydrogen from a methyl group; insertion of ozone into a metal–carbon bond, it is likely that other volatile organometallics are decomposed chemically in the atmosphere this way, and similar mechanisms have indeed been proposed for tetraethyltin.[34]

Thermal homolysis of metal alkyls leading initially to radical species ($R_nM^\bullet$ + $R^\bullet$) produces the radicals in their ground energy states. Photolysis produces products in higher excited energy states and, therefore, more energy is required to decompose compounds than is needed by the thermal route.

Photochemistry is also more effective than thermal decompositions in promoting permanent decomposition as the excited product fragments are less likely to recombine. The additional energy required for photochemical dissociation varies but for relevant organometallic molecules will be of the order of $100-120$ kJ mol^{-1}.[35] Taking D_1 values (Table 1.5) for mercury, tin and lead we might expect energies of the order of 320, 370 and 300 kJ mol^{-1} to be required for the photochemical decompositions of dimethylmercury, tetramethyltin and tetramethyllead respectively. This is easily accomplished by, for example, 240 nm solar radiation (equivalent to 499 kJ mol^{-1}) which penetrates down to below 50 km altitude.

The volatile organometallics considered here are heavy molecules and their mixing ability in the atmosphere and consequent exposure to solar radiation must be considered. Higher diffusion into the atmosphere will give greater exposure to stronger intensities of radiation and faster photolysis. Higher energy radiation at greater height will also produce higher energy fragments which are less likely to recombine. Mixing to cloud base level (varying from 0.5 to 12 km) is a rapid but complex phenomenon dependent on seasonal factors, temperature inversions, maritime influence and the amount of atmospheric pollution in the area. Diffusion to greater heights is a slower process. For a fairly similar mass molecule CCl_3F(MW 137.5 v. 178.7 for tetramethyltin), 30 km altitude concentrations have been shown to have been increasing since about 1968,[36] indicating that diffusion to such heights does occur in parallel with increased production and release. It might be assumed, then, that such diffusion and exposure to high energy radiation could occur also for organometallics if they were to survive in the absence of other processes. CCl_3F is also fragmented by 200 nm radiation and assuming similar photochemical efficiencies broadly similar lifetimes might exist between these and the organometallic molecules. On average it has been calculated that $50-80$ years elapse between the release of CCl_3F at ground level and photochemical destruction at over 25 km altitude. (Boltzman distribution is assumed, leading to varied lifetimes for molecules.) On this basis also, CCl_2F_2 has a lifetime of $90-150$ years on average towards *photochemical* destruction.

It can be concluded that the photochemical processes alone would decompose organometallics into radical species and the metals but that these alone might take decades. However, it should also be considered that other processes are occurring in the atmosphere, particularly where it is polluted. Atmospheric chemistry is complex — a recent source lists 79 reactions that should be considered in modelling stratospheric chemistry.[37] Species involved include O_3, O (triplet^{3}P), $OH^\bullet$, $ClO_x^\bullet$, $HO_x^\bullet$, $NO_x^\bullet$. All of these are highly reactive radicals particularly able to abstract hydrogen atoms from hydrocarbons or organometallics. In addition, particle-based heterogeneous decomposition may occur. Radical or singlet oxygen attack on organometallic species in polluted atmospheres will actually decompose these molecules rather than direct photochemical decomposition alone. It can be concluded that the lifetime of organometallic species in the atmosphere is of the order of hours or days rather than decades, with metals and hydrocarbons being produced. Actual measurements and estimates reinforce these conclusions (see, e.g. Ch. 2).

Actual decay rates will vary widely and have been studied under environmental conditions, e.g. for lead where rates vary from 3 to 4 per cent per hour in winter to 16 to 29 per cent in summer for tetramethyllead. For tetraethyllead the corresponding rates are 17–23 and 67–93 per cent. These are upper limits but it can be seen that organometallics are not persistent in the atmosphere, with direct photolysis accounting for little of the decay. Even in the area of 340 nm where the atmosphere is fully transparent to light the energy associated with the wavelength, 352 kJ mol^{-1}, is sufficient to break most metal–carbon bonds (Table 1.5) to produce ground energy state products. Shorter wavelengths are needed to produce higher energy state products, but normally *chemical* decomposition processes predominate.

For dimethylmercury also, a relatively fast degradation by hydroxyl attack has been deduced, with elemental mercury being produced and the rate constant for hydroxyl attack calculated at 2×10^{11} cm^3 mol^{-1} s^{-1}.[30] Overall it should be assumed that the common organometallic species emitted to atmosphere degrade rapidly and cannot be considered a serious environmental contaminent in most locations owing to their low concentrations (see relevant chapters) and relatively rapid rates of decay.

1.3.5   Stability to oxidation

Organometallic compounds are thermodynamically unstable to oxidation by reason of the large negative $\Delta G_D^\ominus$ values for formation of the metal oxide, carbon dioxide and water products. Equation 1.26 exemplifies this for tetramethyltin.

$$(CH_3)_4Sn_{(g)} + 8O_{2(g)} = SnO_2(s) + 4CO_{2(g)} + 6H_2O_{(g)} \qquad [1.26]$$

$$\Delta H_D^\ominus = -3591 \text{ kJ mol}^{-1}$$

$\Delta G_D^\ominus$ is even more negative in view of the net generation of gaseous molecules creating an overall positive $T\Delta S$ input into $\Delta G^\ominus = \Delta H^\ominus - T\Delta S^\ominus$.

The initial process of oxidation of organometals by O_2 is a rapid charge transfer interaction which occurs, involving electron donation from the organometallic to oxygen.[20] This is shown in equations 1.27–1.29.

$$R_nM + O_2 \rightarrow R_nM^+.O_2^- \rightarrow R^\bullet + R_{n-1}\overset{+}{M}O_2^- \qquad [1.27]$$

$$R^\bullet + O_2 \rightarrow RO_2^\bullet \qquad [1.28]$$

$$RO_2^\bullet + R_nM \rightarrow RO_2R_{n-1}M^\bullet + R^\bullet \qquad [1.29]$$

The species $R_nM^+.O_2^-$ may decay by various routes, and peroxides may be formed or coupling of the alkyl ligands may occur. Organoboron, -mercury and -phosphorus compounds are oxidized by radical chain SH_2 processes as exemplified below in equations 1.30 and 1.31, following the initial charge-transfer process.[20]

$$R_3B + RO_2{}^\bullet \rightarrow R_2BOOR + R^\bullet \qquad\qquad [1.30]$$

$$R^\bullet + O_2 \rightarrow RO_2{}^\bullet, \text{etc.} \qquad\qquad [1.31]$$

Despite this thermodynamic instability, many organometallics (e.g. those of mercury, silicon, germanium, tin and lead) are kinetically stable to oxygen at room temperature and may persist in the natural environment. Others are kinetically unstable, particularly if they possess empty low-lying orbitals, lone electron pairs or polar metal–carbon bonds. These are the factors also making for thermal or light instability. The stability of organometallic species to oxygen under normal conditions is given in Table 1.7. It is likely in some cases, e.g. $(CH_3)_2As^+$, that complexation will render the species stable to oxygen. In the arsenic case we can consider $(CH_3)_2AsO(OH)$ a special case of this, but as the hydride, $(CH_3)_2As^+$ is unstable. For the CH_3Pb^+ case, environmental stabilization by complexation has been tacitly assumed to occur in order to explain stepwise methylation processes but there is no firm evidence for this. It should also be pointed out that unstable species in bulk may decompose only slowly when they are present in extreme dilution (e.g. ng dm^{-3}), and both $(CH_3)_{4-n}SnH_n$ and $(CH_3)_nAs^{(3-n)+}$ species have been detected in anoxic natural environments (see Ch. 3) although they are unstable to oxygen. Such dilution apparently reduces collision possibilities leading to reaction.

Despite the value of $\Delta H_D^\ominus$ given in equation 1.26, tetramethyltin is quite stable in air (discounting atmospheric processes referred to in sect. 1.3.4). However, the coordinatively unsaturated molecule dimethylzinc, which has a less favourable value for $\Delta H_D^\ominus$ (-1920 kJ mol^{-1}), ignites spontaneously in air. This demonstrates the importance of coordination in conferring kinetic stability.

The comparative susceptibilities of some metal–carbon bonds to oxidation to metal oxides are demonstrated in Table 1.8. For main group elements they suggest increasing liability to oxidation as the group is descended.

1.3.6   Stability to water

The first step in the hydrolysis of an organometallic compound is usually the attachment of the lone electron pair on the water oxygen atom to a metal orbital on the organometallic. Hence hydrolytic instability is also connected with empty low lying orbitals on the metal and on the ability to expand the metal coordination number. The rate of hydrolysis is connected with the polarity of the metal–carbon bond; strongly polarized ($M^{\delta+}$—$C^{\delta-}$) bonds are unstable to water. These are found for example in Groups I–II organometals and for those of zinc and cadmium. The influence of polarity is shown for alkylboron compounds which have low polarity and are water stable although unstable to air. Low polarity compounds which cannot easily expand their

Table 1.7 Stability of methylmetals to oxygen*†

Stable	Unstable‡
$(CH_3)_2Hg$	CH_3PbX_3
$(CH_3)_4Si$, $[(CH_3)_2SiO]_n$, $(CH_3)_nSi^{(4-n)+}$, $(CH_3)_6Si_2$	CH_3Tl^+
$(CH_3)_4Ge$, $(CH_3)_nGe^{(4-n)+}$, $(CH_3)_6Ge_2$	$(CH_3)_2Zn$ (CH_3Zn^+ also)
$(CH_3)_4Sn$	$(CH_3)_2Cd$ (CH_3Cd^+ also)
$(CH_3)_4Pb$§	$(CH_3)_3B$
CH_3HgX (C_6H_5 and C_2H_5 also stable)	$(CH_3)_3Al$
$(CH_3)_{4-n}SnX_n$	$(CH_3)_3Ga$
$(CH_3)_3PbX$	$(CH_3)_3In$
$(CH_3)_2PbX_2$	$(CH_3)_3Tl$
$\pi CH_3C_5H_4Mn(CO)_3$§	$(CH_3)_5As$
$CH_3Mn(CO)_4L$¶	$(CH_3)_3As$
$(CH_3)_2AsO(OH)$	$(CH_3)_3Sb$
$CH_3As(O)(OH)_2$	$(CH_3)_3Bi$
$(CH_3)_2S$	$(CH_3)_2AsH$
$(CH_3)_2Se$	CH_3AsX_2
$CH_3HgSeCH_3$	CH_3SbX_2
CH_3COB_{12} (solid state)	$(CH_3)_{4-n}SnH_n$‖
$(CH_3)_3SbO$	$(CH_3)_6Sn_2$ (At RT gives $[(CH_3)_3Sn]_2O$)
$(CH_3)_2SbO(OH)$	$(CH_3)_6Pb_2$ (To methyl lead products)
$CH_3SbO(OH)_2$	$(CH_3)_5Sb$
$(CH_3)_2Tl^+$, $(CH_3)_2Ga^+$	$(CH_3)_3AsO$
$(CH_3)_3S^+$	$(CH_3)_3P$
$(CH_3)_3Se^+$	$(CH_3)_4SiH_{4-n}$
$(CH_3)_3PO$	$(CH_3)_4GeH_{4-n}$

* At room temperature. Assume similar but lesser environmental stability for ethyls.

† That is against rapid (seconds, minutes) oxidation. Table to be read in conjunction with Table 1.9.

‡ Variously unstable because of empty low lying orbitals on the metal, polar metal–carbon bonds and/or lone electron pairs on the metal.

§ Gasoline additive.

¶ To exemplify ligand-complexed transition metal organometallics. Many of these synthetic compounds are oxygen-stable but none have been found in the natural environment.

‖ But apparently stable in dilute form and detected in the environment (Ch. 3).

Table 1.8   Comparative bond enthalpy terms for metal–carbon and metal–oxygen bonds* (in kJ mol^{-1})

Group IV	M—C	M—O	Group V	M—C	M—O	Group VI	M—C	M—O
C	347	358	N	314	214	O	358	144
					632 (NO)			498 (O_2)
Si	320	466	P	276	360	S	289	522 (S=O)
Ge	247	385	As	230	326	Se	247	
Sn	218		Sb	218		Te		
Pb	155		Bi	141		Po		

B—C = 364   C—H = 413   O—H = 464   N—H = 391
B—O = 520   H—H = 436   Cl—H = 432   C—Cl = 346

* From thermodynamic data on the decomposition of the molecules, e.g. C_2H_6. Calculated from thermodynamic cycles.

(Source: Johnson D A 1982 *Some Thermodynamic Aspects of Inorganic Chemistry* (2nd edn). Cambridge University Press, pp. 201–2.) Data gives an assessment of comparative bond strengths. Reproduced with permission.

Definition as in Table 1.3.

Table 1.9   Stability of organometallic species to water

Organometallic	Stability, comments
R_2Hg, R_4Sn, R_4Pb	Only slightly soluble, stable, diffuse to atmosphere. Higher alkyls less stable and less volatile. Species generally hydrophobic and variously volatile
CH_3HgX	Stable, slightly soluble depending on X
$(CH_3)_nSn^{(4-n)+}$	Soluble, methyltin units stable but made hexa- and penta-coordinate by H_2O, OH^-. Species are solvated, partly hydrolysed to various hydroxo species. At high pH polynuclear bridged hydroxo species form for $(CH_3)_2Sn^{2+}$
$(CH_3)_3Pb^+$	Soluble, hydrolysis as methyltins above. Also dismutates to $(CH_3)_4Pb$ and $(CH_3)_2Pb^{2+}$ at 20 °C
$(CH_3)_2Pb^{2+}$	Soluble as for $(CH_3)_3Pb^+$ above. Disproportionates to $(CH_3)_3Pb^+$, Pb^{2+} and CH_3^- slowly. These reactions cause eventual total loss of $(CH_3)_3Pb^+$ and $(CH_3)_2Pb^{2+}$ from water
$(CH_3)_2As^+$	Hydrolyses to $(CH_3)_2AsOH$ then to slightly soluble $[(CH_3)_2As]_2O$
CH_3As^{2+}	Hydrolyses to $CH_3As(OH)_2$ then to soluble $(CH_3AsO)_n$

Table 1.9 (cont)

Organometallic	Stability, comments
$(CH_3)_2AsO(OH)$	Stable and soluble (330 g dm^{-3}). Acidic pk_a = 6.27, i.e. cacodylic acid, dimethylarsonic acid. Detected in oceans
$CH_3AsO(OH)_2$	Stable and soluble. Strong acid pk_1 = 3.6, pk_2 = 8.3 – methylarsinic acid. Detected in oceans
$(CH_3)_3S^+$, $(CH_3)_3Se^+$	Stable and slightly soluble
$(CH_3)_nSiCl_{4-n}$	Hydrolyse and condense but methylsilicon groupings retained
$(CH_3)_nGe^{(4-n)+}$	Stable, soluble, have been discovered in oceans. Hydrolyse but $(CH_3)_nGe$ moiety preserved
$(CH_3)_2Tl^+$	Very stable, soluble, but not been detected as a natural environmental product
$(CH_3)_3AsO$, $(CH_3)_3SbO$	Stable and soluble
$(CH_3)AsH$, CH_3AsH_2	Insoluble, diffuse to atmosphere, air unstable
$(CH_3)SbO(OH)$	Stable and soluble. Detected in oceans
$CH_3SbO(OH)_2$	Stable and soluble. Detected in oceans

Other species

Stable and insoluble – R_4Si, $(R_2SiO)_n$, $CH_3HgSeCH_3$, most C_6H_5Hg derivatives, $(CH_3)_2S$, $(CH_3)_2Se$, $(CH_3)_4Ge$, $(CH_3)_3B$

Unstable – CH_3Pb^+ (has not been detected in the environment), R_2Zn, R_2Cd, R_3Al, R_3Ga, $(CH_3)_6Sn_2$, $(CH_3)_6Pb_2$, $(CH_3)_5Sb$, CH_3Tl^{2+}, CH_3Cd^+, $(CH_3)_2Cd$, $(CH_3)_2Sb^+$, CH_3Sb^{2+}

Solubility here refers to <u>air-free</u> distilled water, no complexing ligands. Range of solubilities is from mg dm^{-3} to g dm^{-3}. Data from references.

coordination number are expected to be water stable. Despite this, most metal alkyls and aryls are thermodynamically unstable to hydrolysis to metal hydroxide and hydrocarbon (equation 1.32). Many, though, are kinetically stable. Examples are given in Table 1.9.

$$R_nM + nH_2O \rightarrow M(OH)_n + nRH \qquad [1.32]$$

1.3.7 Stability of organometallic compounds in biological systems

Full details of the toxicity and fate of organometallic compounds in higher organisms are given in the appropriate chapters. Despite their general toxicity, there are numerous breakdown routes for organometallic species available in organisms. These vary in rate; methylmercury has a half-life in man of about 70 days, triethyllead around 35 days in one human compartment and 100 days in another (Ch. 4). Obviously, half-lives vary between species and organs but mechanisms for the decomposition of the metal−carbon bond do exist in organisms. There are also several common routes linking decay processes for a variety of organometallics by various organisms.

Decay usually takes place by dealkylation. Methylmercury may be decomposed by numerous bacteria, first to methane and mercury(II), and finally to the mercury(0) state (Ch. 2). Arylmercury salts produce the arene and mercury(0). Fully saturated tetramethyl organometallics (e.g. tetramethyllead and tetramethyltin) are easily absorbed but rapid decay by loss of a methyl group to the more stable trimethyl metal form takes place. For tetramethyllead, this conversion occurs rapidly in the liver − in rat liver there is a half-life of minutes. The trialkyl salts decay at various rates, eventually giving the metallic ions at various yields. In physiological saline, $(CH_3)_2PbCl_2$ is broken down to lead(II) with a half-life of about 12 days (Ch. 4). Half-lives for methyl species are usually longer than for ethyl and higher alkyl analogues, owing to the presence in the latter of facile routes to decay.[38] The main route to dealkylation for higher alkyl metals, however, does not lie in direct metal−carbon bond cleavage but usually occurs after initial hydroxylation of the β carbon atom by the liver.

Enzymatic hydroxylation of alkyl groups attached to metals has been widely reported. It occurs for example in the dealkylation of butyl- and cyclohexyltin compounds (Ch. 3) and in the decay of the ethyllead moiety (Ch. 4). The hydroxylation is usually reported at the β carbon but it also occurs at the α carbon; attachment at the metal may also occur, e.g. in the case of silicon where the initial silanol product condenses to a siloxane (Ch. 7). Phenyl groups do not usually undergo enzymatic hydroxylation but it is known for the phenylsilicon grouping (Ch. 7).

For the lower alkyl groups the organic products have been reported as alkanes and alkenes; for ethyllead species the product is reported to be ethanol. The gasoline additive $CH_3C_5H_4Mn(CO)_3$ is also hydroxylated at the methyl position. Hydroxylation is mainly carried out by liver microsomal mono-oxygenase enzyme systems.

From the above it can be seen that the importance of hydroxylation is the labilization it produces in the metal−carbon bond, allowing eventual dealkylation.

By contrast, it has been shown that organic arsenic compounds occurring naturally in seafood are transferred through the body with little change (Ch. 5). The commonest organoarsenic compounds found in the environment, $(CH_3)_2As(O)(OH)$ and $CH_3AsO(OH)_2$, are very stable against biological decay routes in plants and animals. For arsenic, demethylation seems not to occur in animals. Soil bacteria have the ability to demethylate methylarsenic compounds giving CO_2 by an oxidative route. The lack of arsenic demethylation in animals may be seen in the context of organoarsenic compounds being less toxic than their inorganic counterparts. The reverse is the case for most organometals (e.g. those of mercury, lead or tin).

1.3.8   General conclusions on stability

Organometallic compounds that are stable under environmental conditions will have one or more of the following features:

1. Strong metal–carbon bonds of low $M^{\delta+}$–$C^{\delta-}$ polarity.
2. No available low energy empty orbitals or electron pairs, i.e. little ability to increase readily coordination number by accretion of an attacking (destabilizing) species which would promote decay.
3. Organic groups which are non-labile, i.e. do not have β hydrogen atoms (e.g. CH_3, CF_3), or are not susceptible to biological hydroxylation, or are not unsaturated (do not allow electrophilic attack).

It should also be noted that there is an order of relevant organic groups for stability: i.e. $R_F > R_H$ and $C_6H_5 > CH_3 \gg C_2H_5$ (R_F = fluorinated alkyl group). For main group organometallics it should be noted that metal–carbon bond strengths decrease down the group. For transition metal organometals the reverse holds true. Finally, coordinating ligands enhance stability by denying a coordination position to an external attacking ligand which would promote decay (e.g. H_2O). Such enhanced stability may lead to increased retention in organisms and thereby to increased toxicity (e.g. for methylmercury bound strongly to sulphydryl groups in central nervous tissue – see section 2.3).

1.3.9   Sources of data

The stability and decomposition of organometallic compounds have been studied intensively, but mainly under laboratory rather than environmental conditions. There is as a result much knowledge on mechanisms and pathways for decay for these compounds, but this is often of limited help in predicting final products in the complex natural environment. For commercial organometallic products it has often been necessary to determine final end-products after use in the environment to ensure that decay occurs and that it does so to innocuous products. In such cases there may be information about end-products but little about mechanisms. However, it is not unreasonable to assume that organometallic compounds decay in the environment by combinations of the same chemical and biological routes that have been elucidated in laboratory studies.

A number of excellent discussions on the stability of the metal–carbon bond are in existence. There is a useful concise survey in an Open University Course Unit.[39] More detailed accounts are those of Norton,[40] Schrock and Parshall[41] and Davidson *et al.*[42] in the general area of organometallic stability. Photochemical stability has been reviewed in detail by Koerner von Gustorf *et al.*[43] and the cleavage of metal–carbon bonds by biological means is well covered in a recent book by Thayer.[44] Bond energy and enthalpy data for a wide variety of chemical bonds are available in a monograph by Johnson[45] and these make useful comparison with other metal–carbon enthalpy data, e.g. the reviews by Pilcher and Skinner[46] and Connor.[47]

In this work metabolic studies of organometallic compounds and the routes to decomposition available for each metal are discussed in the appropriate chapter and references to decay mediated by biological means are presented there. Brief reference here is made to excellent studies in the decay of methylmercury compounds by Summers and Silver,[48] and by Silver,[49] to work on organotins by Blunden[50,51] and to studies on lead by Harrison[52,53] and by Grandjean and Nielsen.[54]

Other mechanistic studies of organometallic compounds of general importance include the works by Kochi,[55] O'Neill and Wade,[56] and also are incorporated in the standard works on organometallic chemistry.[57-63] Thermochemical data are taken from the above sources and also from data by Kerr[64] and Pilcher.[65]

1.4   General comments on the toxicities of organometallic compounds

The toxicities of the important environmental organometallic compounds are dealt with in the relevant chapters. There is also a detailed discussion of the toxicity of organometallics and their interaction with living organisms in a recent book.[44] The toxicities of individual groups of organometallic compounds are separately discussed in a number of sources,[66-68] with organomercury compounds receiving particularly detailed attention.[69,70] Detailed toxicity accounts exist also for organotin[71,72] -lead,[54] -arsenic,[73] and the other organometalloidal species.[74] Details are further covered here and in the appropriate chapters.

This section will present a summary of the main toxic mechanisms for environmental organometallics and will highlight the main structure–activity relationships as they are known. In general, organometallic compounds are more toxic, often substantially so, than the inorganic metal compounds from which they derive. Mercury, lead and tin obey this general rule; as noted arsenic is an exception. Usually the toxic effects are at a maximum for the formal monopositive cations, i.e. the species derived by loss of one organic group from the neutral fully saturated organometallic, viz. R_3Pb^+, R_3Sn^+, CH_3Hg^+. It should be noted that the toxic effects of the neutral organometallics (R_4Sn, R_4Pb, etc.) usually derive from conversion to R_3M^+, etc. in organisms. In general alkyl groups are more toxic than aryl groups when attached to metals. The most toxic alkyl groups vary from organism to organism, but methyl, ethyl and propyl groups tend to be the most toxic. For tin compounds toxicity of the organotin cations is at a maximum for the trialkyl series. For trialkyltins methyl and ethyl groups are the most toxic to mammals, with higher alkyl groups being more toxic to bacteria (Ch. 3); longer chain alkyl compounds are of lower toxicity to mammals.

Mechanisms of toxicity are varied, but good coordination to base atoms (e.g. S, O, N) on enzyme sites seems to be the main one. Coordination of the organometallic species (usually R_nM^{m+}) to the enzyme blocks the sites and prevents reaction with the biological substrate. Enzymes blocked in this way include lipoic acid (sulphur site), acetylcholinesterase (ester site) and aminolaevulinic acid dehydrase.[44]

The other main result of organometal introduction into higher animals is a diminution of the myelin coating of nerve fibres. In addition, water accumulation and oedema in the central nervous system may occur. Other toxic effects are due to coordination to non-enzyme sites (e.g. thiols, histidine residues of proteins, haemoglobin, cytochrome P-450, cerebal receptors.) Haematopoietic, bone marrow, immune and essential trace metal systems in organisms may be affected by organometallic compounds. Fundamental interference with DNA and protein synthesis has also been reported.

The reason for the enhanced toxicity of organometallics over the inorganic derivatives − usually as cations, R_nM^{m+} − lies with the existence of lipophilic or hydrophobic groups (R) on the same species having also a hydrophilic dipole. This allows transport in aqueous body fluid, and also solubility and transport through fatty tissue and cell walls by diffusion. As Lewis acids (sect. 1.5) similar to inorganic metal species, there is good bonding to Lewis base coordination sites within the organism (e.g. thiol groups). Coordination preferences are discussed in section 1.5.

The main result of organometallic poisoning is central nervous system damage, leading to varied symptoms including coma, ataxia, hyper-activity, varied motor difficulties, speech problems and psychological−attitudinal changes. Within the most prevalent series of environmental organometallic compounds, triorganotin toxicity arises through disruption of mitochondrial functions, membrane damage, disruption of ion transport and inhibition of ATP synthesis. Dialkyltins also inhibit oxygen uptake in mitochondria and inhibit α-ketoacid oxidation (Ch. 3). Acute organic lead poisoning is chiefly focused on central nervous system effects, with physiological changes to the brain being found on autopsy, including neuron destruction and degeneration of nerve tracts. In general, serious but non-specific damage to the nervous system occurs in acute lead poisoning with organic lead compounds, with nerve cells in the hippocampus, reticular formation and cerebellum being particularly sensitive. At lower levels of poisoning incipient anaemia effects have been noted.[6] There is an excellent detailed review on the environmental health aspects of organic lead compounds.[54] Trialkyllead and -tin compounds, together with monomethylmercury, all bind to sulphide residues. Both trialkyllead and -tin compounds may destroy the normal pH gradient across mitochondrial membranes thereby uncoupling oxidative phosphorylation. Despite the binding differences the results of organomercury poisoning with respect to brain function are quite similar to those considered above. Penetration of the blood−brain barrier leads to sensory disturbance, tremor, ataxia, visual and hearing difficulties. Methylmercury is lipid-soluble, rapidly diffuses through cell membranes and once it enters the cell is quickly bound by sulphydryl groups. It rapidly penetrates the blood−brain barrier. Methyl-mercury inhibits protein synthesis and RNA synthesis and causes particular damage to the developing brain (Ch. 2).

A final general point here is that the most subtle and early results of organometal poisoning cannot always be detected, particularly for lead and mercury. Here behavioural and intellectual effects are shown and these are difficult to evaluate especially in children, e.g. for mercury one of the earliest and most sensitive indices of methylmercury poisoning, paraesthesia, is very hard to measure and evaluate in children. For lead, there is of course the debated and incompletely resolved case of low level intellectual effects in children following (mainly) inorganic lead accumulation from organic lead gasoline additives. A particularly subtle effect of such cases is the difficulty of separating genuine toxic effects from other disease or socio-economic symptoms. These second level (or chronic) effects cause different problems in terms of evaluation and action than do acute high-concentration poisonings and are an integral feature in the consideration of the environmental effects of organometallic compounds.

1.5   Coordination preferences for environmental complexation by organometallic species

The general preferences of metal cations for coordination with either small, electronegative bases (e.g. oxides, nitrogen ligands − hard bases) or to larger, more polarizable species of low electronegativity (e.g. sulphides, selenides, iodides − soft bases) has been developed into a general theory (Hard and Soft Acid Base Principle). Hard acid metals are small, have higher oxidation states and are not easily polarizable − in the natural and laboratory environments they bond preferably to hard bases (donors). Soft acid metals are larger, have lower oxidation states and are more polarizable − they are usually found complexed with soft donor bases. The general classification of metals and ligands along these lines is given in Table 1.10.[75−77]

For inorganic metals this principle is well recognized in the natural environment. Hard acids, such as magnesium(II), calcium(II), tin(IV), etc. are found bonded in the earth's crust as oxide or carbonate ores (i.e. hard donors). Softer acid metals such as mercury(II), lead(II) are found as sulphides (a soft donor ligand), although lead also occurs as carbonate. In the organometallic area the complexation of organic groups to metals has only a small effect on the hard−soft properties of metals. Taking formation constants (Log k from Table 1.11[78,79]) as a measure of stability it can be seen that the constants are generally such that methyl ligands appear to move metals into the direction of reduced affinity, but not to a large extent.[79] Methyl groups are more electro-negative than metals and, therefore, generally withdraw electron density from

Table 1.10   Classification of environmental metals (acceptors) and ligands (donors)

Hard acids (metals)	Hard bases (donors)
H^+, Be^{2+}, $Be(CH_3)_2^*$, VO^{2+}, Fe^{3+}, Co^{3+}, Al^{3+}, $Al(CH_3)_3$, Ga^{3+}, In^{3+}, CO_2, RCO^+,	NH_3, RNH_2, H_2O, OH^-, O^{2-}, ROH, RO, R_2O, CH_3COO^-, CO_3^{2-}, NO_3^-, PO_4^{3-},
Si^{4+}, Pb^{4+}, Sn^{4+}, $(CH_3)_n Sn^{(4-n)+}$, RPO_2^+, $ROPO_2^+$, As^{5+}, SO_3, RSO_2^+, $ROSO_2^+$, Mn^{2+}, Ti^{4+}, Cr^{3+}, As^{3+}, $(CH_3)_nAs^{(3-n)+}$, Ge^{4+}, $(CH_3)_nGe^{(4-n)+}$	SO_4^{2-}, F^-, Cl^-
Borderline acids	**Borderline bases**
Fe^{2+}, Co^{2+}, Ni^{2+}, Cu^{2+}, Zn^{2+}, $(CH_3)_3B$, R_3C^+, Bi^{3+}, $C_6H_5^+$, SO_2, $(CH_3)_nPb^{(4-n)+}$, $(CH_3)_nSb^{(3-n)+}$, $R_nSn^{(4-n)+}(R>CH_3)$, Sn^{2+}, Cd^{2+}	$C_6H_5NH_2$, NO_2^-, SO_3^{2-}, Br^-
Soft acids	**Soft bases**
Cu^+, Cd^+, Hg^+, Hg^{2+}, CH_3Hg^+, Tl^{3+}, CH_3Tl^{2+}, $(CH_3)_2Tl^+$, $(CH_3)_3Tl$, RS^+, O_2, M°, RSe^+, RTe^+	Alkyl groups, CN^-, CO, RNC, SCN^-, R_3P, $(RO)_3P$, R_3As, R_2S, RSH, RS^-, $S_2O_3^{2-}$, I^-

* oligomer

Environmentally relevant metals and ligands only are shown. Fuller details are available in Pearson R G 1973 *Benchmark Papers in Inorganic Chemistry, Hard and Soft Acids and Bases*. Dowden, Hutchinson and Ross, Stroudsberg, Pa. USA. (Ref 77)

metals, increasing acidity. This is in spite of methyl groups themselves being generally thought of as soft ligands. Larger aliphatic organic groups are more electron-releasing than the methyl group and they are also more polarizable. Hence larger groups (e.g. butyl, octyl) should tend to move towards increasing the softness of metals to which they are complexed.[80] Overall, however, in terms of environmental complexation the resultant effect will be small and the species tend to remain in the category in Table 1.10 that the inorganic metal species belong to.

Methylmercury complexes have been most thoroughly investigated and coordination trends are shown in Table 1.11. In the environment, as we see above, the coordination preferences of the organometallic species considered in this work will follow those of their inorganic derivatives. Organic mercury will be found almost entirely bonded to sulphur ligands (if they are present – in saline environments chloride donation is found). Organotin and -lead com-

Table 1.11   Formation constants of mercury(II) and methylmercury(II) with organic and inorganic ligands

Ligand L	Donor atom	$\log \dfrac{[CH_3HgL]^+}{[CH_3Hg^+][L^-]}$	Ligand L	Donor atom	$\log \dfrac{[CH_3HgL]}{[CH_3Hg^+][L^-]}$
F^-	F	1.5	CH_3COO^-	O	3.6
Cl^-	Cl	5.3 (6.6)	Acetylcholine	O	2.7
Br^-	Br	6.6 (8.4)	Cl_2CHCOO^-	O	1.1
I^-	I	8.6 (11.9)	β-alanine	O	2.5
OH^-	O	9.4 (10.9)	R_nNH_{3-n}	N	5.0–7.5
CH_3HgO^-	O	2.4	Glycine	N	7.9
SO_4^{2-}	O	0.9	β-alanine	N	7.6
SeO_4^{2-}	O	1.1	4-aminovaleric acid	N	7.5
CO_3^{2-}	O	6.1	Phenylalanine	N	8.3
SeO_3^{2-}	O	6.5	C_5H_5N	N	4.7 (5.0)
$C_6H_5O^-$	O	6.5	$(oC_5H_4N)_2$	N	5.9
CH_3COO^-	O	3.6	N-methylimidazole	N	7.0
HPO_4^{2-}	P	5.0	Uridine	N	9.0
HPO_3^{2-}	P	4.7	Cytidine	N	4.6
CN^-	C	14.1 (17.4)	Adenosine	N	3.0
S^{2-}	S	21.2	Guanosine	N	8.1
CH_3HgS^-	S	16.3	Imidazole	N	7.3 (8.4)
$(CH_3Hg)_2S^*$	S	7.0	NH_2 (histidine)	N	8.8 (10.6)

Ligand		Value
SCN⁻	S	6.1 (8.7)
SO_3^{2-}	S	8.0
$S_2O_3^{2-}$	S	11.0
$SCH_2CH_2OH^-$	S	16.1
RS⁻ (cysteine)	S	15.7
HEDTA³⁻	N	6.2
NH_3*	N	8.4 (8.7)
C_5H_5N*	N	4.8 (5.0)
$NH_2CH_2CH_2NH_2$*	N	8.3
Se²⁻	Se	
SeCN⁻	Se	6.8

Ligand		Value
Im (histidine)	N	6.4 (7.5)
RS⁻ (cysteine)	S	15.7 (20.1)
RS⁻ (histidine)	S	15.9 (20.5)
$HSCH_2CH_2OH$	S	16.1
Glutathione	S	15.9
Cysteine	S	15.7
Mercaptalbumin	S	22.0
Methionine	S	1.9*

General trends of stability
RHgI > RHgBr > RHgCl > RHgF
i.e. soft RHg more stable with softest X

$CH_3HgX > C_2H_5HgX > nC_3H_7HgX > nC_4H_9HgX$
i.e. For X = Cl, Br, OH etc, hardest RHg is most stable with these hard ligands

$CH_3HgI < C_2H_5HgI < nC_3H_7HgI$
i.e. softest RHg is most stable with the soft I ligand

† i.e. Log k (page 32)
* Neutral molecule. Values in parentheses are log $\left(\dfrac{\sqrt{[HgL_2]}}{\sqrt{[Hg]}\,[L]}\right)$.

* showing comparison with inorganic mercury (II) bonded to the same ligands

Data above reprinted with permission from Rabenstein DL 1978 Accounts Chem Res 11: 100–7. Copyright (1978) American Chemical Society.

Table 1.11 (cont)

General affinity trends for metals and organometals (i.e. M^n and R_nM)

1. Fe(III) > Pb(II) > Cu(II) > Zn(II) > Cd(II) > Co(II) > Ca(II) > Mg(II)
   Towards O on a hydrated SiO_2 surface

2. Salicylate (OO):
   Ca(II) > Cd(II) > Mn(II) > Fe(II) > Zn(II) > Cu(II)

   Glycine (ON):
   Ca(II) < Mg(II) < Mn(II) < Cd(II) ~ Fe(II) < Zn(II) ~ Pb(II) < Cu(II) < Hg(II)

   Cysteine (SN):
   Mn(II) < Zn(II) < Fe(II) < Cd(II) < Pb(II) < Hg(II)

   Hydroxide (O):
   Ca(II) < Mg(II) < Cd(II) ~ Mn(II) < Zn(II) < Co(II) < Fe(II) < Cu(II) < Pb(II) < Hg(II)

   Changing stabilities towards different ligands – donor atoms are given in parentheses

3. oxalate < glycine < ethylenediamine < mercaptoethylamine   ligand
   OO   ON   NN   NS   donor atoms

   Stability of cadmium(II) towards different ligands

Data above printed from *Fates of Pollutants* 1978, US Nat. Res. Council, pp 60–8.

plexes will be found to be associated also with nitro and oxo ligands, though sulphide coordination is important in certain environmental processes (Chs 3 and 4). For these elements coordination is weaker than for mercury and their aquatic environmental mobility is greater. For inorganic mercury(II) the formation constants in Table 1.11 are generally slightly larger than those of the corresponding methylmercury complexes. This *difference* increases as the softness of the ligand increases, showing the hardening effect of methyl coordination here.

The unexpected reduction of formation constants shown in Table 1.11 between inorganic mercury(II) and organic mercury(II) has been ascribed to an antisymbiotic effect in which the soft methyl or other alkyl group decreases the affinity of mercury for other soft bases. Other authors have suggested that the effect might mainly be a consequence of the measurement methods originally used.[80] However, Table 1.11 clearly shows the general coordinating preferences of both inorganic- and methylmercury for soft base ligands. Similar trends will apply to other metals depending on their individual hard–soft properties.

In the next section the coordination and other properties of organotin complexes are discussed in detail as a model showing how these determine the analytical and stability properties of typical organometallic species under environmental conditions. Although organotins are used to illustrate the controlling factors, similar arguments apply to other organometallics found in the environment.

1.6   Some basic chemical factors controlling the stability and analysis of organotins, and their relevance to other environmental organometallics

1.6.1   Kinetic and thermodynamic stability of diagnostic tin analytes

The basis for both development of reliable means for molecular characterization and quantitation (i.e. chemical speciation), and a consequent picture of environmental transport and lifetimes, depend on the longevity of both covalent and ionic bonds between tin and certain key ligands. Also important are the oxidation state of the metal, and stability of certain stereochemistries or conformations in rate-determining abiotic and biological transport or co-ordination processes.

The subjects of analysis may include both equilibrium distributions of organotin species in environmental media, as well as identification of just one or several organotin species that may or may not be at equilibrium. A knowledge of more than the species and concentration distributions of tin-containing moieties at or near equilibrium (which require prediction by stability constants[81]), will be required for analysis of transient or unexpected molecules. The available instrumental 'kinetic window' will dictate the utility of any analytical process, but also the rate at which organotin species might redistribute themselves during the course of sampling, preconcentration or derivatization, separation, and detection will vitally affect the validity of results. In this section the term 'kinetic speciation' will be used where appropriate to ensure the time-dependent basis for any measurement is made clear.

The basic chemical and biological information content of an environmental organotin species, though it may be fluxional or redistributive as noted, can be summarized to include:

1. The basic stereochemistry about the tin centre(s).
2. The overall molecular topology,[82] e.g., surface area or molar volume, based upon the number, size, and likely conformation of covalently bound organic ligands.
3. The size and conformation of semicovalently or ionically bound 'labile' ligands, e.g. acetate, chloride, carbonate, sulphydryl, etc. sites, including features of polydentation and local charge.
4. The extent of inner- and outer-sphere solvent interactions, particularly water as the major environmental solvent, and localized surface or phase boundary effects.[83]

Armed with this information, the likely lifetimes for industrial and biogenic organotins in environmental media can now be assessed in terms of those tin−ligand bonds which are most likely to be available and significant in the environment and for analysis.

For example equation 1.33 describes a common aquatic solvation (by H_2O) and recomplexation process (by Y^-) for a triorganotin salt released, perhaps from a marine antifouling paint, into seawater.

$$R_3SnX + H_2O = (R_3Sn^+) + X^- \xrightarrow{\quad H_2O,Y^- \quad} R_3SnY(H_2O) + X^- \quad [1.33]$$

In Fig. 1.6 two stereorepresentations of one of the most common commercial organotin biocidal moieties, tri-n-butyltin, are compared. These were generated by a computer programme PROPHET from nominal bonding parameters,[84] and show that the local distorted tetrahedral (C_3v) stereochemistry around the four-coordinate starting compound, on solution is transformed to a more symmetric trigonal bipyramid (D_3h) chloride-monohydrate as a principal equilibrium species at very low concentrations in seawater.[85] Interestingly, the effective total surface areas (*TSA*) calculated for these two species, shows the allegedly more 'open' pentacoordinate derivative is the smaller molecule, thereby achieving the contracted form energetically more favoured in water[86] (see later).

Based upon reaction kinetics studies conducted by Brinckman *et al.* and others, it seems likely that rates for redistribution between these two entities, and other equilibrands incorporating different anions and more water, are fast. Eigen and Eyring estimated halide and similar anionic exchange rates for aquated metal and organometal ions proceed at rates of 10^6 mol^{-1} s^{-1},[87] whereas bimolecular reactions resulting in cleavage of tin−carbon bonds were found for two related trimethyltin moieties, $(CH_3)_3$ SnClaq and $(CH_3)_3$Sn^+aq, to range between 7.8×10^{-3} and 130×10^{-3} mol^{-1} s^{-1}, respectively.[81] For bulkier alkyl groups in more frequent industrial use, such as the tributyltin species illustrated in Fig. 1.6 electrophilic S_E2 scission of tin alkyl (Sn—C) is expected to be at least 10^3 times slower than that of tin methyl, based on preliminary NMR experiments.[88] (This bimolecular pathway to cleavage of the tin-carbon bond is not, however, the main environmental route to decomposition of these compounds.)

Numerous environmental agents, both chemical or photochemical and biotic, can also cleave tin−carbon bonds. At concentrations below the 10^{-8} mol dm^{-3} often found for organotins in natural media, even ubiquitous naturally occurring hydrogen peroxide or other oxidants[89] will also be at such low concentrations that significant losses of tin−carbon bonding information by redox cleavages occurring even at bimolecular or pseudo-first-order rate constants $>10^6$ times greater than typical measured aqueous electrophilic substitutions by many metal electrophiles, would require several days or more to achieve half-reaction. This same line of reasoning explains the recent discovery of methylstannanes, $(CH_3)_nSnH_{4-n}$ in polluted estuarine surface

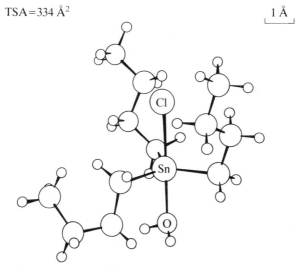

TSA = 343 Å²

Tri-n-butyltin chloride

TSA = 334 Å²

1 Å

Tri-n-butyltin chloride hydrate

Fig. 1.6   Structures for tributyltin chloride species

Computer-generated stereorepresentations of tetracoordinate (C_{3v}) tri-n-butyltin chloride and pentacoordinate (D_{3h}) tri-n-butyltin chloride hydrate illustrate the different degrees of conformational packing expected for these two most likely neutral tin species in seawater at nmol dm^{-3} concentrations.

waters (see Ch. 3), where reactive tin-hydrogen (Sn—H) bonds appear to survive environmental degradation for lengthy periods,[90] as is also seen with biogenic methylarsines, $(CH_3)_nAsH_{3-n}$, released from soils (Ch. 5), despite these species being transitory in the bulk state.[91]

It can be seen that the information content provided by metal–carbon skeletal frameworks may represent useful kinetic speciation tools for both qualitative and quantitative studies of the environmental fate and effects of organometallics. Morever, some rather reactive organometallics, well-known unstable compounds in the laboratory, may still be found to survive at trace concentrations in aggressive natural environments. Normal bulk stability requirements for organometallics (e.g. stability to water, oxidizing agents, air), apply to a lesser extent at the trace levels at which they occur in the natural environment, i.e. unstable organometallics may have ambient but low concentrations in the natural environment.

1.6.2   Isolation, preconcentration and separation of volatile tin analytes

Analysis has long relied upon derivatization schemes that involve placement of new functional groups on existing organometal ligands, stoichiometric displacement of existing ligands on the metal by selected probe ligands, or, most commonly, controlled addition of powerful chelate or polydentate ligands offering stable adducts of known composition and ready measurement. To a great extent kinetic speciation of environmental organometallics at ultratrace ($< ng\ cm^{-3}$) concentrations in complex environmental matrices requires controlled ligation chemistry to achieve desirable separation efficiencies and unequivocal determinations. Almost all modern organometallic speciation methods depend on either a sequence of chromatographic or solvent, i.e. extraction separation and/or concentration steps as outlined in Fig. 1.7. Thus chemical derivatization seeks to increase and stabilize favourable partition coefficients ($P = K_{phase\ 1/phase\ 2}$) selective for each analyte and its matrix, while generating coordinately saturated $R_nR^1_{4-n}Sn$ (R = H or alkyl) or R_nSnL_{4-n} (L = ligand) species of greater volatility and/or uniqueness for specific detectors capable of distinguishing elements or certain functional groups as indicated in Fig. 1.7. All of the methods suggested in Fig. 1.7 are applicable to both gas and liquid chromatographic separation steps, but the volatile analytes are better suited to the first technique.

As an example, Table 1.12 compares a number of recently reported methods that embrace the concepts for kinetic speciation of environmental organotins. In general, they appear valid in so far as retention of the organotin species' original carbon–tin skeletal information is concerned. However, treatment of complex environmental samples with large excesses of strongly basic hydride and carbanion derivatizers may induce misleading redistributive side-reactions, such as those shown to occur during typical gas chromatography (GC) thermal cycles.[92] Morever, even though recoveries of

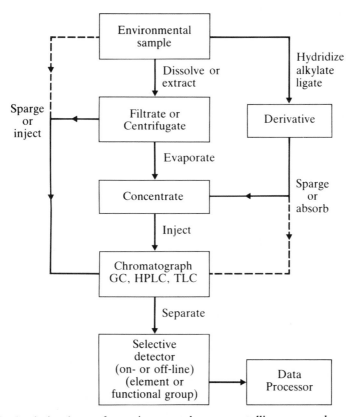

Fig. 1.7   Analysis schemes for environmental organometallic compounds

>90 per cent for the speciated environmental organotins are frequently attained by the methods in Table 1.12, most if not all schemes suffer from the requirement for individual calibration curves for each organotin analyte.

Reasons for this problem reside in both disparate recovery efficiencies in the overall analytical scheme for each organotin species and the wide range of detector sensitivities for individual molecular forms.[90,93,94]

Analogous problems have been reported[93] for the quantitation and speciation of inorganic tin(II) and tin(IV) in aquatic media, where use of hydride generation was expected to allow determination of tin(IV) after separation as stannane in accord with equation 1.34 below.

$$Sn^{4+}aq + 4\,H^-aq = SnH_{4\,gas} \qquad\qquad [1.34]$$

As a practical matter, *both* aquated stannous and stannic species yielded gaseous stannane upon borohyride treatment as described by Schaeffer and Emilius. The reaction was later proposed by Jolly and Drake[95] to proceed by the route shown in equation 1.35 below for tin(IV).

$$4\,HSnO_2^- + 7\,H^+ + 3\,BH_4^- + H_2O = 4\,SnH_{4\,gas} + 3\,H_3BO_3 \qquad\qquad [1.35]$$

Table 1.12 Some speciation methods in current use for organotins

(Substrate(s))* Derivatization, preconcentration	Separation scheme	Detector(s) employed	Tin species: Method, detn. limits, $\mu g\ cm^{-3}$	References
Direct solvent extraction				
(B) benzene, reflux	none	NMR, XRF, GFAA	$(C_8H_{17})_2Sn$; 40	143
(T) HCl + hexane	GC	H-FID	R_4Sn (R = C_2H_5, C_3H_7, C_4H_9; 100	144
(T) R_4NOH + toluene	none	GFAA	total organic, inorg. Sn	153
(M) $CH_3OH + HCl$	CH_3OH/hexane on alumina	GFAA	$(C_4H_9)_nSn$, n = 2, 3; < 0.2	154
Direct hydridization				
(M) sparge w/ or w/o NaBH$_4$ on Tenax-GC	GC	FPD	$(CH_3)_nSn$, n = 0–4; 0.013–0.053 $(C_4H_9)_nSn$, n = 1, 2; 0.018–0.037, C_6H_5Sn; 0.023	145
(T) direct sparge w/ homogenate	GC	MS	$(CH_3)_nSn$, n = 1–4; > 1 ng g^{-1}	152

Extraction with hydridization				
(W) CHCl$_2$ extr, LiAlH$_4$/hexane	GC	EC, FID	(C$_6$H$_5$)$_n$Sn, n = 1–4; 3–15	146
Extraction with aklylation				
(W) benzene/tropolone extr + C$_4$H$_9$MgBr	GC	FAA	(CH$_3$)$_n$Sn, n = 0–3; 0.04 w5dm^{-3} water	147
(W) benzene/tropolone extr + C$_5$H$_{11}$MgBr	GC	FPD	(C$_4$H$_9$)$_n$Sn, n = 0–3; 0.060–0.150	148
Complexometric/spectrophotometric				
(W) 3, 4-dithiol satd polyurethane	LC w/NaOH + acetone	UV at 650 nm	Sn(IV); 0.04	149
(M) silica gel w/CH$_3$OH-HOAc	TLC w/hexane-HOAc	GFAA of spots	R$_n$Sn, n = 1–3; air, 0.1 μgm^{-3} waters, 100 μgdm^{-3} soils, 1 μgg^{-1}	150
Miscellaneous				
(W) direct injection	cation exchange HPLC	GFAA	R$_n$Sn, n = 1, 2 R = alkyl, aryl; 25–150	151

* Matrix speciated: (B) = beverage; (W) = natural waters; (T) = tissues; (M) = mixed matrices, air, water, sediments, dusts.

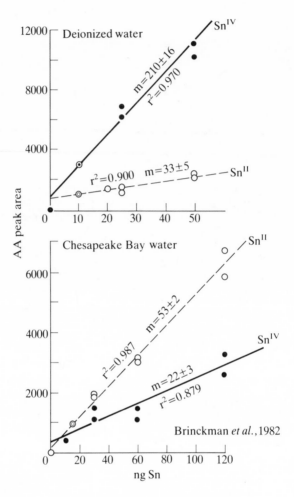

Fig. 1.8 Analysis of inorganic tin species in aqueous media

Calibration curves were obtained by purge and trap gas chromatography with a tin-selective flame photometric detector comparing tin(II) and tin(IV) in deionized Chesapeake Bay water. Equal concentrations of tin(II) and tin(IV) were present before hydridization.

In Fig. 1.8 are graphically summarized results for both deionized and estuarine waters spiked with Sn(II) or Sn(IV) and analysed by purge and trap sparging of samples following hydridization. The stannane produced was collected on a Tenax-GC absorbant preconcentrator and subsequently injected into a gas chromatograph fitted with a sensitive flame photometric detector selective for tin as SnH molecular emission at 610 nm.[94] All the experimental conditions of excess (10[6] molar) borohydride, treatment period, sparging rates, etc. were kept constant since the rate of hydridization is dependent upon both the number and types of organic groups on tin as well as oxidation state.[90,93] Chloride is especially troublesome in the inorganic tin analysis, since these results make it clear that even at trace concentrations an inversion in the method's sensitivity toward the respective two oxidation states of tin is caused by salinity.

Though presumed to predominate in its thermodynamically stable tin(IV) form in aerobic environments,[93,96] it cannot, be inferred from presently available evidence that tin(II) is unimportant in tin's biogeochemical cycle. As the involvement of stannous salts in oxidative methylation by ubiquitous naturally occurring iodomethane represents one possible route for remobilization of precipitated or 'mineralized' tin into overlaying waters,[97] a more accurate speciation method for environmental tin(II) is required. As an alternative to analysis by hydridization, alkylation of inorganic tin substrates to form tetraalkyl R_4Sn analytes suited for GC analysis has been proposed (Fig. 1.7 and Table 1.12). Treatment by Grignard reagents will mainly produce volatile tetraalkylstannanes only from tin(IV), but the potential contribution by disproportionation reactions of tin(II) alkyls, especially under anoxic conditions in saline media remains unclear for quantitative work.

It can be seen that the various derivatization schemes employing reactions drawn largely from conventional organometallic chemistry literature provide means for generating volatile and/or coordinatively saturated organometallic analytes suitable for chromatographic separations and element-selective quantitation. However, not all methods will faithfully preserve the essential information contained in the original organometallic skeleton or in the oxidation state of the metal as these are found in the environmental matrix, particularly for endocellular and macromolecular metal-containing molecules. In these cases the severity of the extraction processes may significantly change the character of the molecule.

1.6.3   Speciation of charged and neutral organotins by liquid chromatography

The solution characteristics of organometallics at interfaces and their relationship to molecular topology and bioactivity are also important in analytical and speciation work. For organotin compounds equation 1.33 defines a common environmental process where both charged and neutral species exist in equilibrium.[81] As further illustrated in Fig. 1.6, either charged or neutral organotin species exhibiting coordination numbers of 4 or 5, as well as 6 also can coexist in solutions.[81,98]

Some of these tin species may be either fragile monomeric or macromolecular species, which are highly susceptible to degradation by chemical or thermal alterations. The question arises whether or not such distinctive molecules can be kinetically speciated in ambient environmental media in ultratrace concentrations by using variations of Fig. 1.7 that do not require excessive rigours of derivatization and thermal volatilization. For example non-destructive spectrophotometric methods operating on very short timescales (10^{-13} s), such as laser Raman, are capable of distinguishing among the symmetry classes represented by nanomolar solutions of R_3Sn^+ [99] in water, but as yet these are insufficiently sensitive and selective for direct environmental work.

Another non-destructive approach depends upon equilibration between well-defined solutions and solid phases as provided by current high performance liquid chromatography (HPLC) methods. Since HPLC is conducted on a very long (minutes) timescale, it is subject to the limitations of bulk derivatization as described for gas chromatographic speciation,[99] that is, only the essential information contained in relatively long-lived tin−carbon frameworks is retrieved in the separation/concentration process. Nevertheless, because very well-defined reactions occur in the liquid−solid chromatographic process which achieve equilibrium under very mild and reversible conditions,[100] it is possible to collect other molecular parameters of profound importance to environmental questions without disruption of metal−carbon structures contained, e.g. even in high molecular weight organometal polymers.

By way of illustration Fig. 1.9 summarizes both homo- and heterogeneous equilibria that govern the liquid chromatographic process for either neutral or

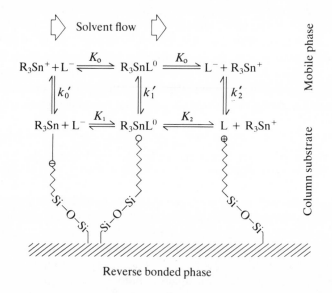

Fig. 1.9 Chromatographic separation of organotins

Liquid chromatographic separation by a combination of ion-exchange and solvophobic (lipophilic) processes is represented by the idealized triorganotic reactants in mobile phase solution and at the reverse bonded-phase column substrate interface, and their respective equilibria. The conventional homogeneous equilibrium constant k_0 defines the reversible dissociation−aquation reaction described in the text. The equilibrium constants k_{0-1} define heterogeneous reactions corresponding, respectively, to capacity factors for the free anionic ligand.

charged organotin species that coexist separately or in mutual reactions.[101] Not depicted in this scheme are the size-exclusion (SEC) or gel-permeation (GPC) chromatographic separation processes which primarily depend on molecular size (molecular weight)-dependent interactions with sized channels or pores in the solid chromatographic substrate,[102] though both ionic and partition (hydrophobic) interactions can participate to a lesser extent, depending on the gel composition and eluate.

Experimentally, from simple chromatographic measurements[103] obtained with pumps, columns, mobile phase(s), and detectors operating at thermal and flow equilibrium, the capacity factor, k' can be determined. This is a useful molecular characteristic[104] for each analyte separated (eluted) by a given HPLC system (see equation 1.36):

$$k_I' = \frac{(t_I - t_O)}{t_O} \qquad\qquad [1.36]$$

Here, t_O is the retention time for a non-retained molecule (equivalent to the column void volume) and t_I is the retention time (equivalent to the retention volume) for the Ith analyte peak.

Table 1.13 provides illustration of the power of the capacity factor in discerning the mechanism of column separation and the nature of ambient organotin species transported into and through the HPLC. Values of k' measured for fresh or aged aqueous and methanolic solutions of a series of tri-n-butyltin salts, (R_3Sn-X), are compared with those obtained for neutral covalent organotins, including bis(tri-n-butyltin) oxide.[101] In accord with the cation exchange mechanism for R_3Sn^+ (defined by equilibrium constants K_0, and k_0' in Fig. 1.9), the retentivity for tributyltin cation is essentially constant, irrespective of the original source counterion X. For the neutral organotins, $(C_4H_9)_3Sn-L)$, $(L = H, C_4H_9,$ or $OSn(C_4H_9)_3)$, only the first two exhibit sufficient longevity of the tin–ligand $(Sn-L)$ bond to undergo chromatographic separation by a non-ionic solvophobic process on the HPLC column in accord with the k_I' equilibrium depicted in Fig. 1.9. Table 1.13 also shows that $(C_4H_9)_3SnH$ uniquely undergoes slow hydrolysis as evidenced by appearance in aged solutions of *two* tributyltin peaks at appropriate retention times, the first eluate being the original neutral hydride (at $k' = 0.58$) and the second its cation hydrolysate (at $k' = 3.37$). For $[(C_4H_9)_3Sn]_2O$, on the other hand, solvolysis occurs rapidly by equation 1.37.[105]

$$[(C_4H_9)_3Sn]_2O + H_2O = 2(C_4H_9)_3SnOH \qquad\qquad [1.37]$$

to yield a chromatogram (single peak at $k' = 3.22$) identical with those obtained for the $(C_4H_9)_3SnX$ salts.

In complex environmental and biological samples, many naturally occurring and pollutant ligands may strongly compete for exchange or solvolysis sites on the analytical HPLC column, or for the dissolved organometallic species, thereby rendering efficient and reproducible separations difficult. By

Table 1.13 Speciation of tributyltin in cations and neutral molecules from various R_3Sn-X sources by HPLC*

$-X$	k'†	Comments‡
Organotin salts		
$-OAc$	3.33 ± 0.05	fresh or saline solutions
$-F$	3.46 ± 0.10	fresh or saline solutions
$-Cl$	3.36 ± 0.04	fresh or saline solutions
$-Br$	3.42 ± 0.11	fresh or saline solutions
$-SO_4(-Sn(n-C_4H_9)_3$	3.15 ± 0.04	fresh or saline solutions
Neutral covalent organotins		
$-O-Sn(n-C_4H_9)_3$	3.22 ± 0.24	fresh or aged aqueous or CH_3OH
$-H$	3.37 ± 0.13	fresh or aged aqueous or CH_3OH
	0.58 ± 0.10	undissociated $(C_4H_9)_3SnH$ peak
$-n-C_4H_9$	0.45 ± 0.08	undissociated $(C_4H_9)_4Sn$ peak
Mean $k' =$	3.33 ± 0.11	excluding neutral peaks
RSD $=$	3.3 per cent	

* Reverse bonded-phase strong cation exchange column run isocratically with 0.06 mol dm^{-3} NH_4OAc in water CH_3OH mobile phase

† Replicate runs $\pm$ average deviation

‡ Solutions containing $1-2$ μg g^{-1} as tin

use of either preliminary sample clean-up procedures or pre- or post-column derivatization with a strong chelator possessing enhanced detection properties, greatly improved HPLC conditions can be achieved. One general method for metal ions uses formation of highly stable β-diketonates followed by their high resolution separation on a reverse bonded-phase (solvophobic) HPLC column.[106] Variations of this approach using traditional polydentate ligands and crown ether cryptands appear promising for organotins and other organometallics.[107] Another recent approach using post-column derivatization with a fluorgenic chelate ('Morin' $= 2'$, 3, $4'$, 5, 7-pentahydroxyflavone) permits highly sensitive and selective fluorescence detection of dialkyltins extracted from tissues.[108]

1.6.4 Correlations between organotin substituent parameters and chromatographic separation

It has long been recognized that the logarithm of the capacity factor, $\ln k'$, is proportional to the free energy change associated with the chromatographic partitioning process, whether this occurs by solvophobic or charged interactions.[109,110] By either mechanism, as defined by the heterogeneous equilibria depicted in Fig. 1.9, it is reasonable to expect that the relative positions of k'_{0-2} for homologous series of molecules will reflect regular changes in the number and kind of substituents bound to the parent molecule's core structure. In organometallic systems, for example, the core structure might be represented by a single central metal atom or a polynuclear metal cluster. Whether or not simple additive relationships exist strongly depends, of course, on the degree of linearity in the sum of steric, electronic, and hydrophobic effects produced by changing substituents in a given solvent.

Taft–Hammett substituent parameters have long been used to rationalize chemical reactivity on the basis of the relative contributions of electronic effects of inductive (σ_I) or π resonance (σ_π) factors which, in turn, depend upon bond hybridization, electronegativities, and steric features.[104,111] Hydrophobic effects generally underly transport properties of both organic and organometallic molecules in environmental and biological systems, where the interaction between a non-polar lipophilic surface and the solute commonly involves water. Mackay and Paterson have employed[112] the physico-chemical term fugacity ('... a thermodynamic quantity related to chemical potential or activity that characterizes the escaping tendency from a phase ...') to describe distribution between phases. A traditional experimental basis for such fugacity capacities is derived from the logarithm of the measured n-octanol–water partition coefficient for a given molecule, usually termed log P, or the hydrophobic structural substituent parameter.[113]

Quantitative structure–activity relationships (QSAR) embodying either electronic, steric, or hydrophobic factors have been extensively developed for a host of organic compounds, and in recent years widely applied to thin layer chromatography (TLC)[114] and HPLC.[104,115] These relationships adopt the basic form given in equation 1.38.

$$\ln k' = m(\text{QSAR}) + \text{constant} \qquad [1.38]$$

where the slope (m) and intercept (constant) of the linear free energy correlation are characteristic of the class of homologous molecules involved and the specific measurement system.

Brinckman *et al.* have demonstrated an application of these concepts to organometals, including both cationic organotin species $R_nSn^{(4-n)+}$ ($n = 2,3$)[101], as well as neutral tetraorganotin molecules.[116] With cation exchange HPLC, a Hammett-type QSAR term σ^ϕ, derived from pK_A values for ionization of a large variety of organophosphonic acids in water,[117] was found to provide the most accurate predictions of k' for both alkyltin and aryltin cationic species in accord with equation 1.38.

However, these two classes of organic tin substituents were not inter-correlated, nor was additivity observed for $n = 2,3$ between the R_2Sn^{2+} and R_3Sn^+ alkyl series (R = CH_3, C_2H_5, n–C_3H_7, n–C_4H_9, etc.). These results make it clear that both inductive and resonance substituent effects contribute non-linearly in this ion exchange chromatographic process (k'_0 in Fig. 1.9) for these particular organotin cations, though steric factors relating to molecular shape or conformation, as discussed below, also may be involved.

In contrast, for the solvophobic separation of neutral tetraorganotins by reverse bonded-phase HPLC (k_1' in Fig. 1.9), this group demonstrated that, by again simply using the tetracoordinate tin atom as the parent 'core structure', completely additive rules are followed (equation 1.39).

$$\ln k_1' = m\,(\text{sum }\pi) + \text{constant} \qquad\qquad [1.39]$$

The π values used for each R (= alkyl) or R' (= aryl) group were taken from the literature[113] and appropriately added for each $R_nSnR'_{4-n}$ molecule chromatographed. As before, with cation exchange HPLC, organotins containing aryl substituents give correlations different from the single one obtained for symmetric and unsymmetric tetraalkyltins combined, though all are linear in accord with equation 1.39. Since the solvophobic separation mechanism chiefly depends upon eluate solubility in the mobile and reverse bonded phase,[104,115] the question of molecular geometry also controlling k_1' is raised. In the next section organotin compounds are used in a study of these stereochemical selectivity phenomena, which bear strongly on predicting both chromatographic retention and biological activity from molecular topology.

To summarize, HPLC offers a precise and economical means for both separating and quantifying organotins and other organometallics at trace concentrations in complex environmental media. Use of sensitive detectors directly coupled with HPLC allows speciation at sub-μg g^{-1} concentrations of tin-containing monomers and polymers, irrespective of fixed or labile substituents and charge, in complex environmental media where conventional detectors are unusable.[118]

Because HPLC analyses can be performed at equilibrium conditions, direct measurements of critical free-energy parameters related to individual or summed substituent effects are possible, and, consequently, chromatographic retention properties of new or unknown organometallic eluates can be realistically evaluated.

1.6.5   Correlation of quantitative structure–activity relationships (QSAR) with organotin molecular topology and bioactivity

In the previous sections it has been shown that liquid chromatography provides an elegant basis for both theoretical and practical assessments of fragmental or additive substituent contributions to molecular solution behaviour, including organotins as trace cations or neutral species. Introduction of a more complete approach, depending on each molecule's entire geometrical or electronic constitution, is even more attractive[119] because it has shown much utility in forecasting biological response in rational organic drug design,[120] and because the rapid advances in molecular structure determinations of organometals[121] provide an invaluable base for estimating their bioactivity.

The latter prospect involves applications of molecular topology (MT), augmented by computer calculations and imaging.[122] Among the leading topology approaches currently under study are the following:

1. Molecular connectivity – a bond-centred summation of electronegativity and electron density that relates to molecular volume and valence states.[123]
2. Branching index – a perturbational analysis of central and adjacent atoms that relates to molecular refraction and volume.[124]
3. Total surface area (TSA) – obtained by summing all Van der Waals' radii for specified atom positions (from bond lengths and distances)[121] in any conformation while minimizing occluded surfaces. It relates to molecular volume and the hydrophobic 'solvent cavity' or 'exclusion volume' in a polar solvent.[122,124]

All of these topological methods give remarkable correlations with those essential physico-chemical properties of organic hydrophobic molecules in water that influence biological activity, such as solubility,[82,124,125] partition coefficient,[126] or solvent polarity.[127] Morever, simple molecular size or shape estimates have proven accurate in predicting chromatographic retention for homologous series of toxic polycyclic aromatics,[128] just as have refined connectivity indices correlated with chromatographic retention for highly polar aliphatic compounds[129] or barbiturates.[130] For all those properties reported in common for the three MT methods listed, correlation matrices indicate good agreement.

In keeping with the basic implications for cellular uptake or depuration of dilute solute molecules by organisms, the bioconcentration factor (K_β) in aquatic organisms has been explained[131] in terms of the octanol–water partition coefficient (log P), or aqueous solubility (S). Attainment of thermodynamic equilibrium is presumed, and such approaches will be rendered invalid if the xenobiotic molecule is significantly degraded by any of the chemical or biological side-reactions prevalent in environmental media. None the less, in practice, bioaccumulation in fish of a wide range of organic compounds, including aliphatic and aromatic hydrocarbons, halides, acids and phosphate pesticides has been linearly correlated[132] for solubility (S) or octanol–water partition coefficient (log P) over many orders of magnitude by the expressions given in equations 1.40 and 1.41.

$$\log K_\beta = m (\log S) + \text{constant} \qquad\qquad [1.40]$$

$$\log K_\beta = m (\log P) + \text{constant} \qquad\qquad [1.41]$$

On the basis of these predictions, it cannot be stated with certainty what may be the specific uptake sites or route of entry into the test organism. The observed strong dependence on log P implies lipophilic transmission, though at sufficiently large molecular sizes, membrane permeation resistance is expected.[84,131,133]

Consistent with this picture, Wong *et al.* showed that a direct relationship exists between the organic substituent (R), partition coefficient and the toxicity of triorganotins towards algae.[134] Prior to the work of Brinckman *et al.*, none of the above MT correlations had been applied to organometallic systems for predicting QSAR, chromatographic retention, or toxicity.[135]

1.6.6   Application of molecular topology to predictions of organotin toxicities

1.6.6.1   Additivity of total surface area (TSA) for aqueous solubility and chromatographic parameters

The utility of TSA has been used by Brinckman *et al.* because of its simplicity and relevance to stereochemical constructs of mono- and polynuclear core structures and diverse organic substituents common to organometallic molecules. Morever, a broad data base of good quality bonding distances, angles and Van der Waals' radii are readily available, and in the organometallic field principally for organotin compounds. In addition to complete TSA calculations for individual organotin molecules in various degrees of coordination or charge and likely conformations (cf. Fig. 1.6),[84] Brinckman *et al.*, have assembled (Table 1.14) mean TSA values for the organic groups (R) and labile inorganic ligands of primary interest to industrial and environmental organometallic chemists.[135] Since these TSA are derived from the full tetra- or pentacoordinated triorganotin(IV) calculations, their use as additive substituent TSA along with a tin core structure TSA value appears potentially subject to error introduced by conformational changes required in new organotin structures. In spite of this limitation, suprisingly good agreement with complete structure computer TSA calculations is obtained, as was also shown for organic molecules.[82]

Recent results show that TSA for a series of methyl elements, $(CH_3)_nE$ (E = Br, I, Hg, C, Sb, and Sn) correlate linearly with aqueous solubilities, in the form of equation 1.40.[136] In Fig. 1.10 are summarized additional solubility data for tetraalkyl derivatives of Group IVA, drawn from the literature.[136,137] Over a solubility range of 10^5, a good linear fit was obtained with summation of the appropriate TSA values (Table 1.14).

Brinckman *et al.*, have also applied TSA values from Table 1.12 to an estimation of the chromatographic retentivity of two series of mixed tetraorganotins (Fig. 1.11). Two excellent linear correlations in the form of equation 1.38 were obtained, but again, as with equation 1.39, two distinct patterns of solvophobic behaviour are observed with the conformationally flexible alkyl groups exhibiting far greater retentivity in the C_{18} reverse bonded-phase column than the rigid, planar phenyl subsituents.[128] Unfortunately, insufficient toxicity data are yet available to test fully the hypothesis that partial rigidity in a mixed alkylaryltin moiety will differentiate its uptake in cellular lipids from a fully flexible alkyltin moiety of comparable TSA. This idea merits further study because design[100,101] of mixed reverse bonded-phase HPLC column substrates with mixed siloxo-bonded organic groups of

Table 1.14 Additive group/atom TSA values*

	TSA (Å^2)	Standard deviation
Organic group		
CH_{2^-}	20.7	1.5
CH_3	32.7	0.02
C_2H_5	55.4	0.01
$n-C_3H_7$	74.4	1.6
$i-C_3H_7$	75.6	2.2
$n-C_4H_9$	95.4	1.8
$i-C_4H_9$	95.6	1.9
$c-C_6H_{11}$	117.9	2.2
C_6H_5	92.3	0.1
Inorganic group		
Sn	17.7	0.9
Cl	27.6	2.3
CO_3	45.4	3.3
OH	20.2	0.05
OH_2	23.4	2.2

* Mean TSA derived from both 4- and 5-coordinate R_3Sn structures (cf. Fig. 1.6)

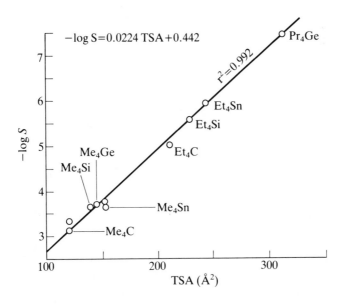

Fig. 1.10 Solubility data for tetraalkyl derivatives of Group IVA

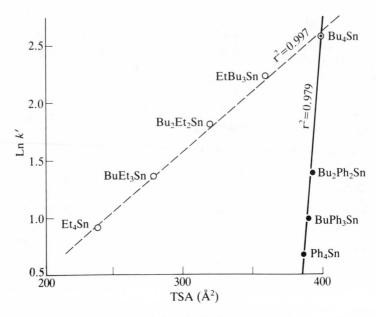

Fig. 1.11   Chromatographic retentivity of mixed tetraorganotins

Very high linear correlations are separately obtained between the experimentally determined *capacity factor*, as in ln k' (see text), and TSA calculated for the two classes of tetraalkyltin and arylalkyltin compounds shown.

different controlled degrees of stererorigidity may offer even better models of cellular behaviour with organotins and other organometals.

1.6.6.2   Molecular topology prediction of organotin toxicities in aquatic organisms to illustrate the use of MT in toxicity prediction

Several groups of mud crab larvae, *Rhithropanopeus harrisii*, were exposed in a study for the duration of zoeal development (10–12 days) to seawater containing spikes of eight structurally distinct triorganotin moieties derived from dissolution of their respective R_3Sn—X parent compounds (X = Br, OH, or O—SnR_3 where R = n-C_3H_7, i-C_3H_7, n-C_4H_9, i-C_4H_9). A similar study was conducted for the analogous R_2Sn^{2+} series.[135] Their survival was quantitated as LC_{50}, expressed in nmol dm^{-3}.[84] The foremost conclusions derived from correlation of the LC_{50} values with TSA values computed for each organotin toxicant and summarized for R_3Sn^+ in Fig. 1.12 were:

1. Highly linear relationships existed between LC_{50} and individually calculated TSA[84] for either the tetra- or pentacoordinate triorganotin solvates as exemplified by Fig. 1.6 or the chloroaquates of $R_2SnCl(H_2O)^+$.
2. Only the mononuclear di- or triorganotin moiety appeared operative as the rate-determining toxic agent, even though half the triorganotins examined were spiked as bis(trialkyltin) oxides whose TSA are ⩾ 500 A^2.

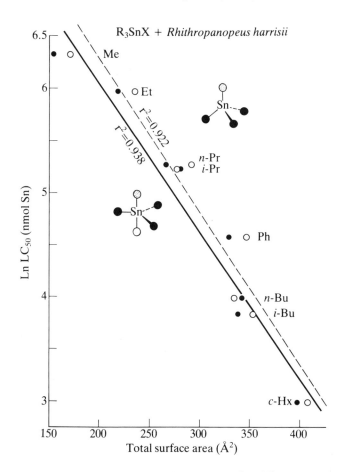

Fig. 1.12 LC_{50} values and TSA for *Rhithropanopeus harrisii*

3. Slightly better fit was obtained for the neutral triorganotin penta-coordinate hydrate forms over the tetracoordinate chlorides (cf. Fig. 1.6), although not significant at the 95 per cent confidence level.

This approach was also applied by Brinckman *et al.*, using approximate additive TSA values provided in Table 1.12, to the study of Wong *et al.*[134] who derived toxicity data for freshwater algal species from Lake Ontario. In Fig. 1.13 are derived regression analyses for the five triorganotin moieties examined by Wong *et al.* (along with a number of inorganic tin agents). As before, excellent correlations were obtained for neutral four- and five-coordinated triorganotin species only, with the latter slightly better at the 95 per cent confidence level. In freshwater, it is unlikely that chlorotin species form, but from inspection of Table 1.12, the distinction between —OH, —Cl, or H_2O probably affects the quality of the linear fit far less that the tin–carbon (Sn—C) bond angles and concomitant number of labile inorganic groups in tin's inner

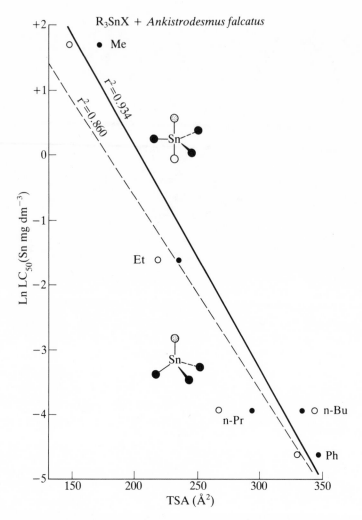

Fig. 1.13 LC$_{50}$ values and TSA for *Ankistrodesmus falcatus*. From Wong *et al.*, 1982 – see Ref. 134.

coordination sphere. In a similar evaluation,[135] one case that reported the toxicity of a trialkyltin series towards spheroplasts obtained from *Escherichia coli*[138] was considered, and again it was found that TSA (from Table 1.12) correlates with the extent of bacterial organotin ($R_3S_n^+$) uptake.

These findings lend strong support to the idea that the utility in kinetic speciation of bioactive organotins lies mainly with those analytical methods which faithfully determine the original information content encoded[139] in the basic Sn—C skeleton.

For all of these applications of TSA to aquatic toxicity, it also appears evident that use of conventional sum π or Taft–Hammett σ is relatively ineffec-

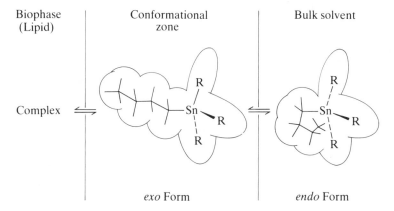

Fig. 1.14 Transfer of species between biophase and bulk solvent

Effective TSA is maximally exerted during solvolysis. When R above = n-C₄H₉, only this part of the molecule exhibits fluxional increase in TSA during membrane solvolysis. The remainder of the molecule is rigid and participates to a much lesser extent.

tive as a structural tool since neither of these can provide recognition of either geometric or conformation details that probably greatly influence the actual mode of transfer from bulk polar hydrophobic solvent through a reorganizational zone then into the membrane complex form prior to transmission into the cell. Figure 1.14 shows such a process where only one flexibile n-alkyl group undergoes transition (with concomitant increase in TSA) while three stereochemically rigid R groups remain unaffected during the uptake process.

Finally it bears noting that not all uptake processes on cells result in transmission of the organotin toxicant. The mentioned study on *E. coli* uptake is relevant. It was also shown that when tri-n-butyltin cation was rapidly adsorbed on to outer membranes of Gram-negative heterotrophic bacteria isolated from the Chesapeake Bay, essential metabolic process continued unabated. Employing HPLC–GFAA[101] it was also shown that the bulk of the $(C_4H_9)_3Sn^+(aq)$ could be recovered unchanged by washing live cells with methanol. Such passive resistance to toxic metal-containing species by metal sorption is well known and has great environmental importance in metal biogeochemical cycles.[140,141] It now remains to be seen whether such non-transmission membrance uptake can also be predicted by the new MT method as was done for *E. coli*, and whether some special geometrical features of the tin centre or conformation of certain organic groups on tin regulate the path and rates of these processes. Ultimately this line of investigation must focus on the modality and prediction of bacterial tin or other metal uptake, retention, and biomethylation, or metal–carbon bond degradation, as well as microbial resistance,[142] all of which underlie most of present-day use and concern for environmental effects of organotins and other organometallic compounds which enter the natural environment.

1.7   Acknowledgements

For this chapter PJC was primary author of sections 1.1–1.5 and FEB primary author of section 1.6.

Part of this chapter was presented in different form (by FEB) at the 4th International Conference on the Organometallic and Coordination Chemistry of Germanium, Tin and Lead, Montreal, Canada, August 1983.

The field of environmental organometallic chemistry is an interdisciplinary one, and FEB is greatly indebted to chemical and biological colleagues who have given much of their time and insights in preparing this review of published work and new results from the NBS Laboratory. In particular FEB thanks Drs Greg Olson, Rolf Johannesen, John Thayer, Roy Laughlin, Hal Guard and Peter Smith who share in much of the latest work on molecular topology predictions of toxicity. Mr Eric Harrell was an outstanding summer student (1983) who ably assisted FEB in assembling additive TSA values, and FEB thanks him.

The biodynamics of environmental organotins has long occupied the NBS Group's attention. They owe a great deal to the sustained and valuable guidance and financial support of several divisions within the US Office of Naval Research which has permitted them to maintain that level of interdisciplinary focus necessary for quantiative work in the field.

PJC also thanks SERC and NERC, United Kingdom, for valued research support over the past years; also NBS Washington, DC; the Royal Society, London; the British Council, London; and Leicester Polytechnic for other support. He also thanks several excellent graduate students including Drs Stephen Morton, Paul Bartlett, Peter Moreton and Spyridon Rapsomanikis.

References

1. Ridley W P, Dizikes L J, Wood J M 1977 *Science* **197**: 329–32
2. Fanchiang Y-T, Ridley W P, Wood J M 1978. In *Organometals and Organometalloids: Occurrence and Fate in the Environment* (eds Brinckman F E, Bellama J M) ACS Symp Ser No 82, pp 59–65
3. Wood J M 1975 *Naturwiss* **62**: 357–64
4. Wood J M, Fanchiang Y-T, 1979. In *Proc Third Euro Symp Vit B_{12} and Intrinsic Factor* Zurich (ed Zagalak B). de Gruyter W F, Berlin and New York, Ch 3, p 539–56
5. Craig P J 1980 Metal cycles and biological methylation. In *Handbook Env Chem* Vol 1, Part A (ed Hutzinger O). Springer Verlag, Berlin; Heidelberg, pp 169–227
6. Craig P J 1983 Organometallic compounds in the environment. In

Pollution: Causes, Effects and Control (ed Harrison R M). Royal Society of Chemistry, London, pp 277–322

7. Craig P J 1982 Environmental aspects of organometallic chemistry. In *Comprehensive Organometallic Chemistry* (eds Abel E W, Stone F G A, Wilkinson G). Pergamon, Oxford, Vol 2, pp 979–1020
8. De Simone R E, Penley M W, Charbonneau L, Smith S G, Wood J M, Hill H A O, Pratt J M, Ridsdale S, Williams R J P 1973 *Biochim Biophys Acta* **304**: 851–63
9. Craig P J, Morton S F 1978 *J Organometallic Chem* **145**: 79–89
10. Bertilsson L, Neujahr H Y 1971 *Biochemistry* **10**: 2805–8
11. Fanchiang Y-T, Wood J M 1981 *J Amer Chem Soc* **103**: 5100–3
12. Cotton F A, Wilkinson G 1980 *Advanced Inorganic Chemistry* (4th edn). Wiley Interscience
13. Pfeiffer P, Heller I 1904 *Chem Ber* **37**: 4619–25
14. Challenger F 1978. In *Organometals and Organometalloids, Occurrence and Fate in the Environment* (eds Brinckman F E, Bellama J M). ACS Symp Ser No 82, p 6
15. Espenson J H, Sellars T D Jr 1974 *J Amer Chem Soc* **96**: 94–7
16. Wood J M 1975, Metabolic cycles for toxic elements in the environment. In *Heavy Metals in the Aquatic Environment* (ed Krenkel P A). Pergamon Press, Oxford, pp 105–12
17. Challenger F 1978 Biosynthesis of organometallic and organometalloidal compounds. In *Organometals and Organometalloids. Occurrence and Fate in the Environment* (eds Brinckman F E, Bellama J M). ACS Symp Ser No 82, Washington DC, pp 1–23
18. Chobert G, Devaud M 1978 *J Organometallic Chem* **153**: C23
19. Cotton F A, Wilkinson G 1980 *Advanced Inorganic Chemistry* (4th edn). Wiley Interscience, pp 1138–40
20. Kochi J K 1978 *Organometallic Mechanisms and Catalysis*. Academic Press, New York, pp 230–2
21. Jarvie A W P, Markall R N, Potter H R 1981 *Environ Res* **25**: 241–9
22. Ref 20 pp 285–8
23. Wardell J L 1982. In Ref 7 vol 2, pp 681–707
24. Bottrill M, Gavens P D, Kelland J W, McMeeking J 1982. In Ref 7 pp 459–64
25. Coates K, Green M L H, Wade K 1967 *Organometallic Compounds* (3rd edn) Vol 2. Methuen, London, pp 220–4
26. Aylett B J 1979. In *Organometallic Compounds* (4th edn) vol 1 (eds Coates G E, Aylett B J, Green M L H, Mingos D M P, Wade K). Chapman and Hall, p 11
27. Balzani V, Carasiti V 1970. In *Photochemistry of Coordination Compounds*. Academic Press, New York, p 281
28. Blunden S J 1983 *J Organometallic Chem* **248**: 155–9
29. Gilroy S M, Price S J, Webster N J 1972 *Can J Chem* **50**: 2639

30. Lindqvist O, Jernelov A, Johansson K, Rodhe H 1984 Mercury in the Swedish environment *Nat Swed Environ Prot Board Rept* SNV pm 1816
31. Edwards S J, Leicester Polytechnic 1984 (Personal communication to PJC)
32. Coyle J D, Hill R R, Roberts D R (eds) 1982 *Light, Chemical Change and Life*. Open University Press
33. Harrison R M, Laxen D P 1978 *Environ Sci Technol* **12**: 1384−92
34. Aleksandrov Y A, Sheyanov N G, Shushunov V A 1969 *J Gen Chem USSR* (Eng trans) **39**: 957−60
35. Fox M F, Leicester Polytechnic 1984 (Personal communication to PJC)
36. Rowland F S 1982 Chlorofluorocarbons and stratospheric ozone. In Ref 32, pp 142−8
37. Phillips D 1982 Kinetics of reactions of stratospheric importance. In Ref 32 pp 139−41
38. Ref 20 pp 246−58
39. The Open University 1977 S304 Science. *The Nature of Chemistry, Units 22−24 Organometallic Chemistry*
40. Norton J R 1979 *Accounts Chem Res* **12**: 139−45
41. Schrock R R, Parshall G W 1976 *Chem Rev* **76**: 243−68
42. Davidson P J, Lappert M F, Pearce R 1976 *Chem Rev* **76**: 219−42
43. Koerner von Gustorf E A, Leenders L H G, Fischler I, Perutz R N 1976 *Adv Inorg Chem Radiochem* **19**: 65−185
44. Thayer 1984 *Organometallic Compounds and Living Organisms*. Academic Press, New York
45. Johnson D A 1982 *Some Thermodynamic Aspects of Inorganic Chemistry* (2nd edn). Cambridge University Press pp 195−207
46. Pilcher G, Skinner H A 1982 Thermochemistry of Organometallic compounds. In *The chemistry of the metal carbon bond*, Vol 1 (eds Hartley F R, Patai S). John Wiley and Sons, Chichester UK
47. Connor J A 1977. Thermochemical studies of organo transition metal carbonyls and related compounds. In *Topics Current Chem* (ed Boschke F L) **71**: 71−111
48. Summers A O, Silver S 1978 *Ann Rev Microbiol* **32**: 637−72
49. Silver S 1984 Life Sci Res Rept. In *Changing Metal Cycles and Human Health* (ed Nriagu J O) Springer Verlag, Berlin, Heidelberg, pp 199−223
50. Blunden S J 1983 *J Organometallic Chem* **248**: 149−60
51. Blunden S J, Chapman A H 1982 *Environ Tech Letters* **3**: 267−72
52. Harrison R M, Laxen D P H 1981 *Lead Pollution: Causes and Control*. Chapman and Hall, London and New York
53. Harrison R, Perry R 1977 *Atmos Environ* **11**: 847−52
54. Grandjean P, Nielsen T 1979 *Residue Rev* **72**: 97−148
55. Ref 20 pp 230−371
56. O'Neill M E, Wade K 1982 Structure and bonding relationships amongst main group organometallic compounds. In Ref 7 Vol 1: 1−43

57. Abel E W, Stone F G A , Wilkinson G (eds) 1982 *Comprehensive Organometallic Chemistry*. Pergamon, Oxford
58. Coates G E, Green M L H, Wade K 1967 *Organometallic Compounds* Vols 1–2. Methuen, London
59. Aylett B J 1979 *Organometallic Compounds* (vol I – *The Main Group Elements* – Part 2, Groups IV and V). Chapman and Hall, London
60. Aylett B J, Lappert M F, Pauson P L (eds) 1984 *Dictionary of Organometallic Compounds*. Chapman and Hall, London and New York
61. Twigg M V (ed) 1983 *Mechanisms of Inorganic and Organometallic Reactions*. Plenum, New York
62. Nriagu J O (ed) 1984 *Changing Metal Cycles and Human Health*. Springer Verlag, Berlin, Heidelberg, New York, Tokyo
63. Collman J P, Hegadus L S 1980 *Principles and Reactions of Organometallic Chemistry*. University Science Books, Mill Valley, Ca, USA
64. Kerr J A 1972–3 Strengths of chemical bonds. In *CRC Handbook of Chemistry and Physics* **64** F176–186
65. Pilcher G 1975 Thermochemistry of organometallic compounds containing metal–carbon linkages. In *MTP Intern Rev Science* Ser 2 *Phys Chem* (ed Skinner H A) Vol 10, Ch 2, pp 45–81
66. Barnes J M, Magos L 1968 Toxicology of organometallic compounds. In *Organometallic Chem Rev* **3**: 137–50
67. Marsh D O 1979. In *Handbook of Clinical Neurology* (eds Vinken P J and Bruyul G W). Elsevier, Amsterdam, Vol 36, pp 73–81
68. Carty A J, Malone S F 1979. In *Topics Env Health* **3**: 433–79
69. Junghans R P 1983 *Environ Res* **31**: 1–31
70. Clarkson T W 1972 *CRC Rev Toxicol*. Chem Rubber Co pp 203–34
71. Smith P J 1979 *Toxicity Data Organotin Compounds*. Intern Tin Res Inst, London, pub no 538
72. Wada O, Manabe S, Iwai H, Arakawa Y 1982 *Japan J Ind Health* **24**: 24–54
73. Tatken R L, Lewis R J (eds) 1983 *Reg Tox Effects Chem Substances*. US Dept Health Human Services, Cincinnati, Ohio, USA
74. Ref 44 pp 51–94
75. Pearson R 1968 *J Chem Educ* **45**: 581–7
76. Pearson R 1968 *J Chem Educ* **45**: 643–8
77. Pearson R 1973 Hard and soft acids and bases. In *Benchmark Papers in Inorganic Chemistry*. Dowden, Hutchinson and Ross, Stroudsberg, Pa, USA
78. Rabenstein D L 1978 *Accounts Chem Res* **11**: 100–7
79. Simpson R B 1961 *J Amer Chem Soc* **83**: 4711–7
80. Hojo Y, Sugiura Y, Tanaka H, 1976 *J Inorg Nucl Chem* **38**: 641–4
81. Jewett K L, Brinckman F E, Bellama J M 1978 *Organometals and*

Organometalloids: Occurrence and Fate in the Environment (eds Brinckman F E, Bellama J M). ACS Symp Ser No 82, Washington DC, pp 158–87

82. Herman R B 1972 *J Phys Chem* **76**: 2754–9. Also see Valvani S C, Yalkowsky S H, Amidon G L 1976 *J Phys Chem* **80**: 829–35
83. Menger F M 1972 *Chem Soc Rev* **1**: 229–40
84. Johannesen R B, Brinckman F E 1981 Unpublished results. Also see Laughlin R B, French W, Johannesen R B, Guard H E, Brinckman F E 1984 *Chemosphere* **13**: 575–84
85. Guard H E 1984 (Personal communication to FEB)
86. Sarma T S, Ahluwalia J C 1973 *Chem Soc Rev* **2**: 203–33
87. Eigen M, Eyring F M 1967 *Inorg Chem* **2**: 636. Also see Simpson R B *J Chem Phys* **46**: 4775–83
88. Jewett K L, Brinckman F E 1984 Unpublished results
89. Cooper W J, Zika R G 1978 *Earth Science Rev* **14**: 97–115. Also see Aleksandrov Y A, Tarunin B I 1973 *Zh Obshch Khim* **44**: 1835–40
90. Jackson J A, Blair W R, Brinckman F E, Iverson W P 1982 *Environ Sci Technol* **16**: 110–9
91. Cheng C N, Focht D D 1979 *Appl Environ Microbiol* **38**: 494–8
92. Burns D T, Glockling F, Harriott M 1980 *J Chromatogr* **200**: 305–8
93. Brinckman F E, Jackson J A, Blair W R, Olson G J, Iverson W P 1983. In *Trace Metals in Sea Water* (eds Wong C S, Bruland K W, Burton J D, Goldberg E D). Plenum, New York, pp 39–72
94. Schwedt G 1981 *Chromatographic Methods in Inorganic Analysis*. Huthig Verlag, Heidelberg, see Ch 3 and pp 216–22
95. Schaeffer G W, Emilius M 1954 *J Amer Chem Soc* **76**: 1203–4. See also Jolly W L, Drake J G 1963 *Inorg Synth* **7**: 34–44
96. Pourbaix M 1966 *Atlas of Electrochemical Equilibria*. Pergamon, New York, p 476
97. Thayer J S, Olson G J, Brinckman F E 1984 *Environ Sci Technol* **18**: 726–9
98. Tobias R S 1978 The chemistry of organometallic cations in aqueous media. In *Organometals and Organometalloids: Occurrence and Fate in the Environment* (eds Brinckman F E, Bellama J M). ACS Symp Ser No 82 Washington DC, pp 130–48
99. Brinckman F E, Parris G E, Blair W R, Jewett K L, Iverson W P, Bellama J M 1977 *Environ Health Perspectives* **19**: 11–24
100. Snyder L R, Kirkland J J 1979 *Introduction to Modern Liquid Chromatography* (2nd edn). Wiley, New York. See also Horvath C (ed) 1980 *High-Performance Liquid Chromatography – Advances and Perspectives*, vol 2. Academic Press, New York
101. Jewett K L, Brinckman F E 1981 *J Chromatogr Sci* **19**: 583–93. See also Liao J C, Vogt C E 1979 *J Chromatogr Sci* **17**: 237–44
102. Yau W W, Kirkland J J, Bly D D 1979 *Modern Size-Exclusion Liquid*

Chromatography: Practice of Gel Permeation and Gel Filtration Chromatography. Wiley-Interscience, New York

103. Ettre L S 1981 *J Chromatogr* **220**: 29–63
104. Kaliszan R 1981 *J Chromatogr* **220**: 71–83
105. Neumann W P 1970 *The Organic Chemistry of Tin*. Wiley, New York
106. Gurira R C, Carr P W 1982 *J Chromatogr Sci* **20**: 461–5
107. Cassidy R M 1981 The separation and determination of metal species by modern liquid chromatography. In *Trace Analysis*, vol 1. Academic Press, New York, pp 121–93
108. Yu T H, Arakawa Y 1983 *J Chromatogr* **258**: 189–97
109. Horvath C, Melander W, Molnar J 1976 *J Chromatogr* **125**: 129–56
110. Martin A J P 1950 *Ann Rev Biochem* **19**: 517–42
111. Sanderson R T 1983 *Polar Covalence*. Academic Press, New York; Burdett J K 1980 *Molecular Shapes*. Wiley-Interscience, New York
112. Mackay D, Paterson S 1982 *Environ Sci Technol* **16**: 654A–60A
113. Hansch C, Lee A *Substituent Constants for Correlation Analysis in Chemistry and Biology*, Wiley, New York
114. Tomlinson E 1975 *J Chromatogr* **113**: 1–45
115. Chen B K, Horvath C 1979 *J Chromatogr* **171**: 15–28. See also Nahum A, Horvath C 1980 *J Chromatogr* **192**: 315–22 and Hammers W E, Meurs G J, De Ligny C L 1982 *J Chromatogr* **247**: 1–13
116. Weiss C S, Jewett K L, Brinckman F E, Fish R H 1981. In *Environmental Speciation and Monitoring Needs for Trace Metal-Containing Substances from Energy Related Processes* (eds Brinckman F E, Fish R H). NBS Spec Pub 616 Washington, DC pp 197–210
117. Mastryukova T A, Kabachnik M I 1969 *Russ Chem Rev* (Eng trans) **38**: 795–9. Mastryukova T A, Kabachnik M I 1971 *J Org Chem* **36**: 1201–5
118. Jewett K L, Brinckman F E 1983. In *Liquid Chromatography Detectors*, (ed Vickrey T M). Marcel Dekker, New York, pp 205–41
119. Hansch C 1969 *Accounts Chem Res* **2**: 232. See also Dagani R 1981 *Chem Eng News* (March 9) pp 26–9
120. Kier L B, Hall L H 1976 *Molecular Connectivity in Chemistry and Drug Research*. Academic Press, New York. See also Froimowitz M 1982 *J Med Chem* **25**: 689–96 and ibid 1982 *J Med Chem* **25**: 1127–33
121. For an interesting example of a triorganotin aquate, see Davies A G, Goddard J P, Hursthouse M B, Walker N P C 1983 *J Chem Soc Chem Comm*: 597
122. Max N L 1984 *J Mol Graphics* **2**: 8–13
123. Kier L B, Hall L H 1981 Derivation and significance of valence molecular connectivity *J Pharm Sci* **70**: 583–9
124. Cammarata A 1979 *J Pharm Sci* **68**: 839–42
125. Hine J, Mookerjee P K 1975 *J Org Chem* **40**: 292–8
126. Yalkowsky S H, Valvani S C 1976 *J Med Chem* **19**: 727–8

127. Kier L B 1981 *J Pharm Sci* **70**: 930–2
128. Wise S A, Bonnett W J, Guenther F R, May W E 1981 *J Chromatogr Sci* **19**: 457–65
129. Kier L B, Hall L H 1979 *J Pharm Sci* **68**: 120–2
130. Wells M J M, Clark C R, Patterson R M 1981 *J Chromatogr Sci* **19**: 573–82
131. Mackay D 1982 *Environ Sci Technol* **16**: 274–8
132. Neely W B, Branson D R, Blau G E 1974 *Environ Sci Technol* **8**: 1113–5. See also Chiou C T, Freed V H, Schmedding D W, Kohnert R L 1977 *Environ Sci Technol* **11**: 475–8
133. Fendler J H 1980 *Accounts Chem Res* **13**: 7–13
134. Wong P T S, Chau Y K, Kramar O, Bengert G A 1982 *Can J Fish Aquat Sci* **39**: 483–8
135. Laughlin R, Johannesen R B, French W, Guard H E, Brinckman F E 1985 *Environ Toxicol Chem* **4**: 343–51
136. Brinckman F E, Olson G J, Iverson W P 1982. In *Atmospheric Chemistry* (ed Goldberg E D). Springer-Verlag, Berlin, pp 231–49
137. De Ligny C L, Van der Veen N G 1971 *Rec Trav Chim Pays Bas* **90**: 984–1000
138. Yamada J 1981 *Agric Biol Chem* **45**: 997–1002
139. Kier L B 1980 *J Pharm Sci* **69**: 807–10 and Crowe A J, Smith P J 1980 *Chem Biol Interactions* **3**: 171–80
140. Blair W R, Olson G J, Brinckman F E, Iverson W P 1982 *Microb Ecol* **8**: 241–50
141. Sterritt R M, Lester J N 1980 *Sci Total Environ* **16**: 55–90
142. Silver S 1983 Bacterial transformations and resistances to heavy metals. In *Changing Biogeochemical Cycles of Metals and Human Health* (eds Hornig O F, Nriagu J O). Springer-Verlag, Berlin, pp 199–223
143. Meranger J-C 1975 *J Assoc Off Anal Chem* **58**: 1143–6
144. Arakawa Y, Wada O, Yu T H, Iwai H 1981 *J Chromatogr* **207**: 237–44
145. Jackson J A, Blair W R, Brinckman F E, Iverson W P 1982 *Environ Sci Technol* **16**: 110–9
146. Soderquist C J, Crosby D G 1978 *Anal Chem* **50**: 1435–9
147. Chau Y-K, Wong P T S, Bengert G A 1982 *Anal Chem* **54**: 246–9
148. Maguire R J, Huneault H 1981 *J Chromatogr* **209**: 458–62
149. Omar M, Bowen H J M 1982 *Analyst* **107**: 654–8
150. Riggle C J, Sgontz D L, Graffeo A P 1978 *Proc 4th Joint Conf Sensing Environ Pollutants*. ACS, Washington DC, pp 761–4
151. Jewett K L, Brinckman F E 1981 *J Chromatogr Sci* **19**: 583–93
152. Means J C, Hulebak K L 1983 *Neurotoxicol* **4**: 37–43
153. Trachman H L, Tyberg A J, Branigan P D 1977 *Anal Chem* **49**: 1090–3
154. Kojima S 1979 *Analyst* **104**: 660–7

Chapter 2

Organomercury compounds in the environment

2.1 Introduction

The impact of organomercury compounds in the natural environment was one of the prime causes of the growth of scientific interest and social concern in environmental matters. This arose from the occurrence of obvious consequences (poisoning epidemics, ecological effects) following misuse or mishaps with organic mercury compounds. In addition in the case of mercury, scientific investigation led to a clearer appreciation of the reactions and pathways of this element in the environment than was available for other metals. The main reasons for interest in organomercury compounds in the environment are summarized in the following sections (sects 2.1.1–2.1.4).

2.1.1 Poisoning outbreaks and ecological problems

There have been a number of large-scale poisonings (e.g. Iraq 1971–72 when 6530 were hospitalized of whom 459·died) in certain populations exposed to food (usually fish or seed corn used as food) which had been contaminated with methylmercury. Other incidents have occurred in Japan (Minamata 1950–70, Niigata 1960s period, Pakistan and Guatemala).[1-7] In the case of Minamata, methylmercury from a vinyl chloride–acetic acid plant entered the Minamata Bay in the effluent, where it was taken up either in the fish food chain and bioconcentrated, or taken up by the fish directly (bioaccumulated) and then consumed by the local populations whose staple food was fish. The other route to methylmercury poisoning epidemics is usually the consumption by subsistence populations at times of food shortage of seed corn treated with methylmercury.

In addition to these obvious human tragedies there have been ecological problems associated even with the correct use of alkylmercury seed dressings. These were first noticed in Sweden and North America in the late 1960s in the avian food chain. Bird populations consuming scattered seed (up to a quarter of the seed may be lost in this way) were found to contain high levels of mercury (viz. $2-120 \, \mu g \, g^{-1}$ range in liver). In addition, reduction in hatching success occurred, owing to eggshell thinning and embryonic mortality, and this is the main cause of population decline.[8] In none of these cases was methylation of inorganic mercury in the environment the origin of the methylmercury. At Minamata the mercury was used as mercuric sulphate catalyst in the conversion of acetaldehyde to vinyl chloride, and the inorganic mercury was partly methylated in the plant. The seed dressings responsible for other outbreaks also contained methylmercury. These incidents led eventually to wide publicity and study of the effects of methylmercury in the environment. The 1971−72 Iraq outbreak was the worst case of a mass poisoning so far on record. In this case the seed had been variously treated with methyl-, ethyl- and phenylmercury.*

2.1.2   Occurrences of methylmercury in fish

It was found that mercury present in fish from Swedish waters was present mainly as methylmercury, usually more than 80 per cent of the total mercury present.[9] This is a general phenomenon, for both freshwater and marine fish, and usually the higher the trophic level and age of the fish the greater percentage of methylmercury in the muscle tissue. Such studies have suggested that mercury uptake by marine fish is a cumulative process resulting in increasing concentrations of methylmercury with age and size, as a consequence of a low rate of elimination from the fish.[10−12]

The historic significance of these obvservations was that they were taken as evidence of a biological mercury methylation process, as in most cases the fish were from waters uncontaminated with methylmercury. Even now details of the process are not totally clear. It is still not known to what extent fish can methylate mercury and it is likely that most methylmercury is absorbed by the fish in the alkyl form from water column and food. However, there are reports of fish gut or liver content being able to methylate mercury,[13,14] although there are negative reports also.[15] Methylmercury is much more efficiently absorbed than inorganic mercury and it is not excreted from fish to any great extent but is stored in muscle tissue.[12] Although numerous pathway studies of mercury in fish have now been undertaken, the original significance of the observations was to suggest that mercury methylation from inorganic mercury substrates did occur in the natural environment. It is interesting to note that natural high levels of methylmercury in marine fish do not seem to lead to obvious toxic symptoms in the fish concerned.

* The above was written before the Bhopal incident in India.

2.1.3   Demonstration of mercury methylation

The first demonstration of the conversion of inorganic mercury to methyl-mercury was given in the now classic experiments of Jensen and Jernelov reported in 1968 and 1969. They showed that mercuric chloride was partially

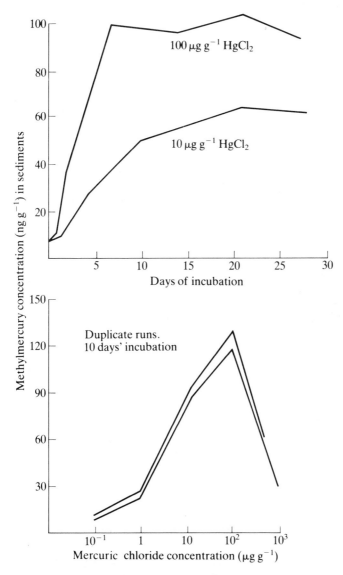

Fig. 2.1   Methylation of inorganic mercury in sediments. Adapted from Jensen and Jernelov, 1969 – see Ref. 16. Reproduced with permission from *Nature* (London) **223**: 753–4. Copyright 1969 Macmillan Journals Ltd.

converted by aquarium sediments to methylmercury, analysable in the sediments.[16] They did not report the presence of dimethylmercury from the sediment but if it had been formed it would have probably diffused out of the sediment, through the water layer and out of the system. The yield of methylmercury was 1.2×10^{-1} per cent and after 20 days a decline in methylmercury levels was observed (Fig. 2.1). As methylmercury was present at only small concentrations in the sediments prior to the addition of mercuric chloride (which did not contain methylmercury) it was assumed that there is a biological or chemical methylating capacity in sediments able to methylate mercury and that there is also an equilibrium between formation and decay. Decay as well as formation seems to be occurring as shown by the tail-off of methylmercury in Fig. 2.1. (This has been confirmed by other experiments – see sect. 2.5.)

2.1.4   Identification of a methylating agent

It was shown in 1968 that a methylcobalamin-utilizing methanogenic bacterium could methylate mercury in sediments.[17] The role of methyl cobalamin (CH_3CoB_{12}) is discussed elsewhere in this chapter and in Chapters 1, 8 and 10. The significance of this 1968 paper was that a naturally occurring biological methyl transfer agent, whose derivatives are known to be present in sediments, was implicated in the mechanism of mercury methylation. This gave credibility ot the idea of biological methylation of metals and provoked a line of research activity still central to methylation studies. It was demonstrated soon after, that CH_3CoB_{12} in aqueous abiotic systems was capable of transferring a methyl carbanion group to mercury(II).[18–23] This view of CH_3CoB_{12} as the sole methylating species in the environment for mercury is now held to be restrictive but the work with CH_3CoB_{12} gave greater understanding to a process that might otherwise have lacked a reasonable and coherent explanation.

2.2   Uses of mercury and organomercury compounds – regulation of organic mercury

2.2.1   Uses of mercury

The uses of inorganic mercury compounds are given in Table 2.1, as loss of these compounds to the environment may be followed by methylation to methylmercury. An estimate of consumption gives an idea of the potential for mercury methylation in the environment and this is also given in Table 2.1. The commercial uses of organomercury compounds are given in Table 2.2. Historic usage is mentioned for completeness and also because removal from use is not uniform through the world as there are disparities in national

Table 2.1 Mercury consumption by use (tonnes)

	1959, USA	1968, USA	1978, USA	1984, USA
Agriculture	110	118	21 (1975)	—
Amalgamation	9	9	<0.5 (1975)	—
Catalytic	33	66	29 (1975)	12
Dental	95	106	18	49
Electrical/batteries	426	677	619	1170
Chlor-alkali	201	602	385	253
Laboratory	38	69	14	7.5
Instruments	351	275	120	98
Paint	121	369	309	160
Paper/pulp	150	14	<0.5	—
Pharmaceuticals	59	15	15 (1975)	—
Metal for inventory/ other	298	298	216	48

From: Bureau of Mines, *Minerals Yearbook* (1984) Mercury Reproduced with permission.

Total world production 1984 = 6000 tonnes. Total USA usage 1984 = 1884 tonnes. Usage in other countries may be gauged from USA figures.

regulatory practices, ranging from outright bans to recommendation and voluntary codes. There are a number of different sources giving usage details for mercury and organomercury compounds.[24,25]

Release of mercury into the environment is now at a lower level than in the period from 1940 to the early 1970s. In Europe the largest single source of mercury discharges is the chlor-alkali industry where liquid mercury is used as a mobile cathode in the electrolysis of brine to produce chlorine and caustic soda. Until recently mercury losses from these plants were up to the order of 150–250 g mercury per 1000 kg chlorine produced.[26] In 1972 UK chlor-alkali plants produced 850 000 tonnes of chlorine and discharged 34 tonnes of mercury (i.e. 40 g per tonne); in 1983 810 000 tonnes of chlorine were produced and 12 tonnes of mercury discharged (14.8 g per tonne).[27] In the UK a heavy investment programme in abatement technology has taken place over the past ten years. In 1973 prices this is calculated at £9.65m but using a net present value calculation involving alternative use of the capital the true cost of abatement to the end of 1980 was £24.28m.[28] In order to bring discharge levels down to 8 g per tonne of chlorine by 1986 the UK chlor-alkali industry in 1980 had budgeted for a cost at present value of another £10.3m for a total cost of £35m present value. Further improvements would be increasingly costly. The cost of installing treatment plants has been estimated at £58 per kg reduction in

Table 2.2   Use of organomercury compounds

Compound	Use	Comments
Compound	Use	Comments
CH_3HgX	Agricultural seed dressing, fungicide	Banned Sweden 1966, USA 1970 as seed disinfectant. Not used today in Europe or USA. Used in laboratories
C_2H_5HgX	Cereal seed treatment	Banned USA, Canada 1970 Used in UK
RHgX	Catalyst for urethane, vinyl acetate production	
C_6H_5HgX	Seed dressings, fungicide, slimicide, general bactericide. For pulp, paper, paints	Banned as slimicide USA 1970. Banned as rice seed dressing Japan 1970. Used in UK
$p-CH_3C_6H_5HgX$	Spermicide	
$ROCH_2CH_2HgX$	Seed dressings, fungicides	Banned Japan 1968. Used in UK
$ClCH_2CH(OCH_3)$ CH_2HgX	Fungicide, pesticide, preservative	
Thiomersal	Antiseptic, C_2H_5Hg derivative	See Fig. 2.2
Mercurochrome	Antiseptic, organomercury fluorescein derivative	See Fig. 2.2
Mersalyl	Diuretic, methoxyalkyl derivative, $RCH_2CH(OCH_3)CH_2HgX$	Little used today R = $oCOOHCH_2OC_6H_4$ $CONH-$
Chlormerodrin	Diuretic, methoxyalkyl derivative, $NH_2CONHCH_2$ $CH(OCH_3)CH_2HgCl$	Little used today
Mercarbolide	$oHOC_6H_4HgCl$	*O*-chloromercuriphenol
Mercurophen	$oNO_2pONaC_6H_3HgOH$	See Fig. 2.2
Mercurophylline	Diuretic	See Fig. 2.2

X = anionic group. Wide range of X known, e.g. OAc^-, PO_4^{3-}, Cl^-, $NHC(NH)NHCN^-$, etc.

emission per year.[28] The emission losses are divided approximately as follows; to air (48 per cent), to water (40 per cent), to products (12 per cent).[29] Overall it can be said that mercury losses to the environment from chlor-alkali plants have been greatly reduced in the past ten years; one plant on the River Mersey estuary, UK, has reduced losses by 95 per cent.[29] The methods used to reduce losses from chlor-alkali plants are outside the scope of this review but are

Table 2.3   Methylmercury in the River Mersey estuary,
UK, 1978 and 1983

Sampling station	1978	1983
15a1	3.2	0.5
15a6	4.8	1.5
15b1	20.1	12.7
15b2	43.3	24.0
15b4	16.0	7.3
15b8	13.1	8.0
15 3*	1.5	11.0

(From Moreton P A 1984 Formation of methyl mercury in the environment, PhD Thesis, Leicester Polytechnic, UK, p 118. Samples are from
surface sediments)

* If this sample is assumed artefactual and ignored, the reduction in
methylmercury levels between 1978 and 1983 is statistically significant.

covered in detail elsewhere.[30] The abatement costs at the plant referred to
above were £20m for capital investment with annual operating costs of £1.3m.
Monitoring costs were £0.7m. There was no significant benefit to the company in terms of recovered mercury.[31] These improvements do, however,
allow the quality standards of the European Economic Community Mercury
Directive to be met with in affected waterways. Direct measurement of
methylmercury levels in sediments adjacent to this plant already suggest a
decline in concentrations.[32] Methylmercury is not a persistent material in
water and sediments and reduction in losses to the environment will eventually
be reflected in lower environmental levels (Table 2.3), although further
methylation of the excess inorganic mercury present may slow this decline.

Of the remaining uses, the incorporation of phenlymercury compounds in
paints might be expected to have the most direct environmental impact, particularly in marine antifouling paints where leaching to harbour sediments
may occur.[33] In marine paints the mercury concentration is between 2 and 5
per cent by weight but in normal domestic emulsion paints its range is
0.001–0.05 per cent. The use of phenylmercury as a slimicide in the pulp and
paper industry has been abandoned in Europe and North America, where it
caused contamination problems in several locations (see sect. 2.6). The other
present-day uses of mercury compounds now cause less environmental difficulty. In UK agriculture about 15 tonnes of organic mercury compounds are
used, mainly phenylmercury and methoxyethylmercury but also small quantities of ethylmercury (1.2 tonnes).[34] Environmental problems arising from
this usage would be expected to be localized rather than widespread and
generalized. Present-day capacity for methylation is most likely to arise from
mercury mineralized or polluted rivers flowing into bacteria-rich polluted

zones, e.g. urban estuaries. Methylation from chlor-alkali effluents is becoming less dominant. Pharmaceutical use of organomercury compounds does not lead to general environmental problems. The UK Department of the Environment has produced a memorandum on the storage, handling and disposal of mercury-bearing wastes.[35] There is still some usage of organomercury antiseptics although they are less common today. Details are given in Fig. 2.2.

2.2.2   Regulation of mercury

The European Economic Community has recently published detailed limit values and procedures for monitoring discharges.[36-39] In general, average concentrations in effluents must not exceed 0.1 mg Hg dm^{-3} of effluent by 1986 and 0.05 mg dm^{-3} by 1989. The limit values correspond to monthly average concentrations or to maximum monthly loads. The EEC has also published a series of quality objectives defining the amounts of mercury allowed to be present in various matrices, e.g. mercury concentrations in fish must not exceed 0.3 μg g^{-1} wet flesh; total mercury concentrations in estuaries affected by chlor-alkali discharges must not exceed 0.5 μg dm^{-3}, in inland surface waters 1 μg dm^{-3} and in coastal seawater 0.3 μg dm^{-3}. Mercury was included on the original EEC Dangerous Substances Directive. The European maximum levels for fish are similar to other national standards defining allowable mercury concentrations in fish, e.g. USA, Canada.[40]

2.3   Toxicity of organomercury compounds

Short-chain alkylmercury compounds are toxic, slow to metabolize and tend to bioaccumulate.[41] The mercury—methyl bond particularly is very stable in most organisms and the addition of the alkyl group confers lipid solubility, allowing penetration of the blood—brain barrier and cell membranes.[42] The solubility and binding ability of the RHg$^+$ moiety to biological ligands results in a large half-life in various organisms for this species, e.g. 60–70 days in man, much longer than for inorganic forms of mercury (3–4 days). Slow excretion, and breakdown to inorganic mercury does occur and intestinal flora will decompose methylmercury.

In addition to the direct toxicity of alkylmercury compounds, a slow decomposition to inorganic mercury(II) ions may lead to secondary toxic effects as for inorganic mercury. Many studies on the metabolism and toxic effects of methylmercury in particular have now been made. Over 90 per cent of methylmercury is absorbed from the gastrointestinal tract in both man and animals, and following such absorption[43] most accumulates in erythrocytes giving red cell : plasma ratios of up to 300 to 1.[44] This allows for efficient

Name	Structure	Other names
Merthiolate	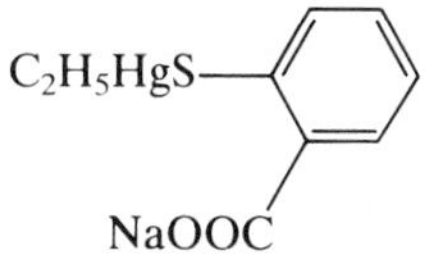 C_2H_5HgS—⟨benzene ring⟩ NaOOC	Thimersol, mercarbolide, mertoxal
Merbromin	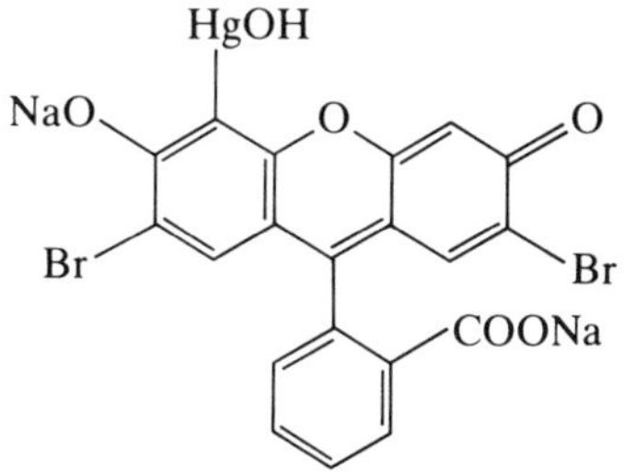 ⟨xanthene structure with HgOH, NaO, Br, Br, O, COONa⟩	Mercurochrome
Nitromersol	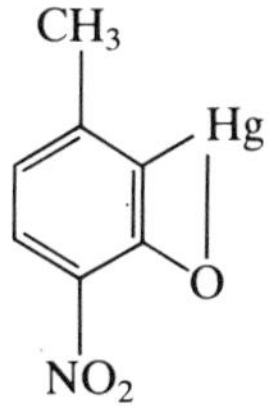 ⟨benzene ring with CH_3, Hg, O, NO_2⟩	Metaphen
Mercurophen	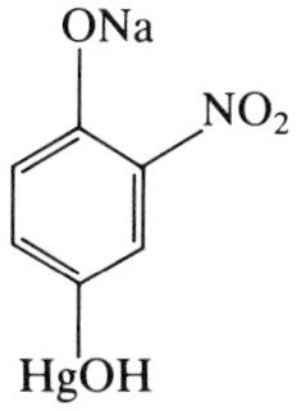 ⟨benzene ring with ONa, NO_2, HgOH⟩	
Mercurophylline	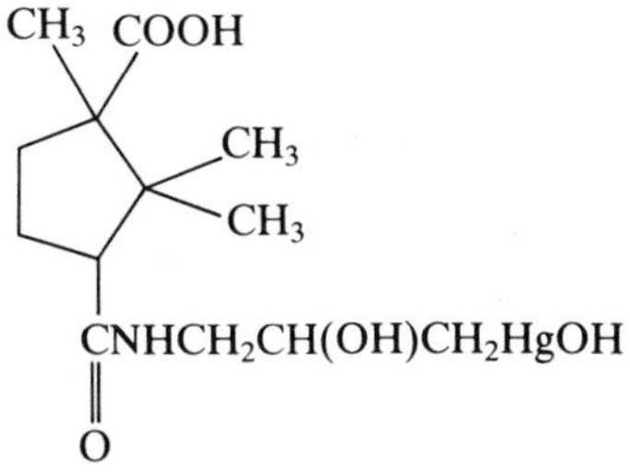 ⟨cyclopentane ring with CH_3 COOH, CH_3, CH_3, $CNHCH_2CH(OH)CH_2HgOH$, O⟩	

Fig. 2.2 Some mercury antiseptics and diuretics

All compounds in this figure are antiseptics, apart from mercurophylline. General formula for diuretics is $RCH_2CH(OCH_3)CH_2HgX$.

transport through the body and results in a generally uniform pattern of distribution in tissues and organs – in blood, kidney and brain concentrations are within a range of one to three by ratio.[45] There is an exceptional ability of methylmercury to pass the blood–brain barrier, and injury to the central nervous system then arises by strong binding of methylmercury to sulphydryl residues and subsequent release of mercuric ions to binding sites in the central nervous system. The slow elimination of methylmercury from the body is a result of the high erythrocyte–plasma ratio.[44] Mercury in plasma can be released from the body relatively easily but in man and higher animals not much methylmercury is present in this compartment.

Methylmercury will accumulate in both cerebellum and also cerebral cortex where it will be tightly bound by sulphydryl groups. Inside the cell methylmercury will inhibit protein synthesis and RNA synthesis.[46,47] The effects are particularly important in the developing foetal and young brain of most animals. Although some puzzling and apparently anomalous results have been found for cats (where 4–9-day-old kittens had greater resistance than adults) and rats, foetal and newborn brain tissue should be regarded as the most sensitive organ.[48] The result of methylmercury poisoning in humans expresses itself mainly in various neurological disturbances, e.g. numbness and

Table 2.4   Toxicity data for some organomercury compounds

Compound	$LD_{50}(\mu g\ g^{-1})$	Notes
CH_3HgCl	58	Rat, oral
	47	Mouse, intraperitoneal
$CH_3HgNHC(NH)$ NHCN	26	Rat, oral
C_2H_5HgCl	50	Rat, oral
	28	Mouse, intraperitoneal
$n-C_3H_7HgCl$	18	Mouse, intraperitoneal
$n-C_4H_9HgCl$	15	Mouse, intraperitoneal
C_6H_5HgOAc	46–50	Rat, oral
	13–70	Mouse, oral
	10	Rat, intraperitoneal
	0.58	LC_{50} (catfish) – 10% product (mg dm^{-3})
$C_6H_5HgNO_3$	27	Mouse, intravenous
C_6H_5HgCl	50–100	Oral, rat
$CH_3OCH_2CH_2HgCl$	30	Oral, rat

Data from References

tingling of extremities, tunnel vision, hearing impairment, speech and intellectual deterioration and coordination problems generally. In an analogy to the problem of alkyllead exposure, mild methylmercury poisoning is hard to detect, particularly at levels leading to slight intellectual and behavioural problems which could have other causes (lead, social, etc.). By contrast, serious poisoning produces effects similar to cerebral palsy, and often only after exposure for several months do symptoms develop. If poisoning is treated early the prognosis is good and treatment is via complexing agents, e.g. D-penicillamine.

The ability of methylmercury to penetrate the placental barrier leads to accumulation in the foetus. The rate of transport across the placental barrier is ten-fold higher than for inorganic mercury. It appears that foetal tissue has a greater binding ability for methylmercury than does the pregnant mother – levels being higher in the newborn baby than for the exposed mother. Exposure via milk is also important for feeding babies. It does appear that pregnant animals may detoxify themselves by transferences to their foetuses. As has been pointed out the developing brain is particularly vulnerable to methylmercury.[48]

To summarize, methylmercury is toxic because of its high affinity for sulphydryl residues with subsequent poisoning of cysteine-containing proteins giving symptoms chiefly to the sensory and motor functions of the central nervous system. Recent calculations suggest a toxicity threshold for methylmercury of 1 mg dm^{-3} in blood, allowing acceptable levels for exposed populations of 0.1 mg dm^{-3}. This corresponds to a body burden of $7-10$ mg, with steady-state intake of 100 μg d^{-1} for an average 50 kg person.[49] However, natural levels in excess of these standards do not seem to have led to toxic effects in the individuals concerned.[6]

Methylmercury toxicity and biochemistry have been closely researched and reviewed during the past ten years and references to the main sources on toxicity are given.[42,49,50-53] Toxicity data for some organomercury compounds are given in Table 2.4.

2.4 Analysis of organomercury compounds in environmental matrices

2.4.1 Organic mercury in air

Currently cold vapour atomic absorption (AA) or plasma atomic emission spectroscopic methods are the main techniques used for the measurement of mercury in the atmosphere. Sample collection over a period is used with separation of particles on a filter. Most methods of collection involve absorption in acid solution or collection on gold foil or charcoal. Mercury(0) amalgamates with the gold and mercury(II) compounds are adsorbed as a monolayer.

Owing to the enrichment procedures used in sampling it is usually not possible to define the original speciation of the mercury compounds in air when analysis finally takes place. It is assumed that the most common mercury constituent of air is mercury(0) but mercury(II), dimethylmercury and methylmercuric hydroxide and chloride also occur. Trapping with gold allows storage times of 1–2 weeks prior to analysis although it may convert some mercury(II) to mercury(0). Analysis indicates that mercury bound to particulates does not make up a large proportion (i.e. 1–10 per cent) of the total airborne mercury. Particulate mercury concentrations are of the order of 0.1 ng m^{-3} or less compared to the normal 0.5–5 ng m^{-3} for total mercury in unpolluted ambient air. Aerosol (i.e. particulate) samples should be collected on filters prior to total atmospheric mercury analysis.

Several recent measurements of mercury concentrations in the atmosphere have been made.[54–56] Details of some specific procedures are now given, but overall it should be noted that although significant quantities of dimethylmercury are emitted to atmosphere, the lifetime of this species is short owing to hydroxyl radical or photochemical attack and degradation to mercury(0). Details of atmospheric decay for organometallic species are given in Chapter 1 (sect. 1.3.4) and the stability of organic mercury in air and water is also discussed in the next section (sect. 2.4.2). Residence time for total mercury in the atmosphere ranges between 0.4 and 3 years.

The main organic mercury compounds of interest in air have been dimethylmercury and methylmercuric chloride or hydroxide. Selective absorption techniques may be used to separate inorganic mercury, e.g. 3 per cent SE-30 on Chromosorb W treated with sodium hydroxide was used to retain methylmercuric chloride.[57] Each was then separately analysed by d.c. discharge emission spectrometry for mercury. Mercury vapour was trapped as a silver film and dimethylmercury on a gold film.

Alternatively gas chromatography (GC) can be used, e.g. air samples have been collected on a short column packed with 20 per cent Tenax and 80 per cent 1 mm glass chips after filtration. The mercury species were desorbed by heating to 200 °C for 30 minutes and then absorbed in benzene. Gas chromatography (GC) separation with AA detection at 253.6 nm was used to distinguish dimethylmercury and methylmercury chloride. Separation of mercury vapour (if present) was difficult but could be achieved by control of the column packing material.[58] This is an important factor in the analysis of atmospheric samples. Elimination of metal surfaces in contact with analytes was an important consideration as they cause decomposition of organomercurials and consequent error if speciation is a target.

A method for the analysis of all forms of mercury present in air has been described. The mercury was collected on gold-coated sand, desorbed at 800 °C and measured by a cold vapour AA technique at 253.6 mm.[59] The detection limit was 0.1 ng and mercury contents of the order of 6 ng m^{-3} were found in outdoor urban environments. However, the method is not species specific. Atmospheric mercury determination has been recently reviewed by Schroeder.[60] Some levels are given in Table 2.8.

2.4.2 Organic mercury in freshwater and seawater

Most methods are based on the original Westoo procedure originated for fish (see sect. 2.4.3). A preconcentration step is required owing to the low levels of methylmercury present − of the order of 2.0 ng dm^{-3}. In a typical method 500 cm^3 of sample are extracted with acidified benzene and this is followed by formation of a water-soluble adduct of methylmercury and cysteine. This is extracted into water, acidified and the liberated CH$_3$HgX (X = Cl in this case) is extracted into benzene followed by electron capture GC.[61] Toluene may be used for the extractions and electron capture replaced by plasma emission detection. Colourimetric or AA detection has also been used.[62] Total extraction efficiency in one procedure was 42 per cent. Methylmercury was about 2 per cent of the total mercury present − of a similar order to the proportion in sediments. In a similar procedure the last cysteine extract was transferred to a dithizone−chloroform solution and the complex was determined by AA methods. This gave a limit of detection of picogram levels of methylmercury in 1 dm^3 of seawater. Similar methods for methylmercury were used in two large-scale Canadian freshwater studies − on the Ottawa River and the Wabigoon − English River (Ontario) (see sect. 2.6).

Ethyl- and phenylmercury have been analysed in a separate study by the benzene−cysteine extraction method followed by electron capture GC, but the extraction from spiked lake water was not quantitative and the species were not found in natural lake water.[63] This method was also used with flameless AA detection and was successful for submicrogram quantities of methylmercury although there were numerous interferents, including high levels of chloride ion. Recovery efficiencies were greater than 90 per cent but the method was not applied to natural non-spiked samples.[64]

Inorganic and organic mercury at ng dm^{-3} concentrations in freshwater have been determined by simultaneous collection on a dithiocarbamate-treated resin on a column followed by elution with an acidic thiourea solution. Mercury(0), mercury(II) and organic mercury were separated by differential reduction followed by cold vapour AA.[65] A similar dithiocarbamate complexation has been used on filtered seawater and the organic and total mercury separated by solvent extraction procedures.[66] Both of these methods yielded limits of detection in the 0.1−0.2 ng dm^{-3} area. Chromatography with dithizone extraction was also used to preconcentrate phenyl-, methyl- and inorganic mercury from 100 to 500 cm^3 aqueous samples prior to a radioanalytical analysis method. Potable, river water, beer, wine and fruit juices were analysed by this method down to a 0.01 μg dm^{-3} level.[67]

Levels of organomercury compounds found in aqueous environmental samples are given in section 2.6 and Table 2.8. It should be noted that the levels accepted as valid for total mercury in dilute samples, e.g. seawater, have fallen steadily in recent years from tens of ng dm^{-3} to 'true' levels around the 1.0 ng dm^{-3} limit today (1985) with detection at the 0.1 ng dm^{-3} limit. The previous levels included a degree of contamination. Some authors suggested that, in view of the levels found still being near detection limits, the actual mercury concentration of seawater is still unknown.[68] A number of reviews of determinations in aqueous media have been made recently and methods of stabilization of the sample during storage have been detailed.[69,70] When water samples do have to be stored for longer periods the mercury present may be stabilized by adding concentrated nitric acid to the sample stored in a sealed clean pyrex or quartz container. Otherwise losses during storage may be serious. Solutions of mercury concentrations greater than about 0.2 μg dm^{-3} may be stored for several days without significant loss. Species of organic mercury present in natural waters include methylmercuric chloride and hydroxide in freshwater. In seawater inorganic HgCl$_4^{2-}$ is the dominant species. In natural waters methylmercury, as in sediments, is bound mainly to sulphide species in complexes or particulate forms. In seawater it has been estimated that methylmercury is present as chloride (92.2 per cent), hydroxide (5.1 per cent) and bromide (2.7 per cent).

Elemental mercury in both air and water is thermodynamically unstable to oxidation, but as discussed in Chapter 1 (sect. 1.3.2) such processes are often slow, and mercury(0) is kinetically stable and may volatilize unchanged from water to atmosphere. In the atmosphere the kinetic stability of mercury(0) is demonstrated by its lifetime in that compartment of up to two years.

A series of stability zones for organic and inorganic mercury in natural waters has been calculated and these are shown in Fig. 2.3. Residence time for mercury in the oceans has been estimated at 2000 years.

2.4.3   Organic mercury in other environmental matrices

Methods used have been based on solvent extraction, differential reduction,[71] difference calculations between total and 'ionic' mercury,[72] and by paper and thin layer chromatography.[73,74] The most commonly used method at the present time is gas–liquid chromatography (GC), although the first practical method for differentiating between organic mercury and inorganic mercury was a colourimetric method originated by Gage.[75] Organic mercury was extracted into benzene and measured spectrophotometrically as the dithizone complex. Inorganic mercury remains in the aqueous phase. The detection limit was 1 μg g^{-1}. An improved differential reduction method using cold vapour AA spectroscopy to measure mercury(0) was described by Magos. The organomercury compounds were reduced to mercury(0) by a stannous chloride–cadmium chloride combination whereas stannous chloride alone releases only inorganic mercury.[76] The basis of most present methods was evolved by Japanese and Scandinavian workers in 1966.[73,77] The procedure separates methyl, ethyl and butyl forms of mercury. Methylmercury in fish was estimated by acidification of the homogenized sample with hydrochloric acid and then by extraction into benzene. The organomercury compounds were then back-extracted into an aqueous cysteine solution. The aqueous solution was acidified and re-extracted into benzene. This double partition

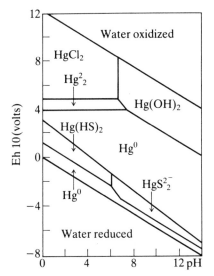

Predominance diagram for soluble mercury species in water containing 10^{-3} mol dm^{-3} Cl$^-$ and 10^{-3} mol dm^{-3} SO$_4^{2-}$ or equivalent total sulphur. The diagram refers to 25 °C and 1 atm pressure. (Benes and Havlik 1979)

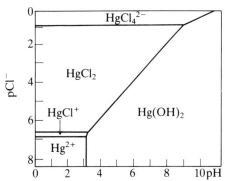

Stability ranges for chloro and hydroxo complexes of divalent mercury in aqueous solution (Benes and Havlik 1979)

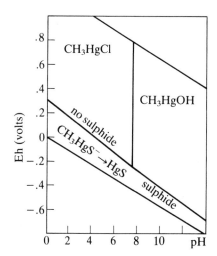

E_h–pH plot of the stability relations among methylmercury species in river water Sambre, Belgium (Wollast, Billen, Mackenzie 1975)

Fig.2.3 Zones of stability for mercury and methylmercury. From: Benes P, Havlik B 1979 in *The Biogeochemistry of Mercury in the Environment* (ed Nriagu J O) Elsevier, North Holland, pp 175–202; Wollast R, Billen G, Mackenzie F T 1975 *Environ Sci Res, Ecol Toxicol Res* **7**: 145–66. Plenum Press, New York. (Ref 139)

method enables removal of interferents in fish analysis (e.g. benzene-soluble thiols). At first the method was unsuccessful in removing mercury bound to sulphur in the sample but addition of mercuric ion was later used to break methylmercury–sulphur bonds.[78,79] The method could be used to detect $0.07-4$ μg g^{-1} methylmercury in fish if electron capture detection was used (which is particularly sensitive to the chloride ion in methylmercuric chloride). Recovery of added methylmercury in hair, food or fish was better than 90 per cent. The method may be used generally for food (eggs, meat, etc.) and fish, and most present-day analytical methods for organic mercury are derivatives of this process. The addition of mercury(II) ions was disadvantageous in that it converted any dimethylmercury present to monomethylmercury and the former was, therefore, not estimated. The use of copper(II) ions to displace organic mercury was later found to avoid the problem of decomposing dimethylmercury.[80] Extractions are also made into toluene rather than benzene for enhanced safety. Overall detection limits in samples after improvements of the method are 0.5 ng g^{-1} from 1 g of sample. It should be noted that use of metal columns leads to decomposition of organomercury compounds,[81] whereas glass columns do not. The original Westoo technique gave poor recoveries of alkoxyalkylmercury compounds but a revised extraction method has been devised.[74] The basic procedure has now been reduced to a semi-micro scale by Uthe *et al.*[82] The initial extraction from fish as bromide into toluene was achieved by grinding in a ball mill and a clean-up procedure used ethanolic sodium thiosulphate. Methylmercury was re-extracted into benzene and determined by electron capture GC as the iodide. Only 1.2 per cent of dimethylmercury present was lost, showing the advantage of copper over mercury in the displacement of organic mercury. The columns were conditioned by injections of aqueous potassium iodide to help prevent decomposition of the organomercury compounds.

Numerous improvements and variations on the Westoo technique have been suggested. Newsome used a filtration step with hydrobromic rather than hydrochloric acid so as to improve the partition ratio.[83] Emission spectrometry in a microwave gas plasma has been used as a detector rather than the more usual electron capture method.[84] Mono- and dimethylmercury have been separated by selective extractions allowing determination of both in a single sample[85] or by use of a microwave discharge detector.[86] Dimethylmercury has also been estimated by combustion to free mercury.[87,88]

Several modifications have also been made for the separation and identification of organic mercury in biological material[89] or blood.[90] Phenylmercury compounds have been analysed by thermogravimetric means and by GC.[91]

Although the main strategy for the extraction of methylmercury is based on the original Westoo technique, methylmercury in fish has been recently identified by a simplified extraction procedure and by AA or high performance liquid chromatography.[92] A standardized procedure for fish and sediments has been evolved by Krenkel and Burrows.[93] A method of nuclear activation analysis allowing determination of the speciation of mercury in fish and blood has also been described.[94] Separation of phenyl- and methylmercury by a radiochemical method was reported by Stary and Prasilova.[95] A method to determine methylmercury in fish using graphite furnace AA spectroscopy has been discussed recently.[96] Acetone extraction is used initially, methylmercury released by cupric bromide and the extraction is to toluene. Addition of dithizone to the extract leads to determination of mercury. Detection limit is 0.08 μg g^{-1}. The procedure was automated. Steam distillation has been used to separate alkylmercury from fish, followed by absorption in potassium persulphate and reduction for cold vapour AA determination in the inorganic form.[97] A new selective reduction method for inorganic and organic mercury in water, fish and hair has been described.[98] Methylmercury has been recently determined by anodic stripping voltammetry in the presence of inorganic mercury but the method was not applied to environmental samples. The detection limit was 0.1 ng dm^{-3}.[99] Such detection limits for solid environmental samples are probably unnecessary but application to aqueous and atmospheric samples would be of considerable interest.

Dichloromethane (CH_2Cl_2) has been used to preconcentrate solvent extracts to volumes down to 0.1 cm^3. Matrices used included fish, water, urine and sediments and GC with an atmospheric pressure active nitrogen detector was used.[100] The method was suitable for dialkyl compounds, as also was an indirect method published by Shariat.[101] Several other variations on methods of analysis for alkylmercury in environmental matrices have been published recently.[102,103]

The main analytical methods for methylmercury in sediments also use the basic Westoo technique but there are several variants. The method of Longbottom *et al*[104] was one of the few directed specifically towards work in sediments. The sediment was extracted with copper(II) bromide and sulphuric acid into toluene in a stoppered centrifuge tube. The toluene was removed and back-extracted as the iodide into benzene prior to chromatography. The method involved relatively large volumes of reagents and it was converted by Uthe *et al* into a semi-micro scale[82] in order to reduce sample and reagent volumes and blank readings. A particular problem with the original larger scale method was the generation of effervescence from the 10 g of sediment used when it was treated with the acidic bromide reagent. The final analyte solutions proved to be more stable if methylmercury was extracted as the bromide rather than the iodide. This method has been discussed in detail by Morton[105] and Bartlett.[106] Steam distillation to separate organic from inorganic mercury compounds followed by reduction and AA analysis has also been used for sediments.[107] It was claimed that this method eliminated the emulsification problems sometimes met with in the extraction methods used prior to chromatography; 2.5 ng g^{-1} could be determined by this technique. Residence time for mercury in sediments has been estimated at 10^6 years but this appears too high in view of the known mobilization routes (see sects 2.5.2 and 2.5.3). Residence time in soils may be 1000 years. Bonding of methylmercury in sediments and biota will be to complex natural sulphur-donating species (see Ch. 1).

2.4.4   Analysis of organomercury compounds – summary

Hundreds of papers on the analysis of organic mercury compounds in environmental compartments have appeared during the past ten years. The clear majority use a variant on the original method reported by Westoo in 1966. It has since been applied to fish, food, seawater, freshwater, blood and urine samples, etc. In this review the essential methodology for organo-mercury analysis in the environment has been given and some emphasis has been placed on more recent analytical developments. Table 2.5 and Fig. 2.4 present details of gas chromatographic methods from various environmental matrices.[108 – 132] There are also several detailed reviews of this area.[133 – 135]

2.5   Formation and decay of methylmercury compounds

During the past fifteen years it has been shown on many occasions that in-organic mercury added to various environmental matrices may be converted to methylmercury. Despite methylation itself being a well-established process, evidence for the mechanism is still largely circumstantial. The main conclusions regarding environmental mercury methylation are given below.

2.5.1   General conditions for methylation and demethylation

In the natural environment methylation of inorganic mercury(II) may occur in sediments, in the water column,[136] in soil[137] and by humic and fulvic materials.[138] The normal pattern is for methylmercury to build up to a maximum level in the matrix and for it then to decline somewhat as a result of demethylation or other removal processes. Hence a measurement of methylmercury in an environmental medium is usually a measurement of an equilibrium level between concurrent methylation and demethylation processes. Methylation may occur in both aerobic and anaerobic environments but maximum rates seem to occur in the oxidizing anaerobic zone where redox potentials exist in the -100 to $+150$ mV range and where there is a wide variety of micro-organisms.[139,140] Under most acid or neutral conditions *mono*methylmercury is formed but *di*methylmercury has been reported under basic conditions. Demethylation may occur to mercury(0) and methane (see later).[141] Factors affecting methylation extent and rates in sediments include total inorganic mercury concentration, organc content of the sediment, pH,

Table 2.5   Application of gas chromatography to the determination of organomercury compounds

Analyte	Matrix	Reference
CH_3HgCl	Hair, fish, aqueous solution, biological and environmental samples	73, 74, 77, 78, 79, 82, 100, 108, 114, 115, 116
CH_3HgCl	Rat liver, fish, vegetables	117, 118
	Rice, hair, aqueous solution	119
CH_3HgDz, C_2H_5Dz, $CH_3OCH_2CH_2HgDz$, $C_2H_5OCH_2CH_2HgDz$ C_6H_5HgCl, $p-CH_3C_6H_4HgCl$	Vegetables, fruit	74
CH_3HgCl	Fish, cereal, grain	83
CH_3HgBr, $(CH_3)_2Hg$, $(C_2H_5)_2Hg$,	Aqueous solution	85, 86, 87, 88, 120, 121
CH_3HgCl, CH_3HgI	Fish, sediments, blood, tissue, hair, aqueous solution	104, 110, 111, 122 – 5, 126 – 131
CH_3HgCl, CH_3HgCl, $CH_3OCH_2CH_2HgCl$	Biological material	132
CH_3HgCl, C_2H_5HgCl, $(CH_3)_2Hg$	Fish, water samples	101, 109
CH_3HgBr, C_2H_5HgBr	Food, biological samples	112, 113
CH_3HgCl, C_2H_5HgCl, $(CH_3)_2Hg)$	Seawater, aqueous solution	61, 63
CH_3HgCl, Hg, $(CH_3)_2Hg$	Air	55

(Adapted and expanded from: Rodriguez-Vazquez J A 1978 *Talanta*, **25**: 299 – 310) Copyright Pergamon Press 1978. Reproduced with permission

Species present in the matrix is RHg; anion X (e.g. Cl, Br, etc.) is usually introduced at analysis. Dz = dithizonate.

redox potential (Eh), temperature, the nature of the micro-organisms present, sulphide levels and the nature of the complexation of mercury by natural ligands. Sulphide is a particularly important controlling factor. When sulphide concentrations in sediments are greater than about 1.8 mg g^{-1} little methylmercury is found in the sediments despite other conditions apparently

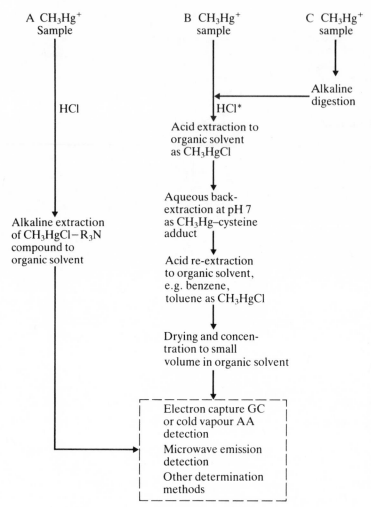

A CH$_3$Hg$^+$
Sample

B CH$_3$Hg$^+$
sample

C CH$_3$Hg$^+$
sample

HCl

HCl*

Alkaline
digestion

Acid extraction to
organic solvent
as CH$_3$HgCl

Alkaline extraction
of CH$_3$HgCl$-$R$_3$N
compound to
organic solvent

Aqueous back-
extraction at pH 7
as CH$_3$Hg–cysteine
adduct

Acid re-extraction
to organic solvent,
e.g. benzene,
toluene as CH$_3$HgCl

Drying and concen-
tration to small
volume in organic solvent

Electron capture GC
or cold vapour AA
detection

Microwave emission
detection

Other determination
methods

*HBr, HI has been used. Process applies to other organomercurials.
A is rapid microwave procedure of Talmi–Refs 109, 110
B is Westoo/Sumino basic method and variations Refs 111–127
C is an alkaline digestion procedure Refs 128–132
Adapted from Rodriguez-Vazquez J A, 1978, *Talanta* **25**: 297–310.
Copyright Pergamon Press 1978. Reproduced with permission

Fig. 2.4 Determination of methylmercury in biological matrices – procedures (Ref 135)

being favourable (e.g. high organic or total mercury content). This is due to formation of intractable mercuric sulphide which is hardly methylated and to removal of any methylmercury by dismutation promoted by sulphide ions (see sect. 2.5.3).[142,143] It has been shown for environments of similar total mercury content but with varying sulphide levels that methylmercury concentrations

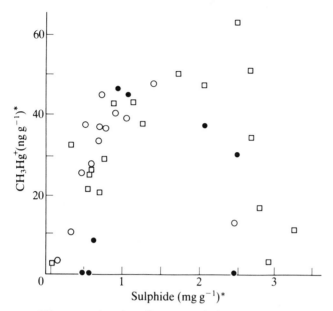

*Concentrations in sediment; symbols represent samples taken on three separate occasions.

Fig. 2.5 Relationship between methylmercury and sulphide in river sediments. From Craig and Moreton, 1983 – see Ref. 142. Permission received from Pergamon Press.

initially increase with increasing sulphide but after 1.8 mg g⁻¹ sulphide they decrease sharply. (Fig. 2.5).[142]

Following methylation, methylmercury does not usually build up in sediments to more than about 1.5 per cent of the total mercury present. This is an approximate equilibrium level between formation and removal.[144] The proportion in biota may be much higher (see sect. 2.6). Some net methylation rates have been estimated: 17–690 ng m⁻² d⁻¹ (river)[145], 15–40 ng g⁻¹ d⁻¹ (sediment),[16] 15 ng g⁻¹ d⁻¹ (sediment)[146] and 137 ng g⁻¹ d⁻¹ (marsh sediment).[147] Rates are slower in saline environments.[148] Photolytic decay in sediments is not expected to account for much loss of methylmercury. Light will not penetrate beyond sediment surfaces although decay by this mechanism in water is possible. Sunlight absorption rate constants for methylmercury have been measured and are low.[149] Despite this methylmercury will decay rapidly under the complex conditions of the atmosphere (Ch. 1).

As nearly all methylmercury is adsorbed either in sediments or suspended particulate matter, it will be seen that photolytic decay is not likely to be important here. In rivers 95 per cent of methylmercury present in the water is bound to suspended solid material (> 45μm). Photolytic decay, as noted, is expected to occur mainly in the atmosphere (Ch. 1, sect. 1.3.4).

It can be seen from the discussion above that demethylation of organic mercury occurs and that methylation and demethylation form an equlibrium process. Demethylation by micro-organisms occurring in water, sediments, soil or the intestine is to mercury(II) then mercury(0).[150−152] Cleavage of the mercury–carbon bond is enzyme catalysed. The genetics of methylmercury resistance is associated with plasmids, their presence leading to demethylation of methylmercury. Details of species able to methylate and demethylate mercury are given in the following section (sect. 2.5.2).

2.5.2   Methylating agents

Numerous pure bacterial strains which may methylate or demethylate mercury are known. These are referred to in Table 2.6.

In several cases no methylation occurred when inorganic mercury was exposed to micro-organisms. Mercury methylation by organisms may be enzymatic, where mercury(II) intercepts a functioning enzyme system in an organism, or non-enzymatic where the abiotic products of active metabolism methylate mercury.[153] Microbes utilizing a number of enzyme systems are capable of methylation, e.g. methionine synthetase, acetate synthetase and methane synthetase. Essentially, mercury is methylated by being present in the proximity of the functioning enzyme and capturing a methyl group destined in principle for transfer elsewhere. In environmental situations it is usually impossible to distinguish this from non-enzymatic processes. Many methyl donor molecules are available in aqueous, particulate or sediment environments and most of these are products of biological processes. Methylation of mercury by such molecules is abiotic or chemical as far as the actual methylation process is concerned. Abiotic methylation in this sense has been observed in amended natural situations, e.g. where the putative donor molecule is added to a mercury-containing sediment and methylmercury is produced. It has also been observed in abiotic, laboratory model experiments, e.g. where the donor molecule is added to mercury(II) in distilled water. Although most evidence about enzymatic methylation is circumstantial it is now known that various naturally occurring molecules will methylate mercury chemically. The most closely studied of these is methyl cobalamin (CH_3CoB_{12}) which has been shown by numerous groups to react with mercury(II).[17−23,154−159] A structure for CH_3CoB_{12} is given in Chapter 1 (Fig. 1.1). Its reaction with mercury(II) in aqueous solution is as follows:

$$CH_3CoB_{12} + Hg^{2+} \xrightarrow{H_2O} CH_3Hg^+ + H_2OCoB_{12}^+ \qquad [2.1]$$

CH_3CoB_{12} also functions with various enzyme systems; it occurs in anaerobic ecosystems and in living organisms (see Ch. 10, sect. 10.4) but it is not known whether environmental methylation with CH_3CoB_{12} occurs

Table 2.6   Methylation and demethylation of mercury by bacteria

Species capable of methylation of $HgCl_2$	Species capable of demethylation of CH_3Hg^+
Pseudomonas fluorescens	*Serratia marcescens*
Microbacter phlei	*Providencia sp.*
Bacillus megaterium	*Pseudomonas fluorescens*
Escherichia coli	*Citrobacter freundii*
Lactobacilli	*Proteus mirabilis*
Aerobacter aerogenes	*Enterobacter aerogenes*
Bifidobacteria	*Enterobacter cloacae*
Enterobacter aerogenes	*Paracolobacterium coliforme*
C. cochlearium	*Achrombacter pestifer*
Aspergillus niger	*Serratia plymuthica*
Scopulariopsis brevicaulis	*Staphylococcus sp.*
Human intestinal bacteria (streptococci, staphylococci, *E coli*, yeasts)	*Pseudomonas aeruginosa*
	Bacillus subtilis
	Flavobacterium marino typicum
	Citrobacter intermedius
	Pseudomonas fragi
	Desulfovibrio desulfuricans (anaerobe)

Data from References

chemically or whether it is also enzymatic.[160] Actual proof that CH_3CoB_{12} is involved at all in mercury methylation in the environment is still lacking, but circumstantial evidence is strong. The methylating factor in tuna fish liver content was shown to be chemically and chromatographically akin to CH_3CoB_{12},[13,14,161,162] and methylation in saline environments where $HgCl_4^{2-}$ is present is slower than where hydrated mercury(II) ions are present.[163] This suggests methylation is a carbanion transfer and it should be borne in mind that CH_3CoB_{12} is the sole methyl carbanion-donating natural methylating agent. As it is known to methylate easily mercury(II) in chemical media, CH_3CoB_{12} still remains the prime candidate for the methyl source in mercury methylation. The other main natural methylating systems transfer methyl groups as methyl carbonium ions (CH_3^+; e.g. *S*-adenosylmethionine and N^5-methyltetrahydrofolate) and would, therefore, be unlikely to transfer to positive mercury(II) ions. In addition to these, naturally occurring methyl donor molecules such as iodomethane (CH_3I), dimethyl sulphide ($(CH_3)_2S$), betaine $((CH_3)_3^+NCH_2COO^-)$ and the trimethylsulphonium cation $((CH_3)_3^+S)$ have been reacted with mercury. None of these will methylate electrophilic mercury(II) since all of them transfer the methyl group essentially as carbonium ion. However, the range of reduction potentials naturally available in certain sediment environments allows reduction of mercury(II) to mercury(0). Therefore, reduced mercury is expected to occur in some locations[139] and might be expected to undergo oxidative methylation, e.g. by methyl carbonium ions. Figure 2.3 shows zones of stability of mercury and methyl species in water or sediments. Methyl carbonium transfer to mercury(0) using iodomethane as an example is shown in equation 2.2.[164]

$$CH_3I + Hg(0) \longrightarrow CH_3Hg^{II}I \qquad [2.2]$$

Such oxidative additions are common in organometallic chemistry where a metal has two stable oxidation states separated by two units. (Ch. 1, sect. 1.2).

In practice none of the above molecules was observed to methylate mercury metal distributed in a natural sediment. Early work showed in fact that light activation was necessary for the methylation with iodomethane and light is only available at the surface of reducing sediments, and sometimes not at all.[164] To the present time, therefore, most attention has focused on CH_3CoB_{12} as the prime mercury methylating molecule and few experiments to test oxidative methylation of mercury(0) have taken place. In locally polluted environments man-made methylating agents may be present (e.g. $(CH_3)_3Pb^+$ from $(CH_3)_4Pb$ decay). These species can methylate mercury(II) in the laboratory in aqueous solution and have been held responsible for methylation in a few natural cases.[165-168] In one case where mutual pollution by mercury and lead alkyls occurred, methylmercury levels of 125 ng g^{-1} and 20-40 ng g^{-1} were found downstream and upstream respectively of the location of the lead source. Transalkylation from lead was suggested as the cause of the excess methylmercury downstream (equation 2.3).[166]

$$(CH_3)_3Pb^+ + Hg^{2+} \longrightarrow (CH_3)_2Pb^{2+} + CH_3Hg^+ \qquad [2.3]$$

This process is, of course, a carbanion transfer.

Other non-cobalamin processes responsible for methylation might also include photolysis of mercury(II) in the presence of acetates (which occur naturally), and photolysis of aliphatic amino acids in the presence of mercury salts.[169-172] Some water-soluble organosilicon compounds, e.g. $(CH_3)_3SiCH_2CH_2SO_3Na$, also transfer a methyl group to mercury(II).[173] Although this may be of no great environmental significance, a number of organosilicon compounds do occur at low levels there and some methylation may occur in this way. It should also be noted that ethylene and acetylene, which are both natural products, may convert $HgCl_2$ to alkylmercury and mercury(0).[174]

It has also been shown that humic and fulvic acids extracted from river sediment and leaf mould methylate mercury(II). Fulvic acid was a much more effective methylating agent than humic acid, and the low molecular weight fulvic acid fractions (MW < 200) are the most active methylators.[175] At pH 3.5 (existing but not very common in the environment) mesitylene, *p*-xylene and especially 2,6-di-t-butyl-4-methylphenol (butylated hydroxytoluene, i.e. BHT) present in humic materials had the capacity to methylate mercury. At pH 7 only BHT did so. The methylations were carried out in darkness, but were only effective at 70 °C. Yields were low, of the order of 4×10^{-3} per cent. Methylmercury production by BHT increased with temperature, inorganic mercury concentration, BHT concentration and varied with pH. Highest methylation rates were at pH 2 and pH 13.5. Abiotic methylation from other low molecular weight compounds in soil has also been reported.[137,176,177]

The very high methylmercury contents found in fish are probably due to absorption and retention from outside more than internal methylation, although intestinal and liver material in fish has methylating properties. Hence the low yields occurring above may still have significant environmental consequences owing to bioconcentration.

2.5.3 Dimethyl mercury

Once formed, methylmercury may be further methylated by CH_3CoB_{12} to dimethylmercury. In aqueous solution the second step is 6000 times slower than the first.[157] Further methylation may also occur in the presence of sulphide ions or hydrogen sulphide (H_2S) by a dismutation process (equation 2.4).[144]

$$2CH_3Hg^+ + S^{2-} \longrightarrow (CH_3Hg)_2S \longrightarrow (CH_3)_2Hg + HgS \qquad [2.4]$$

The presence of light is required for the completion of the process although the first stage may occur in darkness. The dimethylmercury formed may diffuse out of the sediment and through a water layer to the atmosphere. Formation of methylmercury followed by exposure to sulphide may, therefore, be a transport mechanism for organic mercury leading to wide diffusion to the environment, although dimethylmercury is not expected to be very stable in the atmosphere despite low sunlight absorption rates (Ch. 1, sect. 1.3.4). That the above process is practical in natural sediments has recently been demonstrated.[143] The addition of sodium sulphide to a natural sediment containing 58 ng g^{-1} of methylmercury leads to a reduced concentration of 45 ng g^{-1} of methylmercury over 14 days and the generation of dimethylmercury above the sediment layer. The formation of dimethylmercury was such as to suggest an annual loss by this route of at least 12 per cent of the methylmercury present in the sediment if it occurred at a steady rate. This process is likely to be significant in the observed loss of methylmercury from sediments, although other calculations suggest a loss of only 0.10 per cent per annum in the absence of sulphide reactions.[143]

Dimethylmercury is also an important product in soils.[178,179] Interestingly, methylmercury methanethiol (CH_3HgSCH_3) may occur in shellfish and was also found at Minimata in shellfish.[180] The conversion of methyl- to dimethylmercury will lead to greater transportation in the biosphere. Dimethylmercury is volatile and hydrophobic and by diffusion from sediment to atmosphere and subsequent decay there, it will also lead to movement of inorganic mercury.

2.5.4   Mercury methylation and speciation

For *abiotoc* methylation by CH_3CoB_{12} the rates of methylation decline as follows for mercury(II): nitrate > phthalate > acetate > chloride > > > sulphide. These rates are in inverse ratio to mercury−ligand bond strengths and imply carbanion attack on a positive mercury centre.[156,157,181,182] For methylation of added mercury(II) compounds *in sediments*, however, the rates and extend depend less on inherent metal−ligand bond strengths. The important factor seems to be the ease of biological breakdown by micro-organisms present in the sediment. Breakdown produces mercury(II) ions or weak mercury(II) complexes which are the species actually methylated. The order shown in Fig. 2.6 suggests progressively more difficult biological breakdown of the added mercury compounds to species capable of facile methylation.[182]

2.5.5   Mercury methylation and the biogeochemical cycle for mercury

There is a general biogeochemical cycle by which dimethyl- and methylmercury, mercury(II) and mercury(0) may interchange in the atmospheric, aquatic and sediment environments. This accounts for the formation and relatively low persistence of methylmercury and for the presence of methylmercury remote from areas where it has been directly introduced to the environment, but where mercury(II) may have been used. The biogeochemical cycle for mercury is shown in Fig. 2.7.

2.6   Organomercury levels and pathways in the environment

A number of large-scale collaborative efforts to study mercury transport and cycling in the aqueous environment were set up more than ten years ago at the height of concern about mercury pollution. Many of these have now reported their overall conclusions and they demonstrate the close connection between inorganic (mercury(0) and mercury(II)) and organic (methyl) mercury. In the

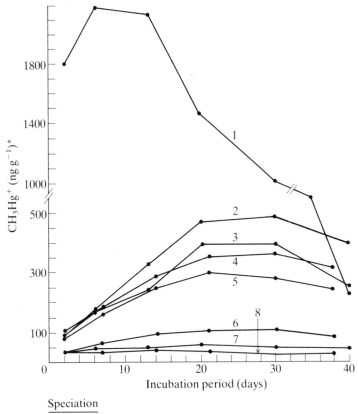

Fig. 2.6 Mercury methylation and speciation. From Craig and Moreton, 1985 – see Ref. 182. Permission received from Applied Science Publishers.

case of the Ottawa River study a total of fifty people were involved, but most studies have been on a smaller scale than this or have involved several independent groups studying a single geographic area.

The conclusions of this large-scale study of mercury transport in the Ottawa River, Canada, were reported in 1978.[183] The group concluded that most of the total mercury (96.7 per cent) and most of the methylmercury (97.8 per cent) is located in bed sediments. Biomass contained only 0.2 per cent of total mercury and 1.7 per cent of methylmercury present, although these components contained high concentrations of mercury. High proportions of

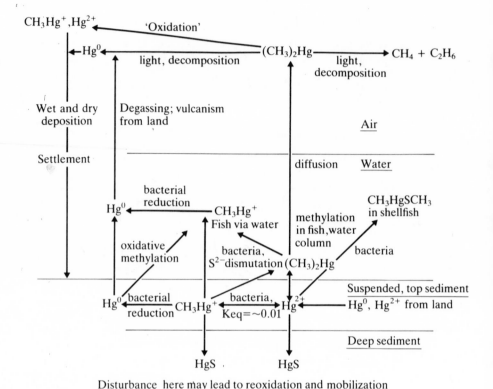

Fig. 2.7 The biogeochemical cycle for mercury

methylmercury were found only in the biological compartments, denoting their importance in toxicity studies. Transport of all forms of mercury in the river was accomplished mainly through movement of suspended sediments (58 per cent) and mercury dissolved in water (41 per cent). The role of bed sediment in movement is, therefore, small (1 per cent). The group also demonstrated the interconvertibility of inorganic mercury and methylmercury in water, sediments, aquatic plants, invertebrates and fish. Table 2.7 shows average mercury concentrations and methylmercury proportions found in the various compartments of the Ottawa River. Although the source of the mercury was uncertain (discharge from the pulp and paper industry or natural weathering from mercury-bearing rock), these results are typical in systems subject to inputs of mercury.

More recent work[184] by the same group has reported a general proportion of methylmercury to total mercury in several river waters of around 30 per cent. In this case, however, the authors do not report filtering the water and this figure may be due to methylmercury complexed to organic matter in the water. A proportion of 30 per cent seems rather high for the compartment.[185]

Table 2.7 Mercury concentrations and distributions in the Ottawa River, Canada

Component	Total mercury concentration (ng g^{-1})	Methyl mercury (%)	Fraction of all mercury present
Filtered water	0.013	< 1	1.3
Suspended matter	1140	< 1	1.8
Bed sediments	80.6*	3.8	96.7
Higher plants	14.2	20	
Benthic invertebrates	223	40	0.2†
Fish	162	85	

* In polluted sediments levels may be typically 100 times greater than this; but the proportion of CH_3Hg^+ is typical here.

† But note implication for bioconcentration and toxicity effects.

(From Kudo A, Miller D R, Akagi H, Mortimer D C, DeFreitas A S, Nagase H, Townsend D R, Warnock R G 1978 *Prog Water Technol* **10**: 329–39) Copyright Pergamon Press 1978. Reproduced with permission (Ref 183)

A similar project has been carried out in Canada on the English–Wabigoon River system in Ontario. This is a highly contaminated river system owing to discharge from a chlor-alkali plant located at Dryden. As an illustration of the effects of controls on mercury discharges it should be noted that discharges between 1962 and 1969 were estimated as of the order of 10 000 kg of mercury; since 1970 they have been 1 per cent of the previous figure. Even so, mercury concentrations in fish in parts of the system are still ten times greater than the legal limit for food consumption.[186–188] The main conclusions of this group at the present time were presented recently.[189] They conclude that most mercury methylation and bioaccumulation occur in the water column and surface sediment layers. Most of the mercury in the total system is buried below the surface sediment layers and does not contribute much to present mercury bioaccumulation. The surface sediment mercury is active, however, as also was the mercury in the water column. Here there was movement between geochemical and biological compartments. Because of this movement substantial amounts of surface sediment mercury and methylmercury were transported down the river lake system. In most fish examined the proportion of methyl to total mercury was over 80 per cent but for crayfish it was less than 30 per cent. This group also showed that methylation occurred in the intestinal content of all fish examined and that mercury *accumulation* in fish is not, therefore, the only cause of observed methylmercury levels.

The pathway of mercury from sediment to fish in the Minamata Bay area in Japan has been described recently.[190] These authors considered conversion from inorganic to methylmercury to occur at the zooplankton stage after release of mercury from the sediment. However, these authors also allowed for the conversion of the mercury in sediments (observed by many groups), followed by dissolution of methylmercury in water and concentration in the various biota leading to fish. The importance of the suspended material as a transport medium has been emphasized by recent studies, e.g. in the Thames estuary, UK where about 98 per cent (a higher figure than the Ottawa River study) of water-borne mercury was found associated with suspended matter following 0.45 μm filtration to define dissolved and particulate forms.[191] A similar proportion was found recently in British Columbia, Canada.[192] In estuaries periods of high chlorinity reduce suspended sediment mercury concentrations.

The origin of methylmercury under normal conditions may be considered to be natural methylation in the water column and sediments. The mercury entering the system is usually inorganic. It has recently been concluded that there is little possibility of the formation of methylmercury in a mercury-containing chlor-alkali plant.[193] For marine systems it has been estimated that similar amounts of methylmercury are formed in the water column and sediments[194] (up to 500 tonnes annually) leading to a total quantity of methylmercury in open ocean waters of 121 000 tonnes, at a concentration of 88 pg dm^{-3}. One group has shown that the reduced rate of methylation of mercury(II) by CH_3CoB_{12} in seawater is due to the bicarbonate rather than the chloride component, although chloride ion in the absence of bicarbonate hinders methylation.[195] Slower methylation in saline environments argues for the process being a carbanion transfer from CH_3CoB_{12}.

After production in sediments, various factors govern the rate of release of methylmercury to the water column. It was found to increase with total mercury content up to a concentration of mercury of $15-20$ μg g^{-1}, to increase with temperature and to increase as nutrients were added.[196] The pH range of natural sediment does not seem to affect mercury binding or methylation rates in the sediments markedly, but it does increase release of methylmercury.[197] Change of pH from 7.0 to 5.0 doubles release. This may account for elevated methylmercury levels in fish from low pH lakes, i.e. in such lakes the ratio of fish mercury to sediment mercury is high. This suggests that the ultimate source of the mercury in fish is the bed sediment − this is reasonable as most of the total mercury is here. Decreasing pH alters the ratio of methylmercury between bed sediment and water column and this change is adequate to explain the increased mercury concentration in fish. Uptake by fish is not pH-dependent but it is mercury-concentration-dependent. It has also been found that the ratio of methyl (fish) to total mercury (sediment) also increases as pH declines;[198] in one case it was 0.84 at pH 5.3, 0.56 at pH 5.7 and 0.41 at pH 6.4. This suggests that the influence of acid rainfall has consequences for organic mercury mobilization as well as for inorganic mercury release from rocks. Even mercuric sulphide is slowly mobilized from sediments under aerobic conditions and taken up by fish, but at slower rates than for mercuric chloride.[199] Total mercury concentration on suspended sediments decreases on exposure to high chlorinity.[200]

These mobilization processes allow absorption of methylmercury and inorganic mercury ultimately by fish and methylation also takes place in fish. Levels of methylmercury in fish may therefore accumulate by these processes to beyond the $0.3-1.0$ μg g^{-1} concentrations allowable in fish for food consumption by the various national regulatory bodies.[201] Table 2.8 gives further details for methyl- and total mercury levels found in various environmental compartments.

Total and organic mercury levels have been measured intensively for a wide variety of environmental compartments during the past fifteen years. In some cases organic levels have not been separately determined as total levels are very low. In general, ambient levels in air are $2-6.0$ ng m^{-3} for total mercury, and in natural uncontaminated waters below around 2.0 ng dm^{-3} g^{-1} but in the ng g^{-1} level in certain mineralized or industrialized areas. In sediments $0-100$ μg g^{-1} of total mercury and $0-100$ ng g^{-1} of methylmercury are often observed in industrial, urban or mineralized areas. In most drinking waters total mercury is less than 2 ng g^{-1} and in natural soils it is in the $0.02-0.4$ μg g^{-1} range. Of twelve food groups monitored by the US FDA since 1971 only the meal, fish and poultry groups contain regularly measurable amounts of mercury.[202] The others contain less than 0.01 μg(Hg) g^{-1}. Total mercury levels in whole blood (0.005 μg g^{-1}), hair (10 μg g^{-1}), kidney ($1-3$ μg g^{-1}) and liver (0.75 μg g^{-1}) are found as noted in unexposed human populations (Table 2.8).

Despite methylmercury levels in sediments declining as emissions of mercury decrease, measurements in Sweden show no apparent change in the mercury content of fish between 1967 and 1979 when mercury discharges were reduced considerably. In addition to increases due to increased mercury loadings, decrease in water pH also causes increased mercury concentrations in fish (see above).

The extent of biomagnification for mercury and methylmercury may be deduced from the ratios shown in Table 2.9.

Overall pathways in the environment are given in Fig. 2.7. From measurements given in Tables 2.8 and 2.9 it can be seen that serious risk of methylmercury poisoning in humans is likely to arise in only two situations. The first is through inadvertent consumption of methylmercury-treated seed and the second arises in communities heavily dependent on fish diets where the fish have high mercury concentrations.

2.7 Methylmercury and selenium

Dietary selenium can act as a protective agent against the toxic effects of certain heavy metals.[203] In fact sodium selenite (Na_2SeO_3) or organoselenium compounds may reverse the effects of methylmercury poisoning in several species.[204] The mechanism appears to be release of methylmercury by selenium complexation from its former linkages to protein sulphydryl groups — the

Table 2.8   Methyl and total mercury in environmental compartments.

Compartment	Total mercury (μg g^{-1})	Percentage as CH$_3$Hg$^+$	Comments
Polluted sediments	1–20	1–3	Normal range, e.g. for polluted rivers; gross pollution gives higher levels
Normal sediments	0.2–0.4	0	
Air (clean)	0.5–5 ng m^{-3}	30	Maximum % for CH$_3$Hg. Mean [Hg] = 0.7 ng m^{-3}
Air (urban)	0.5–50 ng m^{-3}	0–5	Mean = 7.0 ng m^{-3}
Air (polluted – industrial)	20–50 ng m^{-3}	5	Above cinnabar, 20 000 ng m^{-3} known. High levels known above volcanos.
Remote rain	1–10 ng m^{-3}	Unknown	Deposition 5–10 km from point mercury source gives X100 background levels. Dry deposition has similar values
Continental rain	5–30 ng m^{-3}	Unknown	
European rain	10–60 ng m^{-3}	Unknown	
Freshwater fish	0.2–1.0	70–95	Uncontaminated, i.e. from remote location
Contaminated freshwater fish	1.0–7.0	70–95	Some pike in polluted locations have > 10 μg g^{-1}. 23 μg g^{-1} known.
Pike (no direct pollution)	0.36–1.0	70–95	Natural uncontaminated levels in fish are ~ 0.2 μg g^{-1}
Pike (remote, historic)	0.05–0.2	70–95	
Fish in acid lakes	0.7–2	70–95	

Poisoned birds	120	50–100	Poisoned by mercury coated seed – liver concs. Eggs were 3–11 μg g⁻¹
Uncontaminated snow	0.23 ng g⁻¹	Unknown	Greenland snow deposited 1960s
Urban snow	0.002	Unknown	
Snow near chlor-alkali plant	0.0	0	
Non-contaminated river water	0.01–1 ng g⁻¹	0	Including suspended material
Coastal waters	0.01–0.1 ng g⁻¹	Unknown	Uncontaminated; at Minamata Bay levels were about 0.66 ng g⁻¹
Rural surface waters	0.03–0.2 ng g⁻¹	Unknown	
Ocean waters	0.01–0.03 ng g⁻¹	10, conc = ~ 88 pg dm⁻³	
Freshwater	0–0.1 ng g⁻¹	Unknown	Uncontaminated
Agricultural soils	usually < 0.2	Unknown	Mean is 0.1 μg g⁻¹. > 35 μg g⁻¹ known in mineralized areas, etc.
Normal rock	usually < 0.2	0	Mean in some areas > 17 μg g⁻¹ (e.g. Crimea)
Ores	0.5–1.2%	0	
Crustal abundance	0.070	0	
Plants	< 0.1	Unknown	Soils with 10 μg g⁻¹ mercury. Edible portions; not onions. i.e. Little uptake of mercury and no bioconcentration
	0.001–0.02	Unknown	Typical range from non-mercurial soils and no mercury treatment. Some fungi have several μg g⁻¹

Table 2.8 (cont)

Compartment	Total mercury (μg g^{-1})	Percentage as CH$_3$Hg$^+$	Comments
Grain	0.008–0.016	Unknown	From seed wth 12–16 μg g^{-1} Hg
Fruit	< 0.070	Unknown	From treated trees. Average level. Ranges up to 0.36 μg g^{-1}
Zooplankton	0.14		Unpolluted area
Marine fish	0.01–1.5	60–90	Unpolluted. Larger, older fish have higher levels
	50	60–90	In Minamata Bay
Tuna, swordfish	0.3–1	60–95	Natural levels
Shellfish	0.14–0.75	40–90	Unpolluted
	11–40	40–90	Minamata Bay
Seal – liver	5.25	9.1	Non-contaminated
– meat	0.43	62.9	
Whale	0–10	Unknown	
Coastal seabirds	24–48	Unknown	Eiders, UK, 1973, liver concentrations
Pelagic seabirds	< 10	Unknown	Livers
Puffins	2.8–7.7	Unknown	Livers, mean = 2 μg g^{-1} in eggs. UK 1972
Guillemot eggs	4–6	Unknown	UK, 1973. Kittiwake eggs were ~2μg g^{-1}

Gannets	5.9–52.7	Unknown	Livers from dead birds, UK 1972. Possibly poisoned by mercury
Waders	0.3–6/8 range	Unknown	Seasonal range, UK 1973
Deer – heart, liver, meat	90–130 ng g⁻¹	20–45	Taken from remote location
Wood pigeons	0.04–0.43	Unknown	UK, seasonal range. Birds possibly poisoned by treated seed have levels up to 10 μg g⁻¹
Owls	0.18–0.25		UK 1976. Lower than for European equivalents
Human organs – brain	100–400 ng g⁻¹	38.7	Non-occupationally exposed
kidney	1.3	6.0	
heart, liver	140–240 ng g⁻¹	40.0	
spleen, placenta	186 ng g⁻¹ (spleen), 20–40 ng g⁻¹ (placenta)	57.0	

Data from References

Table 2.9   Some bioconcentration factors
for total mercury

Matrix	Concentration range
Freshwater, seawater	1
Algae	10^3
Macrophytes	10^3
Seaweeds	10^4
Fish	$10^4 - 10^5$
Invertebrates	10^5
Oysters	$10^4 - 10^5$
Marine mammals	$10^5 - 10^6$
Seabirds	$10^5 - 10^6$

basic cause of the toxicity.[203,205] Selenium appears to alter the tissue distribution (and therefore effects) of mercury rather than reducing the rate of absorption or increasing the rate of excretion and it does not seem to reduce the half-life of methylmercury in species.[206] The basic cause of the effect is the high affinity of mercury and methylmercury for selenium donor atoms, e.g. the formation constant for the $CH_3HgSeCN$ complex is greater than that for CH_3HgSCN (6.79 versus 6.05), i.e. the mercury−selenium linkage is stronger than that for mercury−sulphur[207] and selenium displaces mercury when it is complexed via sulphur to tissue. Hence selenium appears to operate by complexation with mercury and methylmercury and removal to locations having less deleterious effects. The effect is not due to elimination. Dietary selenium at 5 μg(Se) g^{-1} operates effectively against chronic toxicity of methylmercury in rats. In an experiment, levels of selenium exerting the effect in quail brain tissue were far below stoichiometric 1 : 1 levels and so direct 1 : 1 binding was not occurring.[205] It appears that each molecule of selenium could induce the formation of a large selenoprotein complex which may be able to bind numerous mercury atoms (not just at selenium sites).[208] Restoration of the essential selenium component of glutathione peroxidase − which catalyses removal of harmful peroxides − after its removal by toxic mercury or methylmercury has also been suggested as the mechanism of the mutual antagonism of selenium and mercury.[208,209] Unexpectedly, body retention of mercury is actually higher in selenium-treated animals in spite of the alleviation of toxic symptoms.[210]

Despite the selenoprotein theory outlined above, selenium and mercury are found in 1 : 1 molar ratios in marine animals,[211] tuna,[203] humans,[212] rats and in some cases excess selenium is found.[203] The mechanism of the mercury−selenium interaction is still mysterious and the ideas cited above are speculative. A number of detailed discussions are now available[213−215] and it should also be pointed out that not only antagonism but also synergism is known for selenium.

2.8 Source material on organomercury compounds in the environment – conclusions

The most complete single source describing the environmental effects of total and organic mercury is the book edited by Nriagu.[5] The chemical, rather than environmental, properties of organomercury compounds are discussed in detail in one volume of a recent series.[216] Other chemical information is available in detail in the book of Coates *et al.*,[217] and there is also a book including environmental aspects which is edited by McAuliffe.[218] Several monographs dealing with environmental and health aspects of mercury have been produced within the last ten years.[219-221] Two very recent and detailed reviews of the environmental behaviour of inorganic and organic mercury in the natural environment have appeared.[222,223] In one of these there is a particularly useful discussion of the stability and cycling aspects of mercury.[222] The use of trade names for mercury compounds used in agriculture or as pesticides has caused some confusion, especially to those not intimately connected with the area. There are a number of sources giving trade and/or chemical names for organomercury products.[224-226]

Several conclusions have emerged from the intensive research carried out on mercury in the environment during the past fifteen years. Where poisoning outbreaks have occurred they have been localized, although they may have been severe. Poisonings have occurred in two ways: ingestion of alkylmercury-treated food material or through consumption of a staple diet of mercury-rich fish. Environmental methylation of inorganic mercury is a general process leading to low levels of methylmercury in sediment or water. Ultimately if the food chain conditions are appropriate it may lead to high levels of methylmercury in fish. In this case methylation could be an ultimate cause of a poisoning outbreak in a community heavily dependent on fish. However, the well-known and tragic poisoning cases mentioned in section 2.1.1 did not arise from environmental methylation of mercury. The best documented example of an inorganic mercury pollution problem having potentially serious human problems is that which occurred in the English–Wabigoon River system in Ontario. The high levels of methylmercury in fish were caused by natural environmental methylation of the inorganic mercury effluent of a chlor-alkali plant. Commercial fishing in the region was banned in 1971 and medical surveys of members of local communities reliant on fish in their diet revealed possible, but not unequivocal, symptoms of low level methylmercury poisoning. Methylation of mercury in the environment is not only an academic question; it may give rise to serious practical problems, but in localized areas.

Much research in this area still remains to be carried out. The role of CH_3CoB_{12} in mercury methylation in the natural environment is still not unequivocally established, much less whether or not enzymatic methylation takes place. The cause of the low methyl : total mercury ratios found in sediments is still unknown. Food chain details are now clearer but the extent to which fish themselves methylate mercury is not established.

Finally, let none of us question the extent or resourcefulness of environmental science. Mercury levels in planetary soils on Mars have been estimated[227] and are such that growth of earth-type micro-organisms would be prevented. This suggests that at least we need harbour no lurking doubts on the subject of exobiological mercury methylation!

References

1. Smith W, Smith A 1975 (eds) *Minimata*. Holt, Rinehart and Winston, New York
2. Clarkson T W 1972 *C R C Reviews in Toxicology*. Chem Rubber Co, Cleveland, pp 203–34
3. Clarkson T W, Amin-Zaki L, Tikriti S 1976 *Fed Proc* **35**: 2395–9
4. Bakir F, Damliyi S F, Amin-Zaki L, Murtadha M, Khalidi A, Al-Ravi N Y, Tikriti S, Dhahir H I, Clarkson T W, Smith J C, Doherty R A 1973 *Science* **181**: 230–41
5. Nriagu J O (ed) 1979 *The Biogeochemistry of Mercury in the Environment*. Elsevier North Holland Biomedical Press, Amsterdam
6. Takizawa Y 1979 Epidemiology of mercury poisoning. In Ref 5 pp 325–67
7. Tsubaki T 1968 *Clin Neurol Tokyo, Rinsho Shinkeigaku* **8**: 511–20
8. Fimreite N 1979 Accumulation and effects of mercury on birds. In Ref 5 pp 601–27
9. Westoo G 1966 *Acta Chem Scand* **20**: 2131–7
10. Windom H L, Kendall D R 1979 Accumulation and biotransformation of mercury in coastal and marine biota. In Ref 5 pp 312–23
11. Noren K, Westoo G 1967 *Var Foeda* **19**: 13–7
12. Huckabee J W, Elwood J W, Hildebrand S G 1979 Accumulation of mercury in freshwater biota. In Ref 5 pp 277–302
13. Imura N, Pan S-K, Shimitzu M, Ukita T 1973 In *New Methods Environ Chem Toxicol*. Collected Papers Res Conf New Methods Ecol Chem (ed Coulston F). Int Acad Print, Tokyo, Japan pp 211–6
14. Imura N, Pan S-K, Shimitzu M, Ukita T, Tonamura K 1977 *Ecotoxicol Environ Safety* **1**: 255–61
15. Suzuki T 1979 Dose–effect and dose–response relationships of mercury and its derivatives. In Ref 5 pp 399–432
16. Jensen S, Jernelov A 1969 *Nature (London)* **223**: 753–4

17. Wood J M, Kennedy F S, Rosen C G 1968 *Nature (London)* **220**: 173–4
18. Bertilsson L, Neujahr H Y 1971 *Biochemistry* **10**: 2805–8
19. Yamamoto H, Yokoyama T, Chen J-L, Kwan T 1975 *Bull Chem Soc Japan* **48**: 844–7
20. Craig P J, Morton S F 1978 *J Organometallic Chem* **145**: 79–89.
21. DeSimone R E, Penley M W, Charbonneau L, Smith S G, Wood J M, Hill H A O, Pratt J M, Ridsdale S, Williams R J P 1973 *Biochim Biophys Acta* **304**: 851–63
22. Chu V C W, Gruenwedel D W 1976, *Z Naturforsch Teil C* **31**, 753–5
23. Robinson G C, Nome F, Fendler J H 1977 *J Amer Chem Soc* **99**: 4969–76
24. Nriagu J O 1979. In Ref 5 pp 34–9
25. *Environmental Mercury and Man* 1976 UK Dept Environ, Pollution Paper No 10, HMSO, London
26. Guiruswamy L D, Papps I, Storey D J 1983 *J Common Market Studies* **XXII**: 71–100
27. Ref 26 p 77
28. Ref 26 pp 91–3
29. Ref 25 pp 44–5
30. Beszedits S 1979 Mercury removal from effluents and waste waters. In Ref 5 pp 230–76
31. Edwards P R 1984 *Chem Ind (London):* 506–8
32. Moreton P A 1984 Formation of methyl mercury in the environment. PhD Thesis, Leicester Polytechnic, Leicester, UK
33. Bellinger E G, Benham B R 1978 *Environ Pollut* **15**: 71–81
34. Ref 25 p 47
35. *Mercury Bearing Wastes* 1976 UK Department of Environment, Waste Management Paper No 12, HMSO, London
36. *Official Journal of the European Communities* 1984 No L 74/53–4
37. *Official Journal of the European Communities* 1982 No L 81/32–4
38. *Bulletin of the European Communities* 1984 No 3/43
39. EEC Directive 1976 No 76/464/EEC
40. Takizawa Y 1979. In Ref 5 p 359
41. Gerstner H B, Huff J E 1977 *Clin Toxicol* **11**: 131–50
42. Feltier J S, Kahn E, Salick B, van Natta F C, Whitehouse M W 1972 *Ann Intern Med* **76**: 779–92
43. Berglund F, Berlin M 1969 Risk of methylmercury accumulation in men and mammals and the relation between body burden of methylmercury and toxic effects. In *Chemical Fallout* (eds Miller M W and Berg G C). Thomas, Springfield, Ill, pp 258–73
44. Norseth T, Clarkson T W 1970 *Arch Environ Health* **21**: 717–27
45. Swenson A, Ulfvarson U 1968 *Acta Pharmacol Toxicol* **26**: 273–83
46. Yoshino Y, Mozai T, Nakao K 1966 *J Neurochem* **13**: 397–406
47. Chang L A, Martin A, Hartmann 1972 *Exp Neurol* **37**: 62–7
48. Syversen T L M 1974 *Biochem Pharmacol* **23**: 2999–3007

49. Junghans R P 1983 *Environ Res* **31**: 1–31
50. Friburg L, Vostal J (eds) 1972 *Mercury in the Environment: An Epidemiological and Toxicological Appraisal.* Chemical Rubber Co, Cleveland, Ohio, USA
51. Miller M, Clarkson T (eds) 1973 *Mercury, Mercurials and Mercaptans.* Thomas, Springfield, Ill, USA
52. Report of an International Committee 1969 *Arch Environ Health* **19**: 891–905
53. Doi R, Tagana M 1983 *Toxicol and Applied Pharmacol* **69**: 407–16
54. Ferrara R, Petrosino A, Maserti E, Seritti A, Barghigiana C 1982 *Environ Tech Letters* **3**: 449–56
55. Schroeder W H, Jackson T A 1983 *Spec Conf Measures Monit Non Crit Toxic Contam Air* (ed Frederick E R) 91–100
56. Brosset C 1982 *Water Air Soil Pollut* **17**: 37–50
57. Braman R S, Johnson D L 1974 *Environ Sci Technol* **8**: 996–1003
58. Bzezinska A, Van Loon J, Williams D, Oguma K, Fuwa K, Haraguchi I H 1983 *Spectrochim Acta* **38**B: 1339–46
59. Dumarey R, Heindryckx R, Dams R, Hoste J 1979 *Anal Chim Acta* **107**: 159–167
60. Schroeder W H 1982 *Environ Sci Technol* **16**: 394A–400A
61. Chiba K, Yoshida K, Tanabe K, Haraguchi H, Fawa K, 1983 *Anal Chem* **55**: 450–3
62. Yamamoto J, Kaneda Y, Hikasa Y 1983 *Int J Environ Anal Chem* **16**: 1–16
63. Chau Y K, Saitoh H 1973 *Int J Environ Anal Chem* **3** 133–9
64. Bisogni J J, Lawrence A W 1974 *Environ Sci Technol* **8**: 850–2
65. Minagawa K, Takisawa Y, Kifune I 1979 *Anal Chim Acta* **115**: 103–10
66. Fujita M, Iwasahima K 1981 *Environ Sci Technol* **15**: 929–33
67. Stary J, Havlik B, Prasilova J, Kratzer K, Hanusova J 1978 *Int J Environ Anal Chem* **5**: 89–94
68. Bloom N, Crecelius E 1983 *Marine Chem* **14**: 49–59
69. Jenne E A, Avotins P 1975 *J Environ Qual* **4**: 427–31
70. Yamamoto J, Kaneda Y, Hikasa Y, Takabatake E 1983 *Water Res* **17**: 435–40
71. Clarkson T W, Greenwood M R 1970 *Anal Biochem* **37**: 236–43
72. Doherty P E, Dorsett R S 1971 *Anal Chem* **43**: 1887–8
73. Westoo G, 1969 In *Chemical Fallout*, C.C. Thomas, Springfield, Ill, 75–93
74. Tatton J O'G, Wagstaffe P J 1969 *J Chromatogr* **44**: 284–9
75. Gage J C 1961 *Analyst* **86**: 457–9
76. Magos L 1971 *Analyst* **96**: 847–53
77. Takizawa Y, Kosaka T 1966 *Acta Med Biol Niigata* **14**: 153–61
78. Westoo G 1968 *Acta Chem Scand* **22**: 2277–80
79. Westoo G 1967 *Acta Chem Scand* **20**: 1790–800
80. British Analytical Methods Committee 1977 *Analyst* **102**: 769–76

81. Nishi S, Horimoto Y 1968 *Bunseki Kagaku* **17**: 75–81
82. Uthe J F, Solomon J, Grift B 1972 *J Assoc Off Anal Chem* **55**: 583–9
83. Newsome W H 1971 J Agr Food Chem **19**: 567–9
84. Bachs C A, Lisk D J 1971 *Anal Chem* **43**: 950–2
85. Hartung R, Dinman B D 1972 *Environmental Mercury Contamination*. Ann Arbor Science Publishers, Ann Arbor, USA
86. Grossman W E L, Eng J, Tong Y C 1972 *Anal Chem Acta* **60**: 447–9
87. Dressman R C 1972 *J Chromatogr Sci* **10**: 472–5
88. Longbottom J E 1972 *Anal Chem* **44**: 1111–2
89. Capelli R, Fezia C, Franchi A, Zanicchi G 1979 *Analyst* **104**: 1197–200
90. Goolvard L, Smith H 1980 *Analyst* **105**: 726–9
91. Belcher R, Rodriguez-Vasquez J A, Stephen W I, Uden P C 1976 *Chromatographica* **9**: 201–4
92. Holak W 1982 *Analyst* **107**: 1457–61
93. Krenkel P A, Burrows W D 1975 Tech Report No 34, *Env and Water Resources Engineering Program*. Vanderbilt University, Nashville, Tenn, USA
94. deJong I G, Wiles D R 1976 *J Fish Res Board Can* **33**: 1324–30
95. Stary J, Prasilova J 1976 *Radiochem Radioanal Lett* **26**: 33–8
96. Shum G T C, Freeman H C, Uthe J F 1979 *Anal Chem* **51**: 414–6
97. Collett D L, Fleming D E, Taylor G A 1980 *Analyst* **105**: 897–901
98. Oda C E, Ingle J D Jnr 1981 *Anal Chem* **53**: 2305–9
99. Ireland-Ripert J, Bermond A, Ducauze C 1982 *Anal Chem Acta* **193**: 249–54
100. Rice G W, Richard J J, D'Silva A P, Fassel V A 1982 *J Assoc Off Anal Chem* **65**: 14–9
101. Shariat M 1979 *J Chromatographr Sci* **17**: 527–30
102. Cappon C J, Smith J C 1978 *Bull Environ Contam Toxicol* **19**: 600–7
103. Bye R, Paus P E 1979 *Anal Chim Acta* **107**: 169–75
104. Longbottom J E, Dressman R C, Lichtenberg J C 1973 *J Assoc Off Anal Chem* **56**: 1297–303
105. Morton S F 1977 Some studies in the aklylation and transport of mercury in the environment. PhD Thesis, Leicester Polytechnic, UK
106. Bartlett P D 1979 Studies in the distribution, mobility and methylation of mercury in the environment. PhD Thesis, Leicester Polytechnic, UK
107. Floyd M, Sommers L E 1975 *Anal Letters* **8**: 525–35
108. Kitamura S, Tsukamoto T, Hayakawa K, Sumino K, Shibata T 1966 *Med Biol Tokyo* **72**: 274–81
109. Talmi Y, 1975 *Anal Chem Acta* **74**: 107–17
110. Talmi Y, Norvell V E 1976 *Anal Chem Acta* **85**: 203–8
111. Sumino K 1974 *Water Technology VII*. Pergamon, New York, pp 35–45
112. Schafer M L, Rhea U, Campbell J E 1976 *J Agr Food Chem* **24**: 1078–81
113. Schafer M L, Rhea U, Peeler J T, Hamilton C H, Campbell J E 1975

J Agr Food Chem **323**: 1079–83

114. Johansson B, Ryhage R, Westoo G 1970 *Acta Chem Scand* **24**: 2349–54
115. Kamps L R, McMahon B 1972 *J Assoc Off Anal Chem* **55**: 590–5
116. Tanaka K, Shimada T, Fukui S, Kanno S 1972 *Eisei Kagaku* **18**: 256–63
117. Sumino K 1968 *Kobe J Med Sci* **14**: 115–30
118. Sumino K 1968 *Kobe J Med Sci* **14**: 131–48
119. Nishi S, Horimoto S 1970 *Bunseki Kagaku* **19**: 1646–51
120. Longbottom J E, Dressman R C 1973 *Chromatogr Newslett* **2**: 17–19
121. Gonzalez J G, Ross R T 1972 *Anal Lett* **5**: 683–94
122. Ohkoshi S, Takahashi T, Sato T 1973 *Bunseki Kagaku* **22**: 593–5
123. Freeman H C, Horne D A 1973 *J Fish Res Board Can* **30**: 454–6
124. Holden A V 1973 *J Food Technol* **8**: 1–25
125. Zelenko V, Kosta L 1973 *Talanta* **20**: 115–23
126. Von Burg R, Farris F, Smith J C 1974 *J Chromatogr* **97**: 65–70
127. Hall E T 1974 *J Assoc Off Anal Chem* **57**: 1068–73
128. Giovanoli-Jakubczak T, Greenwood M R, Smith J C, Clarkson T W 1974 *Clin Chem* **20**: 222–9
129. Takeuchi M, Ebato K, Ito K, Amemiya T, Harada H, Totani T 1974 *Tok Tor Eisei Kenk Kenkyu Nempo* **25**: 133–9 (see *Chem Abstr* 1975, **82**: 150028)
130. Ueta T, Yamazoe R, Yamanobe H, Totani T 1975. In Ref 129, **26**: 110–4 (see *Chem Abstr* 1976, **84**: 160202)
131. Hobo T, Ogura T, Suzuki S, Araki S 1975 *Bunsekei Kagaku* **24**: 288–93
132. Ealy J A, Shults W D, Dean J A 1973 *Anal Chem Acta* **64**: 235–41
133. Uthe J F, Armstrong F A J 1974 *Toxicol Environ Chem Rev* **2**: 45–77
134. Fishbein L 1970 *Chromatogr Rev* **13**: 83–162
135. Rodriguez-Vasquez J A 1978 *Talanta* **25**: 299–310
136. Topping, G, Davies I M 1981 *Nature* **290**: 243–4
137. Rogers R D 1977 *J Environ Qual* **6**: 463–7
138. Nagase H, Ose Y, Sato T, Ishikawa T 1984 *Sci Total Environ* **32**: 147–56
139. Wollast R, Billen G, Mackenzie F T 1975 *Environ Sci Res 7 (Ecol Toxicol Res)*: 145–67
140. Bartlett P D, Craig P J 1981 *Water Research* **15**: 37–47
141. Spangler W J, Spigarelli J L, Rose J M, Miller H M 1973 *Science* **180**: 192–3
142. Craig P J, Moreton P A 1983 *Mar Poll Bull* **14**: 408–11
143. Craig P J, Moreton P A 1984 *Mar Poll Bull* **15**: 406–8
144. Craig P J, Bartlett P D 1978 *Nature (London)* **275**: 635–8
145. Langley D G 1973 *J Water Poll Control Fed* **45**: 44–51
146. Spangler W J, Spigarelli J L, Rose J M, Flippin R S, Miller H M 1973 *Appl Microbiol* **25**: 488–93

147. Windon H, Gardner W, Stephens J, Taylor F 1976 *Est Coastal Mar Sci* **4**: 579–83
148. Compeau G, Bartha R 1983 *Bull Environ Contam Toxicol* **31**: 486–93
149. Baughman G L, Gordon J A, Wolfe N L, Zepp R G 1973 *US EPA Ecol Res Ser* EPA-660/3-73-012
150. Silver S 1984. In *Changing Metal Cycles and Human Health* (ed Nriagu J O). Dahlem Konferenzen 1984. Springer-Verlag, Berlin; Heidelberg; New York; Tokyo, pp 199–223
151. Summers A O, Silver S 1978 *Ann Rev Microbiol* **32**: 637–72
152. Misra T K, Silver S, Mobley H L T, Rosen B P 1984. In *Molecular and Cellular Approaches to Understanding Mechanisms of Toxicity* (ed Tashjian A H Jr). Harvard School of Public Health, pp 63–81
153. Bisogni J J 1979 Kinetics of methyl mercury formation and decomposition in aquatic environments. In Ref 5 pp 211–27
154. Hill H A O, Pratt J M, Ridsdale S, Williams F R, Williams R J P, 1970 *J Chem Soc Chem Commun*: 341–2
155. Lewis J, Prince R H, Stotter D A 1973 *J Inorg Nucl Chem* **35**: 341–51
156. Agnes G, Bendle S, Hill H A O, Williams F R, Williams R J P 1971 *J Chem Soc Chem Commun*: 850–1
157. Tauzher G, Dreos R, Costa G, Green M 1974 *J Organometallic Chem.* **81**: 107–10
158. Imura N, Sukegawa E, Pan S-K, Nagao K, Kim J-Y, Kwan T, Ukita T 1971 *Science* **172**: 1248–9
159. Schrauzher G N, Weber J, Beckham T M, Ho R K Y 1971 *Tetrahedron Letters* 275–7
160. Lindstrand K 1969 *Nature (London)* **204**: 188–9
161. Hidemitsu S K P, Imura N 1981 *Eisei Kagaku* **27**: 184–6
162. Imura N, Pan S-K, Ukita T 1972 *Chemosphere* **1**: 197–201
163. Blum J E, Bartha R 1980 *Bull Environ Chem Toxicol* **25**: 404–8
164. Maynard J L 1932 *J Amer Chem Soc* **54**: 2108–12
165. Beijer K, Jernelov A 1979 Methylation of mercury in aquatic environments. In Ref 5 pp 203–10
166. Jewett K L, Brinckman F E, Bellama J M 1975. In *Marine Chem Coastal Environ* (ed Church T C). ACS Symp Ser No 18, Washington D C, pp 304–18
167. Huey C, Brinckman F E, Grim S, Iverson W P 1974 *Proc Int Conf Trans Persist Chem Aquat Ecosystems*. NRC, Ottawa, Canada, pp 11: 73–8
168. Jewett K L, Brinckman F E 1974 *ACS Preprints Div Environ Chem* **14**: 218–25
169. Akagi H, Takabatake E 1973 *Chemosphere* **3**: 131–3
170. Akagi H, Fujita Y, Takabatake E 1975 *Chem Lett* **1**: 171–6
171. Hayashi K, Kawai S, Ohno T, Maki Y 1977 *J Chem Soc Chem Commun*: 158–9
172. Takabatake E, Fujita M, Akagi H, Hashizume K 1972. In Ref 162

pp 199–210
173. De Simone R E 1972 *J Chem Soc Chem Commun*: 780–1
174. De Filippis L F, Pallaghy C K 1975 *Bull Environ Contam Toxicol* **14**: 32–7
175. Nagase H, Ose Y, Sato T, Ishikawa T 1984 *Sci Total Environ* **32**: 147–56
176. Rogers R D 1977 *U S EPA Ecol Res Ser* EPA-600/3-77-007
177. Rogers R D 1976 *J Environ Qual* **5**: 454–8
178. Johnson D L, Braman R S 1974 *Environ Sci Technol* **38**: 1003–9
179. Lofroth G 1969 *Ecol Res Bull*, No 4. Swedish Nat Res Council
180. Fujita E 1969 *Kumamota Igakkai Zasshi* **43**: 47–62
181. Chu V C W 1976. Ph D Thesis, University of California, Davis, California, pp 91–6
182. Craig P J, Moreton P A 1985 *Environ Poll Ser B* **10**: 141–58
183. Kudo A, Miller D R, Akagi H, Mortimer D C, DeFreitas A S, Nagase H, Townsend D R, Warnock R G 1978 *Prog Water Technol* **10**: 329–39
184. Kudo A, Nagase H, Ose Y 1982 *Water Res* **16**: 1011–5
185. Armstrong F A J, Hamilton A L 1973. In *Trace Metals and Metal–Organic Interaction in Natural Waters* (ed Singer P C). Ann Arbor Science Publishers, Ann Arbor, Mi, USA, pp 131–56
186. Parks J W, Hollinger J D, Almost P M 1984 Mercury Pollution in the Wabigoon–English River System of Northwestern Ontario. Report available from OME, Thunder Bay, Ontario, Canada
187. Rudd J W M, Turner M A, Townsend B E, Swick A L, Furutani A 1980 *Can J Fish Aquat Sci* **37**: 848–57
188. Rudd J W M, Turner M A, Furutani A, Swick A L, Townsend B E 1983 *Can J Fish Aquat Sci* **40**: 2206–17
189. Rudd J W M, Furutani A, Turner M A 1980 *Appl Environ Microbiol* **40**: 777–82
190. Nishimura H, Kumagai M 1983 *Water Air Soil Pollut* **20**: 401–11
191. Nelson L A 1981 *Environ Tech Letters* **2**: 225–32
192. Wallace G T, Seibert D L, Holzknecht S M, Thomas W H 1982 *Estuarine Coastal Shelf Sci* **15**: 151–82
193. Inoko M, Matsuno T 1984 *Environ Pollut Ser B* **7**: 7–10
194. Topping G, Davies I M 1981 *Nature (London)* **290**: 243–4
195. Compeau G, Bartha R 1983 *Bull Environ Contam Toxicol* **31**: 486–93
196. Wright D R, Hamilton R D 1982 *Can J Fish Aquat Sci* **39**: 1459–66
197. Miller D R, Akagi H 1979 *Ecotoxicol Environ Safety* **3**: 36–8
198. Landner L, Larsson P O 1972 IVL Report B115 Swedish Inst Water Air Pollution Res, Stockholm (in Swedish)
199. Gillespie D C, Scott D P 1971 *J Fish Res Board Can* **28**: 1807–8
200. Rae J E, Aston S R 1982 *Water Res* **16**: 649–54
201. Hildebrand S G, Strand R H, Huckabee J W 1980 *J Environ Qual* **9**: 393–400

202. DeCarlo V J 1977 U S EPA Report No. 560/6-77-031 Batelle Columbus Laboratories, Columbus, Ohio, U S A
203. Genther H E, Goudie C, Sunde M L, Kopecky M J, Wagner P, Oh S-H, Hoekstra W G 1972 *Science* **175**: 1122−4
204. Chang L W, Reuhl K R, Lee G W 1977 *Environ Res* **14**: 414−23
205. Stoewsand G S, Bache C A, Lisk D J 1974 *Bull Environ Contam Toxicol* **11**: 152−6
206. Carty A J, Malone S F 1979 The chemistry of mercury in biological systems. In Ref 5 pp 433−80
207. Rabinstein D L, Tourangeau M C, Evans C A 1976 *Can J Chem* **54**: 2517−25
208. Chang L W 1979 Pathological effects of mercury poisoning In Ref 5 pp 519−80
209. Taylor T J, Reiders J, Kocsis J J 1978 *Toxicol Appl Pharmacol*
210. Eybl V, Sykora J, Mertl F 1969 *Arch Toxicol* **25**: 296−305
211. Koeman J H, Peeters W H, Koudstaal-Hol C H M, Tijoe P S, de Goeiff J J M 1973 *Nature (London)* **245**: 385−6
212. Kosta L, Byrne A R, Zelenko V 1975 *Nature (London)* **254**: 238−9
213. Groth D H, Stettler L, Mackay G 1976 In *Effects and Dose−Response Relationships of Toxic Metals* (ed Nordbergf G F). Elsevier, Amsterdam, pp 527−43
214. Cappon C J, Crispin Smith J 1981 *J Anal Toxicol* **5**: 90−8
215. Magos L, Webb M 1979 Synergism and antagonism in the toxicology of mercury. In Ref 5 pp 581−600
216. Wardell J L 1982. In *Comprehensive Organometallic Chemistry* (eds Abel E W, Stone F G A, Wilkinson G). Pergamon Press, Oxford, pp 863−978
217. Coates G E, Green M L H, Wade K 1967 *Organometallic Compounds*, Vol 1 Methuen, London, pp 147−77
218. McAuliffe C A (ed) 1977 *The Chemistry of Mercury*. MacMillan, London
219. *Mercury and the Environment. Studies on Mercury Use and Emission, Biological Impact and Control* 1974 Organisation for Economic Cooperation and Development, Paris
220. *Mercury Contamination in Man and his Environment* 1972 International Atomic Energy Agency, Vienna
221. *Environmental Health Criteria 1, Mercury 1976* World Health Organisation, Geneva
222. Lindquist O, Jernelov A, Johansson K, Rodhe H 1984. In *Mercury in the Swedish Environment Global and Local sources*. Report of the National Swedish Environmental Protection Board. No. snv pm 1816.
223. Clarkson T W, Hamada R, Amin-Zaki L 1984. In *Changing Metal Cycles and Human Health* (ed Nriagu J O) Life Res Report No 28. Dahlem Konferenzen, Springer Verlag

224.   D'Itri, F M 1972 *The Environmental Mercury Problem* Ch III: 20−4
225.   Anonymous 1982 *Chemical Compounds Used as Pesticides* U K Dept of Environ (Series of Recommendations Sheets)
226.   *The Agrochemicals Handbook* 1983 Royal Society of Chemistry, UK
227.   Siegal B Z, Siegal S M 1979 *Life Sci Space Res* **17**: 87−97

Chapter 3

Organotin compounds in the environment

3.1 Introduction

During the last decade, interest in many countries in the environmental chemistry of organotin compounds has risen dramatically, and several articles have already been written[1-9] reviewing certain areas of this topic. The reason for the increased awareness of the environmental aspects of organotins is probably due to the fact that the world-wide production of organotin chemicals has risen, over the last thirty years, from under 5000 tonnes in 1955 to at least 35 000 tonnes[10] at the present time. This value represents approximately 7 per cent of the total consumption of tin metal. The increased production and consumption is due primarily to the wide range of industrial applications discovered for organotin chemicals. In fact tin has a larger number of its organometallic derivatives in commercial use (Fig. 3.1) than any other element. However, before these industrial applications are reviewed, a brief description of the toxicological pattern of organotin chemicals will be given, since, although inorganic tin compounds are basically not harmful, some organotin species are toxic to both animal and vegetable life, and this relates to many of the uses and environmental parameters discussed later.

3.2 Toxicological patterns for organotin compounds

3.2.1 Toxicology and mode of action

The toxicological pattern of organotin compounds has been found to be very complex.[11] However, in general, progressive introduction of organic groups at the tin atom in any R_nSnX_{4-n} series produces a maximum biological activity when $n = 3$, i.e. for the triorganotin compounds, R_3SnX.[12] The nature of the

Fig. 3.1   Some organotin-containing products (Photograph courtesy of International Tin Research Institute)

X group in an R_3SnX derivative generally has very little effect on the biological activity,[12] unless, of course, X is itself biologically active, in which case the activity of the compound may be enhanced. It has recently been shown,[13] however, that in complexes where the X group results in the formation of a polymeric structure, or a five coordinate chelated monomer, a significant reduction in activity may occur. Within any R_3SnX series, the species towards which the triorganotin compound is most active is determined by the three organic groups (Table 3.1). For example, in the tri-n-alkyltin series, the trimethyltins have a high toxicity to insects and mammals,[14-16] the triethyltins are most toxic to mammals,[15,16] the tripropyltins to Gram-negative bacteria[17] and the tributyltins to Gram-positive bacteria and fungi.[17] Further increase in the n-alkyl chain length produces a sharp drop in biological activity and the trioctyltin compounds are essentially non-toxic to all living organisms.[16] Additionally, an effect of particular importance to the environment is the high toxicity of tributyl-, triphenyl- and tricyclohexyltin derivatives to fish (Table 3.2).[17-22] A bibliography of the toxicity of organotins to aquatic animals has recently been compiled.[23]

The tetraorganotins, R_4Sn, have been found to exhibit a delayed toxic action, which may be due to their conversion to a triorganotin compound, R_3SnX, in the organism, since *in vitro* and *in vivo* studies have shown this process to occur, particularly in the liver.[24,25]

Table 3.1   Species specificity of triorganotin compounds, R_3SnX

Species	R in most active R_3SnX compound
Insects	CH_3
Mammals	C_2H_5
Gram neg. bacteria	$n\text{-}C_3H_7$
Gram pos. bacteria, fish, fungi, molluscs, plants	$n\text{-}C_4H_9$
Fish, fungi, molluscs	C_6H_5
Fish, mites	$cyclo\text{-}C_6H_{11}$ $C_6H_5(CH_3)_2CCH_2$

The underlying cause of the broad spectrum of acute toxicity shown by the triorganotin derivatives is thought to be due to derangement of mitochondrial functions, which can be brought about:[26-28]

(a)   by interaction with mitochondrial membranes to cause swelling and disruption;

(b)   by secondary effects derived from their ability as ionophores to derange mitochondrial function through mediation of chloride–hydroxide ion exchange across the lipid membrane; and

(c)   by their ability to inhibit the fundamental energy conservation processes involved in the synthesis of adenosine triphosphate, in which living organisms share many common features, e.g. triorganotins inhibit mitochondrial oxidative phosphorylation and, in chloroplasts, inhibit photosynthetic phosphorylation.

The biological activity of triorganotin compounds is believed to be due to their ability to bind to certain proteins,[27,29,30] e.g. yeast mitochondrial membrane,[31] and although the exact nature of the binding site is not known, studies of triethyltin compounds on cat haemoglobin have suggested that cysteine residues and histidine are involved.[30,32] The toxic effects arising from mild cases of triorganotin poisoning have generally, in both animals and man, been found to be reversible.[33] However, it has recently been observed that trimethyltin compounds can cause selective and irreversible neuronal destruction in the brain.[34-36]

The biological effects of the diorganotin compounds, R_2SnX_2, are also due to their ability to inhibit oxygen uptake in mitochondria. However, the mechanism of their toxic action differs, since the diorganotins have been shown to inhibit α-ketoacid oxidation,[27,28] probably by combining with enzymes or coenzymes, possessing vicinal dithiol groups. Therefore, in this

Table 3.2 Toxicity of organotin compounds to fish

Compound	Fish species	LC_{50}* (mg dm^{-3})	Reference
$((C_4H_9)_3Sn)_2O$	*Leuciscus idus melanotus* (Golden orfe)	0.05 (48 hr)	3
$(C_4H_9)_3SnOAc$	*Lebistes reticulatus* (Guppy)	0.026 (48 hr)	18
$(C_4H_9)_3SnF$	*Alburnus alburnus* (Bleak)	0.06–0.08 (96 hr)	19
$(C_4H_9)_3Sn$ (naphthenate)	*Leuciscus idus melanotus* (Golden orfe)	0.07 (48 hr)	3
$(C_4H_9)_4Sn$	*Leuciscus idus malenotus* (Golden orfe)	10.0 (48 hr)	3
$(C_4H_9)_2SnCl_2$	*Leuciscus idus melanotus* (Golden orfe)	1.0 (48 hr)	3
$(C_4H_9)SnCl_3$	*Leuciscus idus melanotus* (Golden orfe)	> 45 (48 hr)	3
$(C_6H_5)_3SnOH$	*Salmo gairdneri* (Rainbow trout fry)	0.015 (96 hr)	20
$(C_6H_5)_3SnOAc$	*Lebistes reticulatus* (Guppy)	0.034 (48 hr)	19
$(C_6H_5)_3SnF$	*Alburnus alburnus* (Bleak)	0.40 (96 hr)	19
$(c\text{-}C_6H_{11})_3SnOH$	*Microterus salmonoides* (Large mouth bass)	0.06 (24 hr)	21
$(c\text{-}C_6H_{11})_3Sn$—— —$\overline{N.C=N.C=N}$ | | H H	*Carassius auratus* (Goldfish)	0.01–0.1 (96 hr)	22

* LC_{50} is the initial concentration of a compound killing 50 per cent of the test organisms during the time denoted in parentheses. (mg dm^{-3} = p.p.m.). See Ref. 17.

case the nature of the X group may affect the toxicity. For example, comparison[16] of the acute oral toxicities, to rats, of $(CH_3)_2SnCl_2$ and $(CH_3)_2Sn(SCH_2CO_2\text{-i-Oct})_2$ shows that the former compound, which would be expected to react readily with dithiol moieties, is moderately toxic, whereas the latter, in which two Sn—S bonds are already present in the molecule, is essentially non-toxic. As with the trialkyltins, the mammalian toxicity of the dialkyltins generally decreases with increasing length of the alkyl chain.[16] However, one particular effect of certain dialkyltins, noted by Seinen,[37,38] is their toxicity towards lymphocytes and the suppression of thymus-dependent

Table 3.3 Carcinogenicity studies on organotin chemicals

Compound	Species	Type of test	Result	Reference
$(C_6H_5)_3SnOAc$	Mice (male and female)	18 month feeding study	Negative	39
$(C_6H_5)_3SnOH$	Mice (male and female)	18 month feeding study	Negative	40
$(C_6H_5)_3SnOH$	Rats (male and female)	18 month feeding study	Negative	40
$(C_4H_9)_3SnF$	Mice (male)	6 months dermal study	Negative	41
$(c\text{-}C_6H_{11})_3SnOH$	Rats (male and female)	2 year feeding study	Negative	42
$(C_4H_9)_2Sn(OAc)_2$	Mice (male and female)	18 month feeding study	No conclusive evidence for carcinogenicity	43
$(C_4H_9)_2Sn(OAc)_2$	Rats (male)	18 month feeding study	No conclusive evidence for carcinogenicity	43
$((C_6H_5(CH_3)_2CCH_2)_3Sn)_2O$	Mice (male and female)	18 month feeding study	Negative	44

immunity in rats. This appears to be more marked for compounds of longer chain length and is in contrast to the general pattern of acute toxicities, expressed by LD_{50} values.

The monoorganotin compounds, $RSnX_3$, do not appear to have any important toxic action in mammals,[14] and tin in its inorganic form is generally accepted as being non-toxic[14] because, at physiological pH, the metal does not react and the oxides are insoluble.

It is of interest to note that no carcinogenic effects have yet been demonstrated for any of the organotin compounds tested to date (Table 3.3). A similar study of bis(tri-n-butyltin) oxide is currently under way.[45]

3.2.2   Metabolism

A knowledge of the metabolic fate of organotin compounds in mammals is obviously of considerable importance. It was first demonstrated by Cremer[24] that tetraethyltin is metabolized *in vivo* in rats to a triethyltin derivative. It has also been shown that diethyltin species are broken down *in vivo* in rats to monoethyltin salts, which are rapidly eliminated from the body.[46] The metabolism *in vivo* in rabbits of a series of tetraalkyltin compounds has been studied by Wada and his co-workers[11,47−49] and it was found that the extent of conversion of $R_4Sn \rightarrow R_3SnX$ decreased as the length of the alkyl chain increased.[47]

Freitag and Bock have examined the metabolism of triphenyltin chloride ($(C_6H_5)_3SnCl$) in rats, using a [113]Sn labelled compound, and, by analysis of the degradation products in urine, found that di- and monophenyltin compounds were formed, along with inorganic tin.[50] Additionally, Fish and his co-workers[25] have studied the breakdown of [113]Sn labelled triphenyltin acetate ($(C_6H_5)_3SnOAc$) in rats, and found a similar breakdown pattern of this triaryltin compound. *In vivo* studies in rats and mice using tricyclohexyl-,[51] triethyl-[49] and tributyltin[49,52−54] compounds have indicated that they are excreted essentially intact from the animals and there appears to be no problem of long-term accumulation. However, Blair reported[55] that tricyclohexyltin hydroxide is metabolized *in vivo* in rats to di- and mono-organotin and inorganic tin species, Wada *et al.*[56] recently demonstrated, in rats, that tributyltin fluoride, once transported to the liver, is dealkylated, and Ward *et al.*[57] found that bis(tributyltin) oxide was metabolized in sheepshead minnows, *Cyprinodon variegatus*, to form dibutyl-, monobutyl- and inorganic tin products.

In vitro animal studies have demonstrated, using rat liver microsomal mono-oxygenase enzyme systems, that the primary metabolic reaction for butyltin compounds is not Sn—C bond cleavage, but carbon hydroxylation of the butyl groups, where the α- and β-carbon hydrogen bonds are found to be susceptible to hydroxylation.[25] Similarly, results indicating hydroxylation of the organic group have been obtained with tricyclohexyltin hydroxide,[51] but this process does not appear to occur with triphenyltin species.[25,58]

In the majority of studies carried out to date on the metabolism of organotin compounds, there is very little information on the dealkylation or dearylation products derived from the organic moieties. Recent work by Prough *et al.*[59] on the NADPH- and oxygen-dependent microsomal metabolism of the ethyltin series $(C_2H_5)_nSnX_{4-n}$ $(n = 2-4)$, has demonstrated that the major organic metabolite is ethylene and the minor product, ethane. Similarly, Fish and his co-workers[25] detected 1-butene from the butyltin compounds $(n-C_4H_9)_4Sn$ and $(n-C_4H_9)_3SnX$ (where X = Cl, OAc and OSn $(n-C_4H_9)_3)$.

3.3   Industrial applications of organotin compounds

The wide range of industrial applications of organotin chemicals, and the specific compounds used are listed in Table 3.4.

At the present time, the biocidal uses[60,61] of the triorganotin compounds (approximately 8000 tonnes world-wide) are exceeded by the non-toxic applications[62,63] of the di- and monoorganotin derivatives (approximately 27 000 tonnes world-wide). The tetraorganotins do not have any large-scale commercial outlets,[64] at the present time, but are important intermediates in the manufacture of other organotins, R_nSnX_{4-n}, from $SnCl_4$.[65]

Some of the major industrial applications of tri-, di- and monoorganotins are described below.

3.3.1   Organotins in agriculture

Triphenyltin acetate and triphenyltin hydroxide have both been widely used,[66-68] since the early 1960s, to combat a range of fungal diseases in various crops, particularly potato blight, leaf spot on sugar beet and celery, rice blast and coffee leaf rust. It is interesting to note[69] that these triphenyltin fungicides also possess antifeedant properties, in that they deter insects from feeding, and, additionally, they may act as insect chemosterilants.

Tricyclohexyltin hydroxide was introduced in 1968 as an acaricide for the control of mites on apples, pears and citrus fruits.[66,70] More recently, bis(trineophyltin) oxide[66,71] (i.e. 2-methyl-2-phenylpropyl = neophyl) and 1-tricyclohexylstannyl- 1, 2, 4-triazole[22,66] have been introduced as acaricides (see Table 3.4). The organotin fungicides and acaricides function mainly through a prophylactic action and tend not to be absorbed by the plant.[68]

Tributyltin compounds are not used in agriculture, due to their relatively high phytotoxicity, and, although trimethyltin derivatives, e.g. $(CH_3)_3SnSn(CH_3)_3$,[72] possess a high insecticidal activity, their use in the field has been precluded by their high mammalian toxicity.

Table 3.4   Industrial applications of organotin compounds

Application	Compound
	$\underline{R_3SnX}$
Agriculture (fungicides)	$(C_6H_5)_3SnX$ (X = OH, OAc)
(antifeedants)	$(C_6H_5)_3SnX$ (X = OH, OAc)
(acaricides)	$(c\text{-}C_6H_{11})_3SnX$
	(X = OH, $-N.C=N.C=N$) with H, H
	$(C_6H_5(CH_3)_2CCH_2)_3Sn)_2O$
Antifouling paint biocides	$(C_6H_5)_3SnX$ (X = OH, OAc, F, Cl, SCS.N(CH_3)_2, OCOCH_2Cl, OCOC_5H_4N\text{-}3)
	$(C_6H_5)_3SnOCOCH_2CBr_2COOSn(C_6H_5)_3$
	$(C_4H_9)_3SnX$ (X = F, Cl, OAc)
	$((C_4H_9)_3Sn)_2O$
	$(C_4H_9)_3SnOCOCH_2CBr_2COOSn(C_4H_9)_3$
	$(C_4H_9)_3SnOCO(CH_2)_4COOSn(C_4H_9)_3$
	$(-CH_2C(CH_3)(COOSn(C_4H_9)_3)-)_n$
Wood preservative fungicides	$((C_4H_9)_3Sn)_2O$
	$(C_4H_9)_3Sn(naphthenate)$
	$((C_4H_9)_3Sn)_3PO_4$
Stone preservation	$((C_4H_9)_3Sn)_2O$
Disinfectants	$(C_4H_9)_3SnOCOC_6H_5$
	$((C_4H_9)_3Sn)_2O$
Molluscicides (field trials)	$(C_4H_9)_3SnF$
	$((C_4H_9)_3Sn)_2O$

3.3.2   Preservation of materials

Since the work of van der Kerk and Luijten[12] proposing the use of organotin compounds as biocides in wood preservation, a number of formulations have become commercially available. Bis(tributyltin) oxide has been used as the fungicidal constituent in many of these formulations since 1958,[73] and, more

Table 3.4 Cont'd . . .

Application	Compound
	R_2SnX_2
Heat and light stabilizers for rigid PVC	$R_2Sn(SCH_2COO\text{-}i\text{-}C_8H_{17})_2$ $(R = CH_3, C_4H_9, C_8H_{17},$ $(C_4H_9)OCOCH_2CH_2)$
	$(R_2SnOCOCH{=}CHCOO)_n$ $(R = C_4H_9, C_8H_{17})$
	$(C_4H_9)_2Sn(OCOCH{=}CHCOOC_8H_{17})_2$
	$(C_4H_9)_2Sn(OCOC_{11}H_{23})_2$
	$(C_4H_9)_2Sn(SC_{12}H_{25})_2$
Homogeneous catalysts for RTV silicones, polyurethane foams and transesterification reactions	$(C_4H_9)_2Sn(OCOCH_3)_2$ $(C_4H_9)_2Sn(OCO^iC_8H_{17})_2$ $(C_4H_9)_2Sn(OCOC_{11}H_{23})_2$ $(C_4H_9)_2Sn(OCOC_{12}H_{25})_2$ $((C_4H_9)_2SnO)_n$
Precursor for forming SnO_2 films on glass	$(CH_3)_2SnCl_2$
Anthelmintics for poultry	$(C_4H_9)_2Sn(OCOC_{11}H_{23})_2$
	$RSnX_3$
Heat stabilizers for rigid PVC	$RSn(SCH_2COO\text{-}i\text{-}C_8H_{17})_3*$ $(R = CH_3, C_4H_9, C_8H_{17},$ $C_4H_9OCOCH_2CH_2)$
	$(C_4H_9SnS_{1.5})_4$
Homogeneous catalysts for transesterification reactions	$(C_4H_9Sn(O)OH)_n$ $C_4H_9Sn(OH)_2Cl$
Precursor for SnO_2 films on glass	$C_4H_9SnCl_3$ CH_3SnCl_3*

* These compounds are used in combination with the corresponding R_2SnX_2 derivatives.

recently, tributyltin naphthenate and tris(tributyltin) phosphate have been introduced.[74,75] One disadvantage of the tributyltin wood preservatives is their low aqueous solubility and they are, therefore, usually applied as a 1–2 per cent solution in an organic solvent, by dipping, spraying, brushing or double-vacuum impregnation methods. Obviously, the use of an organic solvent may give rise to toxicity problems, constitute a fire hazard and increase application costs. One method, at present, of overcoming this problem is to form an aqueous dispersion of bis(tributyltin) oxide with a suitable quaternary ammonium salt, $R_4N^+X^-$. Formulations of this type have been shown to be effective for the eradication of moss, algae and lichens on stonework,[76] and for treating cotton textiles to prevent fungal attack. Subsequent work[77] demonstrated their effectiveness as aqueous wood preservatives and they are now used for timber treatments in Sweden.[78] An alternative approach has been the development of discrete water-soluble tributyltin biocides and, although not yet in commercial use, the tributyltin alkanesulphonates, $(n-C_4H_9)_3SnOSO_2R$ (R = CH_3, C_2H_5) have been found[79-81] to possess an aqueous solubility in the range 2−3 per cent. They have an excellent activity against wood-destroying fungi when applied to a wood substrate in aqueous solution.[82,83]

The use of organotins in wood preservation has recently been reviewed by Evans and Hill.[84]

3.3.3   Disinfectants

Since tributyltin compounds are active against Gram-positive bacteria,[12,17] their combination with a second chemical which is active against Gram-negative bacteria produces a highly effective disinfectant which may be used on open areas posing a risk of infection, such as hospital floors. Examples of such formulations are a mixture of tributyltin benzoate and formaldehyde[85] and some bis(tributyltin) oxide/quaternary ammonium halide dispersions.[86]

3.3.4   Antifouling paints

The attack of timber-hulled boats by marine borers, such as the Teredo worm, and the attachment of barnacles, sea grass, hydroids and other marine organisms to all types of ships hulls can seriously impede the running of the vessel.

The use[87] of marine paints containing tributyl- and, to a lesser extent, triphenyltin compounds, as toxic additives has been found to be very effective in eliminating this fouling problem (Fig. 3.2). The organotin-based marine paints usually contain up to 20 per cent by weight of a suitable tributyl- or triphenyltin toxicant (Table 3.4) which is slowly leached into the surrounding water in the immediate vicinity of the hull. The active lifetime of these paints is usually 1−2 years, after which time the vessel must be repainted.[88]

In order to increase the performance life of antifouling coatings, organotin polymers have been developed, in which the triorganotin compound is chemically bound to the polymer chain, and these have been formulated into

Fig. 3.2 Sea water immersion tests are often used in assessing the efficiency of anti-fouling paints in preventing the attachment of marine organisms. The panels which have not been protected with an organotin-based paint are heavily encrusted with marine growth

paints. Such polymers typically have the chemical structure shown in structure A,[87]

$$
\left[\left[-CH_2-\underset{\underset{CO_2SnR_3}{|}}{\overset{\overset{CH_3}{|}}{C}}-\right]_x\left[-CH_2-\underset{\underset{CO_2CH_3}{|}}{\overset{\overset{CH_3}{|}}{C}}-\right]_y\left[-CH_2-\underset{\underset{CO_2SnR_3}{|}}{\overset{\overset{CH_3}{|}}{C}}-\right]_z\right]_n \quad [A]
$$

where x, y, z and n represent repeating monomeric units. The mode of action of these polymer-based paint systems is for the triorganotin moiety to hydrolyse in seawater, releasing the active species; the depleted surface layer of the paint then becomes water-swollen and is eroded away.[87] Thus, a fresh surface layer of triorganotin acrylate polymer is exposed and the process repeated. The advantages claimed for the organotin polymer-based antifouling paints include a constant delivery of the triorganotin toxicant with respect to time, there is little depleted paint residue to remove and dispose of, and the surface is self-cleaning. The active lifetime of the organotin polymer antifouling systems is at least four years.[89] Other aspects of tin and related metallic polymeric systems in the environment are covered in Chapter 9.

An alternative antifouling system used on low surface area applications, e.g. sea bottom or ship mounted sonar equipment, is based on covering the surface with a rubber coating, in which the triorganotin biocide is dispersed. The release mechanism of the organotin compound involves diffusion–dissolution processes, whereby the elastomer remains intact while the toxicant is released from the surface on a molecular scale. Release rates are adjusted by specific selection of polymers, compounding ingredients, vulcanization conditions, coating thickness and toxicant loading.[87] The rubbers are applied in sheet form, and as well as providing a matrix through which the triorganotin biocide can diffuse, also act as a corrosion barrier.

3.3.5   PVC stabilizers

Poly(vinyl chloride) (PVC) has a tendency to degrade upon heating (during processing at 180–200 °C) or on prolonged exposure to light, due to loss of HCl from the polymer, leading to yellowing of the plastic and severe embrittlement. The addition of certain chemicals known as stabilizers, to the PVC before processing inhibits this breakdown and protects it during its service life. In fact, some of the most effective PVC stabilizers known are di- and monoorganotin compounds[66] (Table 3.4). The relatively high cost of the organotin stabilizers is offset by the excellent long-term transparency conferred to the plastic, and, on a tonnage basis, this is the principal use of organotins, accounting for some 66 per cent (approximately 23 000 tonnes) of all production world-wide.

The stabilizers are added to the PVC before processing at a level of approximately 5−20 g kg⁻¹.[90] Individually, the diorganotin compounds are generally the more effective stabilizers, but, in practice, up to 60 per cent of the corresponding monoorganotin derivative is added, as this combination is found to give a synergistic improvement in the stabilizing activity.[91,92] Those compounds which contain Sn—S bonds provide good stability to heat, while those with Sn—O bonds, in general provide long-term stability to light.

The mode of action of the di- and monoalkyl tin stabilizers is probably due to a combination of a number of effects, which include:[29]

(a) inhibition of dehydrochlorination, by exchanging anionic groups, X, with reactive chlorine sites in the polymer;
(b) scavenging of the hydrogen chloride which is produced and which would induce further elimination;
(c) production of the compound HX, and possible inhibition of other side-reactions;
(d) action as antioxidants, and prevention of breakdown of the polymer by atmospheric oxygen.

Organotin-stabilized PVC finds many applications, including potable water piping, drainage piping, food contact grade PVC (di- and monooctyltin compounds), roofing and glazing materials and, more recently, window frames. Environmental aspects of polymeric tin-containing PVC materials are further discussed in Chapter 9.

3.3.6 Precursors for forming SnO₂ films on glass

Dimethyltin dichloride, monomethyltin trichloride and monobutyltin trichloride are all used as alternatives to stannic chloride for coating glass with a thin film of stannic oxide. The organotin vapour is brought into contact with the hot glass surface (at 500−600 °C), where decomposition and oxidation occur. This treatment renders the glass scratch resistant, lustrous or electro-conductive, depending on the thickness of the SnO_2 film deposited (*c.* 10 nm to > 1000 nm).[93]

3.3.7 Homogeneous catalysts

Dibutyltin dilaurate, and, to a lesser extent, dibutyltin diacetate and dibutyltin di(2-ethylhexoate) are used for the room temperature vulcanization (RTV) of silicones.[94] Room temperature addition of the organotin catalyst to the liquid silicone oligomer brings about cross-linking of the silicone, due to the reactive carboxylate groups, and produces a flexible elastomeric solid. The same dibutyltin compounds are also used commercially to catalyse the addition of alcohols to isocyanates producing polyurethanes.[95] Two monobutyltin compounds, n-C₄H₉Sn(OH)₂Cl and (n-C₄H₉Sn(O)OH)ₙ, have been introduced as esterification catalysts, e.g. in the reaction of phthalic anhydride with octanol to form dioctylphthalate.[96]

The use of organotin compounds as homogeneous catalysts constitutes an annual consumption of approximately 2000 tonnes.

3.4   Modes of entry of organotin compounds to the environment

As a result of the wide range of industrial applications of organotin chemicals, just described, there are a variety of pathways which can be envisaged for their entry into the environment (Table 3.5).

The principal commercial use of organotins is as PVC stabilizers, and the

Table 3.5   Possible direct modes of entry of organotins into the environment

Medium	Species	Source
Air	R_3SnX	Agricultural spraying Volatilization from biocidal treatments Antifouling paint sprays
	R_3SnX, R_2SnX_2 and $RSnX_3$	Incineration of organotin treated or stabilized waste materials
	R_2SnX_2 and $RSnX_3$	Glass coating operations − spraying of organotins on to glass at high temperatures to give SnO_2 films
Soil	R_3SnX	Agricultural applications Wood preservation
	R_3SnX, R_2SnX_2 and $RSnX_3$	Burial of organotin-containing waste materials
Water	R_3SnX	Antifouling coatings Molluscicides Overspray from agricultural applications Land run-off from agricultural usage Industrial processes, e.g. slimicides in paper manufacture
	R_2SnX_2 and $RSnX_3$	Leaching from organotin-stabilized PVC

possible routes by which these stabilizers may enter the environment are (a) leaching/weathering, (b) land burial and (c) incineration of waste material. Of these routes, incineration is probably not significant since this method of disposal is not as common as land filling and, additionally, the organotins are likely to undergo thermal decomposition to inorganic tin compounds at the combustion temperatures. It has been estimated that a typical level of inorganic tin released into the environment from the incineration of domestic waste is approximately 1 μg per gram of suspended particles.[97]

Leaching of organotin stabilizers from PVC has been studied in some depth, due to their use in food contact applications, and this subject has been reviewed.[98] Studies with potable water piping[99,100] and food contact packaging[98,101] have shown that the leaching rate is generally very low, although the actual rate of loss depends on the nature of the leachant (pH and aqueous or organic character),[99,101] the alkyl chain length of the organotin compound present[99,101] and the level of plasticizer.[102] It is of interest to note that although leaching rates are very low, it has been found that organotin-stabilized PVC is not suitable for certain medical applications, due to localized adverse tissue reactions towards the organotins.[103,104]

With regard to the disposal of PVC waste by land burial, little is known about the environmental fate of the organotin stabilizers, which are the less toxic di- and monoorganotin compounds. It is reasonable to assume that leaching by aqueous media continues to occur, but the relationships between chemical environment, soil adsorption, leaching rates and degradation are unknown. However, it can be inferred that the PVC waste would be evenly distributed in domestic rubbish and so the expected levels of organotin compounds would be very low indeed.

The use of mono- and dialkyltins as homogeneous catalysts results in their incorporation into the finished product at low levels, especially in terms of the final volume. A typical concentration of organotin catalysts, e.g. dibutyltin dilaurate, in a flexible polyurethane foam formulation, is of the order of 0.2 g kg^{-1}.[95] Therefore, the methods of release of these compounds will be similar to those encountered with the PVC stabilizers.

From the foregoing discussion it can be concluded that, although PVC stabilizers and homogeneous catalysts together comprise the largest consumption of organotin compounds, it is unlikely that these contribute significantly to the levels found in the environment.

Although the biocidal uses of organotin chemicals comprise only approximately 30 per cent of the total world consumption, they probably give rise to the largest proportion of free organotins in the environment, due to their direct introduction into natural media. One example is of the use of triorganotin compounds as non-systemic agrochemicals ($\sim$ 2000 tonnes annually world-wide), where because of the manner in which they are applied (often aerial spraying), there is the possibility that not only the soil, but also the air and adjacent waterways, could be contaminated by airborne sprays.

Volatilization of triorganotin biocides, after application, is unlikely to be a

Table 3.6   Vapour pressure data on organotin compounds

Compound	Vapour pressure (mmHg)	Reference
$(C_4H_9)_3SnOAc$	2.7×10^{-3} (20 °C)	105
$(C_4H_9)_3SnO.CO.C_6H_5$	1.5×10^{-6} (20 °C)	106
$((C_4H_9)_3Sn)_2O$	7.5×10^{-6} (20 °C)	106
$((C_4H_9)_3Sn)_2O$	1.2×10^{-4} (20 °C)	105
$((C_4H_9)_3Sn)_2O$	1.1×10^{-5} (25 °C)	107
$((C_4H_9)_3Sn)_2O$	6.4×10^{-7} (20 °C)	108
$1\text{-}(c\text{-}C_6H_{11})_3Sn(1,2,4\text{-triazole})$	$<5 \times 10^{-5}$ (25 °C)	25
$(C_6H_5)_4Sn$	9.2×10^{-11} (25 °C)	109
$(C_6H_5)_4Sn$	1.3×10^{-11} (25 °C)	110
$(C_6H_5)_4Sn$	1.2×10^{-11} (25 °C)	110
$(C_6H_5)_6Sn_2$	3.2×10^{-15} (25 °C)	110
$(C_6H_5)_6Sn_2$	2.7×10^{-15} (25 °C)	110

problem, due to the general involatility of these compounds, as reflected by their very low vapour pressures (Table 3.6). However, volatilization by co-distillation from moist surfaces has been observed for tricyclohexyltin hydroxide,[7] which is essentially non-volatile in the dry state, although it is not known to what extent this occurs under field conditions. A low vapour pressure is particularly important in wood preservation, where the active tributyltin fungicide must remain for many years in the treated timber.

Triorganotin compounds have been shown to adsorb strongly on to soil. Barnes et al.[111] found that [^{14}C]triphenyltin acetate could not be leached with water from a 25 cm layer of an agricultural loam over a period of 6 weeks, and subsequent analysis of the soil for radioactivity showed that over 75 per cent of the organotin compound had remained in the top 4 cm. The adsorption behaviour of the same radiolabelled triphenyltin derivative in two types of hops soil has been studied by Suess and Eben,[112] and of ^{113}Sn labelled triphenyltin chloride in clay, by Leeuwangh et al.[113] Strong adsorption of the triphenyltin compounds to the soil was again found, and, in the latter case,[113] it was established that the ratio of Ph_3SnCl in the sediment to that in water was approximately 20 : 1. The adsorption characteristics of bis(tributyltin) oxide and tributyltin fluoride in several soils (clay, sand, topsoil and silt) have been examined[114,115] and it was observed that, over a 16-week period, no significant leaching of the tributyltin compounds occurred. Strong affinity of soil particles for other triorganotin derivatives has also been demonstrated.[116,117] It is likely that, after spray application of triorganotin biocides, most of these will be removed by adsorption on to the soil before they can enter surface waters. Additionally, it has been shown[118] that up to 100 μg g^{-1} of bis(tributyltin) oxide has no adverse effect on soil micro-organisms or on their ability to maintain soil fertility. The interaction of the methyltin chlorides, $(CH_3)_nSnCl_{4-n}$, with humates, peat and related materials, using the ^{113}Sn labelled compound, has recently been studied.[119]

Direct entry of organotins into the aqueous environment is primarily due to their use in agriculture and in antifouling paint systems. In agricultural applications, contamination could result from run-off water and overspray, as mentioned previously, whereas in marine antifouling paints and in antifouling rubber coatings, which utilize some 3000 tonnes of tributyl- and triphenyltin derivatives annually, the triorganotin compounds are released directly from the paint or rubber matrix into the water in the vicinity of the ship's hull and subsequently dispersed. Consequently, the release of the triorganotin biocide in a static environment, e.g. harbours, marines and bays, is likely to be more significant than its loss when the vessel is under way in the open sea. In accord with this, recent studies[120-124] have shown that elevated levels of butyltin compounds may occur in seawater, in particular, in areas of high pleasure craft activity.[124] These studies[120-124] have given rise to some concern, since it has been found[125-130] that tributyltin compounds cause reduced growth and shell deformity in the commercially cultivated Pacific oyster, *Crassotrea gigas*. It is possible that increased future use of the tributyltin polymer antifouling systems, rather than tributyltin-containing paints, may help to overcome this problem, since the organotin biocide is released when the vessel is under way, and so the concentrations encountered in harbours and estuaries may be significantly reduced.

A further related mode of entry of tributyltin compounds into the fresh water environment is their probable future use in slow release rubber pellets for the control of the snail vectors of schistosomiasis.[131-133]

Other possible modes of entry of organotins into the environment (Table 3.5) result from low tonnage uses, e.g. anthelmintic treatments for poultry, disinfectants, etc., and are not considered to have a significant impact on environmental levels.

3.5   The aqueous chemistry of organotin species

Studies of the nature of organotin compounds in aqueous solution have mainly been limited to methyltin derivatives, since other organotins do not have a sufficiently high solubility readily to permit spectroscopic investigations. It has been demonstrated[134] that trimethyltin compounds in aqueous solution exist at pH less than 5, primarily as the trimethyltin cation, $(CH_3)_3Sn^+$, and, at higher pH, as $(CH_3)_3SnOH$. The trimethyltin cation has been shown to be hydrated and to have a trigonal bipyramidal tin atom geometry with axial water molecules (structure B; R = CH$_3$)[135]

(B)

Speciation of dimethyltin compounds in aqueous solution has also been in-vestigated.[134] At pH less than 4 the predominant species is the dimethyltin ca-tion, (CH$_3$)$_2$Sn^{2+}, which is hydrated and has an octahedral tin atom geometry with *trans*-methyl groups (structure C; R = CH$_3$).[136,137]

(C)

However, at environmental pH (6−8) the main species is (CH$_3$)$_2$Sn(OH)$_2$.[134]

Very few studies have been made of the hydrolysis products of monoalkyltin derivatives, and it has been suggested[134] that these exist in solu-tion only as the hydrated oxides. Recent studies[138] with methyltin trichloride have shown, however, that at pH 1.4, a concentration-dependent equilibrium exists (equation 3.1). At low concentrations CH$_3$SnOH^{2+} was found to be the main species, which is undoubtedly hydrated and is presumably in equilibrium with other hydroxy derivatives (equation 3.2). As the pH is increased the species to the right are increasingly favoured, and, from Tobias's work[134] with tri- and dimethyltin compounds, it might be expected that at environmental pH, CH$_3$Sn(OH)$_3$ will predominate.

$$CH_3Sn(OH)Cl_2.2H_2O \rightleftharpoons CH_3Sn(OH)_2Cl.nH_2O \rightleftharpoons CH_3Sn(OH)^{2+} \quad [3.1]$$

$$CH_3Sn^{3+} \rightleftharpoons CH_3Sn(OH)^{2+} \rightleftharpoons CH_3Sn(OH)_2^{+} \rightleftharpoons CH_3Sn(OH)_3 \quad [3.2]$$

In general, the aqueous solubility of tri- and di-n-alkyltin species decreases as the alkyl chain length increases (Tables 3.7 and 3.8). However, exceptions to this are known, e.g. (n-C$_4$H$_9$)$_3$SnSO$_3$R,[79−81] in which the X radical is highly hydrophilic. A recent ^{119}Sn NMR investigation[80] of 0.5 mol dm^{-3} aqueous solutions of (n-C$_4$H$_9$)$_3$SnSO$_3$R (where R = CH$_3$, C$_2$H$_5$) revealed that the organotin species present in solution was the tributyltin cation, ((n-C$_4$H$_9$)$_3$Sn(H$_2$O)$_2$)$^{+}$, which has been demonstrated[149] crystallographically to have structure B (R = n-C$_4$H$_9$). In the case of triphenyltin compounds, measured solubilities have been of the same order of magnitude as for tributyltin compounds, and Soderquist and Crosby suggested[146] that the solubility of triphenyltin species should be independent of counterion when the bonding is primarily ionic (e.g. Cl^{-}, Br^{-} or OAc). It has been shown[142−144] that the presence of chloride from seawater inhibits the solubility of tributyl- and triphenyltin compounds (Table 3.7), probably by association with the cation to form the covalent organotin chloride.[142]

Guard *et al.*, however, propose[150] that, in seawater, tributyltin compounds exist in an equilibrium between the hydrated tributyltin cation, tributyltin chloride, bis(tributyltin) carbonate and tributyltin hydroxide. It may be seen from Table 3.7 and 3.8 that, in some cases, there is a wide variation in the published figures for solubilities of organotin compounds in water. This reflects the difficulties involved in determining the solubility and is discussed in section 3.8.

Finally, it is of interest to describe the nature of inorganic tin species in aqueous solution, since these could be relevant to the transformations of tin compounds in the environment, which are discussed in the next section. Pettine *et al.* studied[151] the hydrolysis of stannous salts in seawater, from 0.1 to 1.0 mol dm^{-3} at 20 °C, and found that, at pH 8.1, the predominant species is $Sn(OH)_2$. In a similar investigation of stannic nitrate,[152] at a level of 8 μg ion dm^{-3}, Nazarenko *et al.* concluded that, above pH 3, the inorganic tin(IV) hydroxo-complex was primarily $Sn(OH)_4$.

Hence, it may be concluded that, at environmental pH, tin compounds, R_nSnX_{4-n} (n = 0–3), will exist in aqueous solution as simple neutral hydroxides.

3.6 Transformation of organotin species in the environment

Since the early work on the biomethylation of arsenic and selenium by Challenger,[153–156] there has been considerable interest in the possible routes for the environmental methylation of other metalloids and heavy metals. Developments in analytical techniques for inorganic tin and organotin compounds have enabled a number of groups to report the determination of methyltin species, $(CH_3)_nSnX_{4-n}$ (n = 1–4) in environmental media (e.g. seawater, lake water, rainwater and sediments) at concentrations typically of a few parts per billion.[120,121,157–160] Obviously, these methyltin species could have resulted from the biological or chemical methylation of inorganic tin. However, it is of interest to note that, in the USA, di- and monomethyltin derivatives are used ($\sim$ 3000 tonnes in 1981) as stabilizers for PVC potable water pipe, and some of the methyltin species could have arisen from leaching of the organotin.

The first laboratory studies on the environmental methylation of tin were carried out by Brinckman and his co-workers,[161] using pure cultures of tin-resistant *Pseudomonas* bacteria from Chesapeake Bay. Inoculation with $SnCl_4.5H_2O$ led to the production of what was believed to be a dimethyltin species. Later experiments by the same workers[121] showed that *in vitro*

Table 3.7   Aqueous Solubilities of Triorganotin Compounds at Room Temperature*

R_3SnX	Solubility (mg dm^{-3})	Reference
$(C_2H_5)_3SnOH$	~35 000	139
$(C_2H_5)_3SnOCOCH_3$	7500	140
$(C_4H_9)_3SnCl$	50	140
$((C_3H_7)_3Sn)_2O$	50	140
i-$(C_3H_7)_3SnCl$	25	140
$(C_4H_9)_3SnF$	6	141
$(C_4H_9)_3SnF$	4.5	142
$(C_4H_9)_3SnCl$	50	140
$(C_4H_9)_3SnCl$	17†	143
$(C_4H_9)_3SnCl$	16	143
$(C_4H_9)_3SnCl$	5.4	142
$(C_4H_9)_3SnOCOCH_3$	256†	144
$(C_4H_9)_3SnOCOCH_3$	65†	143
$(C_4H_9)_3SnOCOCH_3$	50	140
$(C_4H_9)_3SnOCOCH_3$	16	143
$(C_4H_9)_3SnOCOCH_3$	6.4	142
$(C_4H_9)_3SnOCOCH_3$	5	144
$((C_4H_9)_3Sn)_2O$	19.5†	144
$((C_4H_9)_3Sn)_2O$	18†	143
$((C_4H_9)_3Sn)_2O$	8	143
$((C_4H_9)_3Sn)_2O$	8–10	145
$((C_4H_9)_3Sn)_2O$	3	140
$((C_4H_9)_3Sn)_2O$	1.4	144
$((C_4H_9)_3Sn)_2S$	<1	140
$(C_4H_9)_3SnSO_3CH_3$	~31 000†	80
$(C_4H_9)_3SnSO_3C_2H_5$	~29 000†	80
$(C_6H_5)_3SnF$	1.2	141
$(C_6H_5)_3SnCl$	5.2	142
$(C_6H_5)_3SnCl$	<1	140
$(C_6H_5)_3SnOH$	1.2†	141

Table 3.7 (cont)

R_3SnX	Solubility (mg dm^{-3})	Reference
$(C_6H_5)_3SnOH$	1.2†	146
$(C_6H_5)_3SnOH$	<1	140
$(C_6H_5)_3SnOCOCH_3$	<3.3†	111
$(C_6H_5)_3SnOCOCH_3$	2.9	147
$(C_6H_5)_3SnOCOCH_3$	<1	140
$((C_6H_5)_3Sn)_2O$	<1	140
$(C_6H_5)_3SnSO_3C_2H_5$	1000†	80
$(C_6H_5)_3SnSO_3C_6H_5$	1000†	80
$(C_8H_{17})_3SnCl$	<1	140
$(c\text{-}C_6H_{11})_3SnOH$	<1	21
$(c\text{-}C_6H_{11})_3Sn\text{-}\overset{\displaystyle\frown}{N}.C{=}N.C{=}N$ $\qquad\quad\underset{H}{\mid}\quad\underset{H}{\mid}$	<1	22
$((C_6H_5)(CH_3)_2CCH_2)_3Sn)_2O$	<0.005	148

* Solubility values marked † refer to distilled water; unmarked refer to seawater.

Table 3.8   Aqueous Solubilities of Diorganotin Compounds at Room Temperature*

R_2SnX_2	Solubility (mg dm^{-3})	Reference
$(CH_3)_2SnCl_2$	20 000	140
$(C_4H_9)_2SnCl_2$	92†	143
$(C_4H_9)_2SnCl_2$	50	140
$(C_4H_9)_2SnCl_2$	6–8	145
$(C_4H_9)_2SnCl_2$	4	143
$(C_4H_9)_2Sn(OCOCH_3)_2$	6	140
$(C_4H_9)_2Sn(OCO\text{-}n\text{-}C_{11}H_{23})_2$	3	140
$(C_4H_9)_2Sn(OCO\text{-}i\text{-}C_8H_{17})_2$	6	140
$((C_4H_9)_2SnOCOCH{=}CHCOO\text{-})_n$	6	140
$(C_8H_{17})_2SnCl_2$	<1	140
$(C_6H_5)_2SnCl_2$	50†	4

* Solubility values marked † refer to distilled water; unmarked refer to seawater.

biomethylation of inorganic tin(IV) by the same bacterial strain produced methylstannanes, $(CH_3)_nSnH_{4-n}$, where $n = 2-4$.

The formation of methyltin compounds in sediments has been described:[162-164] both tin chlorides, $SnCl_2$ and $SnCl_4$, were found to be methylated, and, in the study by Hallas,[163] the species formed from $SnCl_4.5H_2O$ were the hydrides, $(CH_3)_2SnH_2$ and $(CH_3)_3SnH$ (on the basis of gas chromatographic retention times and mass spectral data). Trimethyltin chloride[164] and hydroxide[162] are converted to tetramethyltin in sediments and this process appears to be slow. Since $(CH_3)_4Sn$ was also produced from $(CH_3)_3SnOH$ in sterile sediments, the tetraalkyltin derivative could be formed via a chemical pathway.[164] This may be due to disproportionation of the initially formed bis(trimethyltin) sulphide[165] (equation 3.3) from naturally occurring sulphides

$$3(CH_3)_3SnSSn(CH_3)_3 \rightarrow ((CH_3)_2SnS)_3 + 3(CH_3)_4Sn \qquad [3.3]$$

However, a study of tricyclohexyltin hydroxide in soils under aerobic and anaerobic conditions, indicated an absence of methylated products.[166]

There have been a number of laboratory investigations using methylcobalamin (CH_3CoB_{12}), the methyl coenzyme of cyanocobalamin (vitamin B_{12}), or methyl iodide, which is produced by certain seaweeds and algae,[167] as chemical methylating agents for tin(II) and tin(IV) compounds. Further details on the action of CH_3CoB_{12} are given in Chapters 1, 2 and 8.

Wood and his co-workers[168,169] have reported that CH_3CoB_{12} is demethylated by $SnCl_2$ in aqueous HCl solution, in the presence of an oxidizing agent, to form a monomethyltin species. Thayer found[170,171] that finely divided tin(IV) oxide, SnO_2, reacts with CH_3CoB_{12} in aqueous HCl solution to form methyltin derivatives, but the reaction appears to be very slow. This may have been due to the particle size of the tin(IV) oxide since it has recently been shown by the same author,[172] that when a solution of $SnCl_4.5H_2O$ dissolved in nitric acid it slowly neutralized, a very finely divided precipitate was gradually formed, and this reacted more rapidly with CH_3CoB_{12}.

An even more rapid reaction occurs[172] upon the addition of liquid tin tetrachloride $(SnCl_4)$ to aqueous CH_3CoB_{12}, producing methylstannanes $(CH_3)_nSnH_{4-n}$ ($n = 1, 2$) of which CH_3SnH_3 predominated. Similar products were formed[172] when tin-containing sediments were treated with CH_3CoB_{12}. In addition, Thayer[170,173] studied the reaction of $(CH_3)_3SnOAc$, $(C_2H_5)_3SnOAc$, $(C_6H_5)_3SnOAc$ and $(CH_3)_2SnCl_2$ with CH_3CoB_{12} although the products were not identified.

Chau and his co-workers[164] found that methyl iodide could methylate inorganic tin(II) salts, in an aqueous medium to form monomethyltin(IV) species, along with small amounts of di- and trimethyltin derivatives, whereas tin(IV) compounds did not react. More recently, Brinckman *et al.* have demonstrated the oxidative methylation of SnS by methyl iodide in aqueous solution, under mild anaerobic or aerobic conditions, to form CH_3SnI_3.[174]

Therefore, it may be concluded that both tin(II) and tin(IV) compounds appear to be biologically and chemically methylated under simulated environmental conditons. Thus, with regard to the question of methylation in the environment, there appears to be circumstantial but as yet no direct unequivocal evidence for this phenomenon.[164,175] The significance of methyltin compounds in the environment cannot be readily assessed. However, the levels so far detected do not justify undue concern, and it is possible that a steady-state situation exists, in which their entry into the environment is equal to their removal by, for example, degradation, which is discussed in the next section.

3.7   Degradation of organotin compounds

Owing to the widespread industrial uses of organotins, along with their biological activity and the possibility of biomethylation, their fate in the environment is of considerable interest. Consequently, work has been carried out over a number of years to investigate the possible ways in which organotin compounds may degrade in the environment.

The degradation of an organotin compound may be defined as the progressive removal of the organic groups from the tin atom (equation 3.4)

$$R_4Sn \rightarrow R_3SnX \rightarrow R_2SnX_2 \rightarrow RSnX_3 \rightarrow SnX_4 \qquad [3.4]$$

This stepwise loss of organic groups from the tin atom is attractive to potential users of organotin compounds because it is accompanied by a progressive lowering in biological activity (sect. 3.2), and, as such, the use of these compounds would be unlikely to lead to environmental pollution.

Early studies[176-178] of the fate of organotins emphasized the 'disappearance' of the compounds, either after having been applied to crops or on exposure to ultraviolet light, and it is only recently that attempts have been made to establish the mode of degradation and to identify the breakdown products.

Degradation involves the breaking of a Sn—C bond and this can occur by a number of different processes. These include:

1. Ultraviolet (UV) irradiation.
2. Biological cleavage.
3. Chemical cleavage.
4. Gamma (γ) irradiation.
5. Thermal cleavage.

Of these processes, γ-irradiation will have little effect on environmental degradation, due to its negligible intensity at the earth's surface. It is worth noting, however, that the importance of studying the effects of γ-irradiation of organotin compounds may increase because of its possible use for the sterilization of food, some of which is packaged in organotin-stabilized PVC, and Dunn and Oldfield[179] have shown that γ-irradiation will degrade tributyltin compounds. Thermal cleavage is also unlikely to be of environmental significance, because the Sn—C bond is reported to be stable at temperatures up to 200 °C.[1] Therefore, only processes (1), (2) and (3) are considered below.

3.7.1   Ultraviolet degradations

The light emitted by the sun, reaching the earth's surface, consists mostly of wavelengths above 290 nm, although it has been estimated[180] that the quantum flux density below 290 nm is about 10^{16} photons cm^{-2} $month^{-1}$. Therefore, with regard to the degradation of organotins, the shorter wavelengths may have a long-term effect. Ultraviolet (UV) light of wavelength 290 nm possesses an energy of approximately 300 kJ mol^{-1} and the mean bond dissociation energies for some Sn—C bonds are found[181] to be in the range 190–220 kJ mol^{-1}. Consequently, provided that absorption of the light takes place, Sn—C bond cleavage could occur, and, in fact, the maximum absorption wavelength of organotin compounds is generally within the UV region.

Many studies have been made of various photochemical reactions of organotins and these have been discussed previously.[182–184] Therefore, it is only those studies that are of direct environmental interest that are reviewed here.

Cenci and Cremonini[185] investigated the rate of disappearance of triphenyltin acetate and triphenyltin hydroxide in different types of soil when exposed to UV light, without identifying the nature of the breakdown products. Getzendaner and Corbin[186] studied the UV degradation of tricyclohexyltin hydroxide, suggesting that dicyclohexyltin, monocyclohexyltin and inorganic tin species were produced, but no further details were given. Akagi and Sakagami[187] irradiated several triphenyl-, dibutyl- and dioctyltin compounds on a watch glass and concluded that they all decomposed to inorganic tin. The breakdown of the triphenyltin compounds was suggested to occur stepwise (equation 3.5).

$$(C_6H_5)_3SnX \rightarrow (C_6H_5)_2SnX_2 \rightarrow C_6H_5SnX_3 \rightarrow \text{inorganic tin} \qquad [3.5]$$

Barnes *et al.*[111] showed that triphenyltin acetate was degraded to inorganic tin by UV light, while Massaux[188] irradiated triphenyltin chloride on a watch glass with light of wavelengths of 254 nm and 350 nm, showing that decomposition occurred faster at the shorter wavelength. A similar study was made by Chapman and Price[189] for triphenyltin acetate. In this case, irradiations were carried out using light of wavelengths greater than 235 nm and greater than 350 nm, and diphenyl-, monophenyl- and inorganic tin species were identified and quantitatively determined. Again, degradation was shown to occur at a faster rate with the shorter wavelength light. In a similar study[190] with bis(tributyltin) oxide and tri-n-butyltin acetate, the pattern of degradation was found to follow that observed for the triphenyltin compounds. The UV-induced decomposition of bis(tributyltin) oxide has also been studied by

Klötzer[191] under a variety of conditions of temperature, irradiation intensity and on different matrices. The same author found[192] that the relative rates of breakdown of tributyltin chloride and bis(tributyltin) oxide were similar, while dibutyltin dichloride broke down faster than dibutyltin oxide. These studies were, however, only concerned with the decomposition of the starting compounds and did not suggest possible products or follow the reaction through to inorganic tin. The UV breakdown of bis(tributyltin) oxide has also been reported by Komora and Popl[193] and Woggon and Jehle.[194]

The study of the UV degradation of organotins in aqueous solution has received less attention to date, presumably because their low aqueous solubility makes the analysis particularly difficult. In a paper reporting the rates of hydrolysis of organotins, Mazaev *et al.* noted[195] that irradiation of aqueous solutions of diethyltin dicaprylate and dibutyltin bis(S, S'-isooctyl-thioglycollate) (Table 3.4) produced an increase in the rate of disappearance of the compounds compared to non-irradiated solutions. Soderquist and Crosby reported[146] the action of ultraviolet light on triphenyltin hydroxide in water, showing that, under simulated environmental conditions, a diphenyltin species was produced which underwent further breakdown, possibly forming a polymeric monophenyltin derivative $(C_6H_5SnO_xH_y)_n$. It was not, however, demonstrated whether or not this polymeric species would eventually degrade to inorganic tin. The gradual disappearance of triphenyltin acetate in freshwater, seawater and sewage water has been noted by Odeyemi and Ajulo,[196] but the breakdown products were not identified. Maguire *et al.*[108] studied the UV degradation of tributyltin species in water, and found that the process involved sequential debutylation through to inorganic tin. A similar reaction was shown to occur slowly in natural sunlight. Sherman and his co-workers[197] reported the breakdown of tributyltin chloride and tributyltin fluoride in water to a dibutyltin species, but the reaction was not pursued in order to establish whether complete breakdown to inorganic tin occurred. A study[198] of the UV degradation of the methyltin chorides, $(CH_3)_nSnCl_{4-n}$ ($n = 1-3$), in aqueous solution showed the final product of the reaction to be hydrated tin(IV) oxide, and established approximate relative rates of breakdown.

3.7.2 Biological cleavage

The question of biodegradation of organotins is particularly important in situations where, for example, the compounds are not directly exposed to light, e.g. in the soil or on the sea bed.

It has been shown by Barnes *et al.*[111] that ^{14}C-labelled triphenyltin acetate in soil is broken down to inorganic tin. Since carbon dioxide was evolved and breakdown did not occur in sterile soil, it was concluded that degradation was due to the ability of certain micro-organisms to metabolize the organotin compound. The degradation of [14C]triphenyltin acetate in soil has also been studied by Suess and Eben.[112] Barug and Vonk[115] have shown that bis(tri[1 − [14]C]butyltin)oxide undergoes breakdown in soil due to the action of micro-organisms and Barug[199] demonstrated that the Gram-negative bacteria, *Pseudomonas aerugenosa* and *Alcaligenes faecalis*, will dealkylate, bis(tributyltin) oxide. In contrast, certain other Gram-negative bacteria have been found[199] not to degrade bis(tributyltin) oxide, and Brinckman *et al.*[200] have recently demonstrated the exclusive uptake of the tributyltin cation $(C_4H_9)_3Sn(H_2O)_2^+$, by a number of Gram-negative tin-resistant heterotropes, without degradation. Sheldon reported[201] the breakdown of ^{14}C-labelled bis(tributyltin) oxide, tributyltin fluoride and triphenyltin fluoride in soil, showing that degradation occurred faster under aerobic than anaerobic conditions, and Stein and Küster[202] studied the biodegradation of bis(tributyltin) oxide to inorganic tin by biological sewage treatment. In addition, it has been reported that tricyclohexyltin hydroxide will degrade in soil.[55]

The degradation, in wood, of bis(tributyltin) oxide[203−212] and tributyltin naphthenate[205−207,209] has been examined and di- and monobutyltin species have been detected as breakdown products. This process could be due to fungal degradation, since bis(tributyltin) oxide has been shown to be degraded by various individual microfungi, such as *Coniophora puteana*,[199,213] *Coriolus (Trametes) versicolor*,[199,203] *Chaetomium globosum*,[199] *Aureobasidium pullulans*[213,214] and various *Phialophora* species,[215] as well as fungal culture filtrates of *Coniophora puteana*,[216] *Coriolus versicolor*[216] and *Sistotrema brinkmannii*.[216] There is, however, the additional possibility that organotin degradation in wood is due to acid cleavage, because both formic and acetic acids may be present in some types of timber.[217]

3.7.3   Chemical cleavage

The Sn—C bond is capable of polarization in either direction $(Sn(\delta+)—C(\delta-)$ or $Sn(\delta-)—C(\delta+)$,[218] and is, therefore, susceptible to attack by both nucleophilic and electrophilic reagents. Hence, for reactions of the type

$$\geqslant Sn—C\leqslant + A—B \rightarrow \geqslant Sn—A + \geqslant C—B$$

A—B may be one of a wide variety of compounds, e.g. mineral acid, carboxylic acid, alkali, etc. In addition, the trimethyltin cation has been shown to transmethylate with various hydrated metal cations, e.g. Tl^{3+}, Pd^{2+}, Au^{3+}, Hg^{2+}, forming a dimethyltin species and the corresponding monomethyl metallic derivative.[6] Free radical processes can cause homolytic Sn—C bond fission, the Sn—C bond being a fairly good radical trap.[218]

Consequently, due to the very wide range of chemical reactions that result in Sn—C bond cleavage, a discussion of purely chemical aspects is outside the scope of this review. However, these reactions are described elsewhere.[218,219]

3.7.4   Summary of degradation properties

From the breakdown studies that have been reviewed here and elsewhere[9,220,221] it may be concluded that, within a generally consistent pattern of behaviour, organotins will degrade in natural media, and this has been demonstrated for triphenyltin,[50,222] tributyltin[223] and tricyclohexyltin[186] compounds.

In one of the aqueous degradation studies,[146] inorganic tin was not detected and it was thought that a monoorganotin species might be the end-product of the breakdown process. However, the rate of breakdown of monoorganotins has been suggested,[1] in some cases, to be slower than that of tri- and diorganotins, and it is possible that the timescale was not adequate for inorganic tin to be detected. With regard to the degradation studies in wood[203-212] only di- and monoorganotin species have so far been positively identified as breakdown products. The failure to, as yet, detect inorganic tin may be due to the problems associated with extracting the total tin content from the wood sample. For example, Plum and Landseidel[204] claim to have extracted only 65−85 per cent of the total tin content from the wood sample tested. In addition, it has been observed that as the degradation of bis(tributyltin) oxide proceeds with time, so the amount of an unextractable tin residue increases.[212]

No mention has been made of the timescales involved in the degradation processes, since although some of the studies have reported half-lives, they relate only to the specific laboratory experimental conditions and so are not directly comparable. Additionally, it can be misleading to relate experimental timescales to an environmental situation, where the actual rate of breakdown will be dependent on many factors, e.g. intensity of sunlight, concentration of suspended matter in waterways, etc.

The effect of the anionic radical, X, in an R_nSnX_{4-n} species, has not been discussed since very little is known about how this affects the breakdown. However, at the levels that the organotins are present in the environment, they will either exist as, or will rapidly be converted to, oxides or hydroxides,[111] carbonates[224,225] or hydrated cations.[134,135,138]

3.8   Analysis of organotin species in the environment

The earlier work of the 1960s and early 1970s has already been reviewed.[226] However, in the past five or six years there has been a vast increase in the number of papers published on the determination of organotin compounds in environmental samples. This increase has been due to the advances in the development of analytical techniques for the determination of metals at nanogram levels that have taken place during the past six to seven years, particularly in the fields of high performance liquid chromatography (HPLC), gas chromatography (GC) and electrothermal atomization (ETA) for atomic absorption spectroscopy (AA).

In this section, an attempt will be made to review the more interesting of these techniques for the quantitative determination of tin compounds, and examples of specific applications of these methods will be given. Theoretical aspects of the analysis of organometallic compounds are discussed further in Chapter 1 (sect. 1.6) using organotin species as examples.

3.8.1   Methods of determination of organotin compounds

Atomic absorption and emission spectroscopies are by far the most frequently used techniques for the determination of tin, either in the organometallic form or after oxidation to inorganic tin. In most cases, a prior step of separation for speciation or concentration will have to be carried out. For example, Brinckman and his co-workers[227] used a commercially available flameless (graphite furnace) atomic absorption (GFAA) spectrometer as a detector for an HPLC system to provide element-specific separation and detection of organometallic compounds at nanogram concentrations. The HPLC columns were later[223,228] modified to enable size exclusion chromatography (SEC) or reversed phase-bonded chromatography (RPBC) to be performed before the final detection with GFAA.

Vickrey et al.[229] also used HPLC coupled with GFAA to analyse mixtures of organotin compounds and found that treatment of the surface of the graphite cuvette with zirconium eliminated some of the interferences observed with non-treated cuvettes. Chau et al.[158] used GC to separate butylated methyltin compounds before determination by GFAA.

Where speciation has not been required, concentration of the organotin compound from aqueous solution by extraction with methylisobutyl ketone has been used,[230] prior to determination of the tin by GFAA. However, it should be noted that solvent extraction only gives 100 per cent recovery for R_3SnX compounds, unless carried out in the presence of complexing agents, such as tropolone.[145]

An alternative approach is to reduce the organotin compounds with sodium borohydride, followed by cryogenic trapping and separation of the species, based on the boiling points of the organotin hydrides, and determination by AA.[120] If speciation is not required, the aqueous samples may be treated with sulphuric acid–potassium permanganate solution, followed by reduction to stannane with sodium borohydride.[231]

Other workers have used GC with conventional detectors, e.g. flame photometric detector (FPD),[121,232,233] flame ionization detector (FID),[234,235] electron capture (EC)[235] or mass spectrometry (MS)[145] to separate and determine the organotin species.

Solvent extraction − either prior to derivatization and GC separation[145,232] where speciation is required, or where only a single compound or total tin content is sought — with a spectrophotometric or spectrofluorimetric determination has been used. In this case, measurement of the organotin complex with a suitable reagent is used,[236-239] or the compounds are first wet-ashed in sulphuric acid with nitric acid or hydrogen peroxide to give inorganic tin which then forms the coloured species with the reagent.[240-243]

Electrochemical determination of tin, either as inorganic tin or as an organotin compound is possible, but for environmental samples final determination of tin in the inorganic form appears to be the generally used procedure.[194] However, pulse polarographic studies of triphenyl- and tributyltin compounds in aqueous extracts of paint films have been carried out.[244]

3.8.2 Applications

By far the largest number of published works in this field have been on the determination of organotin compounds in water, either 'naturally' occurring, or in leachates from the testing of antifouling paints. Some of the methods used are described below.

3.8.2.1 Methyltin compounds in aqueous systems

Braman and Tompkins[157] have reported the first ambient low concentration speciation method for tin, and applied this method to the determination of inorganic, mono-, di- and trimethyltin compounds in a range of environmental samples, such as saline and estuarine waters, freshwater lake samples, rain- and tap-water, as well as human urine and shell (marine and fowl) samples. Their procedure is to treat the tin compounds in aqueous solution, at pH 6.5 (adjusted if necessary with 2 mol dm^{-3} Tris-HCl buffer solution), with sodium borohydride, to convert them to the corresponding volatile hydride. The hydrides are cryogenically trapped in a U-tube containing silicone oil (type OV-3) on chromosorb. On warming the U-tube, the hydrides separate according to their boiling points and are determined in a hydrogen-rich, hydrogen–air flame emission type detector, with a detection limit of approximately 0.01 ng (SnH band). It was observed[157] that, in the determination of inorganic tin, high blank values were obtained originating from impurities in the reagents and it was noted that the sodium borohydride used should be of at least 99 per cent purity.

Very soon after this, Hodge *et al.*[120] reported a similar procedure for the determination of organotin compounds but used a conventional AA as the detector. Their initial treatment differs from Braman and Tompkins[157] in that the sample (100 cm^3) is acidified with 1 cm^3 of 2 mol dm^{-3} acetic acid before treatment with the sodium borohydride. Again, the hydrides are separated on the basis of their boiling points, before being burned in a hydrogen–air flame for the determination of the tin compounds by AA. The limit of detection is of the order of 0.5 to 1 ng. Hodge *et al.*[120] also noted that the sodium borohydride was a source of high blanks for inorganic tin and recommended that it should be purified before use, by electrolysis between carbon rod electrodes at 3 V, while bubbling helium through the solution.

Jackson *et al.*[121] have also determined methyltin compounds in natural waters. Their procedure is based on a GC separation of the stannanes, followed by detection with an FPD similar to that of Braman and Tompkins.[157] The sample, with or without the addition of sodium borohydride, is passed into a purge and trap (P/T) sampler containing 60/80 mesh Tenax-GC, followed by chromatographic separation using columns co-packed with 3 per cent SP-2401 and 10 per cent SP-2100 on 80/100 mesh supelcoport. For methyltins the chromatograph was maintained isothermally at 30 °C. The separated compounds were determined by FPD modified to permit tin-selective detection of SnH-emission in a hydrogen-rich flame (cf. Braman and Tompkins[157]); the limit of detection appears to be about 0.3 ng. These authors[121] noted that deposits from combustion products which collected in the detector seriously reduced the detection sensitivity, although this could be restored by repeated injections of a fluorine-containing organic compound, e.g. m-$C_6H_4F_2$. It was reported[121] that naturally occurring methyltin hydrides in water could be determined by running the sample through the analytical procedure without the addition of the sodium borohydride. Jackson *et al.*[121] also observed that tetramethyltin did not react with sodium borohydride and so would not be detected by the P/T–GC–FPD system. They, therefore, used a GC–MS system to determine this compound. The chromatographic separation is carried out on a column of 80/100 mesh supelcoport, as previously described, operated isothermally at 30 °C with helium as the carrier gas.

Jewett and Brinckman[223] studied the speciation of di- and triorganotin compounds in water by ion-exchange HPLC–GFAA. The ion-exchange column is a Whatman 10 μm Partisil (SCX), the eluant is 0.03 mol dm^{-3} ammonium acetate in methanol–water (70 : 30), the HPLC is interfaced with a GFAA; the limit of detection is 5 to 30 ng (as Sn) depending on the compound. Chau *et al.*[158] used solvent extraction with benzene containing tropolone to extract methyltin species, and found that the organotin could be extracted from 5 to 10 litres of water with 100 cm^3 of benzene containing 0.5 per cent tropolone in the presence of sodium chloride (30 to 40 g NaCl per 100 cm^3 water). The extract is evaporated to low volume on a rotary evaporator and butylated by treating with butylmagnesium chloride reagent. The butylated species are separated by GC on a column packed with 3 per cent OV-1 on chromosorb W (80/100 mesh). The separated compounds are determined by AA modified to take a quartz cuvette. The detection limit on a 5 litre sample was 0.04 μg dm^{-3}. It was concluded[158] that the only tetravalent elements that could interfere would be germanium(IV) and lead(IV) species. However, lead(IV) did not give a signal in their AA system and it was thought that there was little chance of germanium(IV) compounds being present in natural water, nor did they expect any spectral interference.

A more recent paper on the determination of methyltin species in water is by Andreae and Byrd,[245] who treated the sample with sodium borohydride to form the organotin hydrides which are cryogenically trapped and eluted in order of increasing boiling point. The tin is determined either by AA with a quartz cuvette or a graphite furnace or by flame photometric detection based on the molecular emission of the SnH species formed in the highly reducing hydrogen/air flame (cf. Braman and Tompkins[157] and Jackson *et al.*[121]). It was concluded[245] that AA with a quartz cuvette was unreliable because of the presence of spurious peaks eluted along with each methyltin species and so the quartz cuvette was abandoned in favour of the graphite furnace. Perhaps the most important observation in this work is that it was found that the FPD is subject to serious interference from germanium compounds, especially methylgermanium derivatives, which are found[245] in many natural samples, making the differentiation of methyltin and methylgermanium peaks virtually impossible with a single channel system. These authors,[245] therefore, developed a dual channel system with the ability to determine both tin and germanium simultaneously, although they observed that the optimum reaction condition for the methyltin species were not the same as those for methylgermanium compounds. As a result of these findings it is possible that the reported levels of methyltin species in the environment (sect. 3.6), determined by the flame emission method, may have been artificially high due to interference from germanium and/or sulphur.[246]

3.8.2.2   Butyltin compounds in aqueous systems

The use of tributyltin compounds in antifouling paints led to a need to determine these derivatives in leachates when evaluating paint formulations containing these biocides. The early work involved the use of extraction techniques, followed by spectrophotometric determination. Chromy *et al*[236] measured the tributyltin/dithizone species and Pettis *et al.*[237] extracted the tributyltin compound into methylisobutyl ketone and used haemetin − an oxidation product of haematoxyltin − to complex the tributyltin species. Neither of these photometric methods is particularly sensitive, the range covered being 1 to 200 mg dm^{-3} in the former and 0.1 to 10 mg dm^{-3} in the latter. However, Mor *et al.*[243] extracted the triorganotin compounds with carbon tetrachloride without a complexing agent, the extract being wet-ashed with sulphuric and perchloric acids to give inorganic tin, which was determined photometrically with phenylfluorone. These authors[243] concluded that the method is suitable for the range 0.01 to 0.2 mg dm^{-3} organotin using a 50 cm^3 sample. The work on butyltin compounds discussed so far was concerned with the determination of either tributyltin compounds or 'total' tin in the samples and no attempt was made to speciate.

Woggon and Jehle[194] were one of the first groups to publish an attempt to separate the butyltin species in water. They initially separated and concentrated the tributyltin compounds by steam distillation from an acidified sample (30 cm^3 of hydrochloric acid dm^{-3}); the di- and monobutyltin species were then extracted from the remaining solution by a suitable solvent (e.g. chloroform). The tributyltin derivative in the distillate was extracted into 5 cm^3 of cyclohexane and an aliquot (500 μl) subjected to thin layer chromatographic separation. The plate was sprayed with quercetin and the spots eluted with 12 per cent hydrochloric acid followed by exposure to UV light for 60 minutes. The inorganic tin species formed by UV degradation were determined by anodic stripping voltammetry (ASV). The di- and monobutyltin compounds, after evaporation of the chloroform, were subjected to a similar separation and the inorganic tin species formed determined by ASV.

Meinema *et al.*[145] also speciated butyltin compounds in water using an extraction technique but found it necessary to extract in the presence of tropolone (0.05 per cent) and hydrobromic acid (5 to 20 cm^3 dm^{-3} of sample), to ensure complete extraction of all the $(n\text{-}C_4H_9)_nSnX_{4-n}$ (n = 1 to 3) derivatives. It was observed[145] that tributyltin compounds were, in general, totally recovered by solvent extraction but that mono- and dibutyltin compounds were not, unless the extraction was carried out in the presence of tropolone and hydrobromic acid. Their method was to take the sample (500 cm^3), acidify with hydrobromic acid and extract with 2 $\times$ 25 cm^3 of 0.05 per cent tropolone in benzene. The benzene extract is submitted to Grignard methylation, to convert the butyltin species to $(n\text{-}C_4H_9)_nSn(CH_3)_{4-n}$, followed by separation by GC and determination of the organotin species by MS. The sensitivity of the method is 0.01 mg dm^{-3} on a 1 litre sample. Some workers using oxo and sulphur ligands with organotin complexes have observed rearrangement of the organic groups on tin.[5]

Hodge *et al.*[120] and Jewett and Brinckman[223] have applied the procedures for methyltin compounds already described (3.8.2.1) to the determination of butyltin compounds. Maguire and Huneault[232] used an extraction system similar to that of Meinema *et al.*[145] but derivatized the butyltin species by reacting with n-pentylmagnesium bromide, since they considered that the methylated derivatives were of similar volatilities to solvents such as hexane and benzene. It was found[232] that appreciable quantities of the methylbutyltin derivatives were lost during routine concentration procedures such as rotary or vortex evaporation of solvents. After pentylation the compounds are separated by GC and determined using a modified FPD. The optimum concentration of bis(tributyltin) oxide appeared to be 10 mg dm^{-3} on a 25 cm^3 sample. However, in a later paper,[122] the method was applied to the determination of organotin compounds in lakes and rivers and by using 8 litres of sample they were able to detect $(n\text{-}C_4H_9)_nSn(n\text{-}C_5H_{11})_{4-n}$ species at 0.01 μg dm^{-3}.

3.8.2.3 Phenyltin compounds in aqueous systems

As with the determination of butyltin compounds in environmental samples, the early work on this class of organotins tended to be limited to the determination of either 'total' tin, or, at most, identification of triphenyltin species and their determination as inorganic tin. The method of Mor *et al.*[243] described above (sect. 3.8.2.2) is one such procedure, no attempt being made to identify the presence of triphenyltin compounds. However, Freitag and Bock[240] studied the analysis of mixtures of tri-, di- and monophenyltin compounds by solvent extraction techniques. They studied a number of different schemes, a typical one being based on the sequential extraction of the compounds as follows: the solution containing 2 to 100 μg of the phenyltin compounds (as chlorides) is treated with an equal volume of citrate–phosphate–EDTA buffer solution at pH 8.5 and extracted twice with 10 cm^3 of dichloromethane. The triphenyltin species in the organic layer are determined directly with dithizone. The aqueous layer is extracted with dithizone or α-benzoinoxime in chloroform or PAR in isoamyl alcohol (all 10^{-3} mol dm^{-3}) and the diphenyltin derivative determined directly by spectrophotometry. The monophenyltin compounds remaining in the aqueous layer are extracted with tropolone in chloroform and the tin is determined photometrically with dithiol, after wet-oxidation.

Blunden and Chapman[238] used a spectrofluorimetric method for the determination of triphenyltin compounds in water using 3-hydroxyflavone in toluene as the reagent. It was found[238] that, if the triphenyltin/3-hydroxyflavone solutions were shaken with a saturated solution of sodium acetate then at least a ten-fold excess of di- and monophenyltin compounds and a fifty-fold excess of inorganic tin had no effect on the determination of the triphenyltin species; the method was found to be suitable for the range of 0.004 to 2 mg dm^{-3} in water.

Soderquist and Crosby[235] developed an analytical procedure for the determination of phenyltin compounds in water, based on the extraction of a 200 cm^3 sample with dichloromethane in the presence of an acetate buffer (2.0 mol dm^{-3} at pH 4.7). The extract is divided into three equal portions each of which is concentrated to about 0.1 cm^3 by evaporation with a stream of nitrogen at 40 °C. Total extractable tin is determined on one of the portions, after wet-ashing in sulphuric acid, followed by determination of the tin in the inorganic form with catechol violet/cetyltrimethyl-ammonium bromide. The second extract is treated with hexane and LiAlH$_4$ and the resulting phenyltin hydrides are separated by GC and determined using an ECD. Any tetraphenyltin present is determined in the third extract by treating this after concentration, with hexane and passing the solution through a Florosil microcolumn, eluting with hexane, concentrating the first 2.5 cm^3 of eluate to 0.3 cm^3 by evaporation and analysing by GC with FID. The authors claim a sensitivity of 0.003 mg dm^{-3} for a 200 cm^3 sample for mono-, di- and triphenyltin compounds and later[146] applied the method to a study of the degradation of triphenyltin hydroxide in water. Schaefer *et al.*[247] also used this method to determine triphenyltin hydroxide residues in rice field water and soil after treatement of growing rice with this fungicide.

Some of the methods for the determination of organotins in water reviewed in this section are summarized in Table 3.9.

Table 3.9   Methods for determining organotin compounds in water

Compound investigated	Method	Approximate detection limit	Reference
R_3SnX ($R = CH_3$, C_2H_5, C_4H_9) R_2SnX_2 ($R = CH_3$, C_2H_5, C_4H_9) $RSnX_3$ ($R = CH_3$, C_2H_5, C_4H_9)	Conversion to hydride with $NaBH_4$, collection in a trap, separation by boiling point, detection by AA	1 ng dm^{-3}	120
R_3SnX ($R = CH_3$, C_2H_5, C_3H_7, C_4H_9, C_6H_5, $c\text{-}C_6H_{11}$)	HPLC separation on strong cation exchange columns with determination by GFAA (graphite furnace AA)	$5-30$ ng as Sn	223
$(CH_3)_nSnH_{4-n}$ ($n = 1-4$)	Automatic purge trap sampler followed by GC separation and determination by FPD (flame photometric detection)	1 μg dm^{-3} on a 10 cm^3 sample	121
$(CH_3)_nSnX_{4-n}$ ($n = 0-3$)	Conversion to hydride with $NaBH_4$, separation by boiling point and determination by FED (flame emission detection)	<1 ng dm^{-3} on a 100 cm^3 sample	157
$(CH_3)_nSnX_{4-n}$ ($n = 1-3$)	Conversion to hydride with $NaBH_4$ separation by boiling point and determination by FPD using a dual-detector system to eliminate interferences	$20-50$ pg as Sn	245

Compound	Method	Detection limit	Reference
$(CH_3)_nSnX_{4-n}(n = 0-3)$	Extraction with benzene, followed by butylation and separation by GC and determination by AA	0.04 ng dm⁻³ on a 5 dm³ sample	158
$(C_4H_9)_nSnX_{4-n}(n = 1-3)$	Solvent extraction, followed by methylation and separation by GC and determination by MS	10 µg dm⁻³ on a 1 dm³ sample	145
$(C_4H_9)_nSnX_{4-n}(n = 1-3)$	Solvent extraction, followed by pentylation and determination with a modified FPD	10 mg dm⁻³ on a 25 cm³ sample	232
$(C_6H_5)_nSnX_{4-n}(n = 0-3)$	Separation by solvent extraction of the individual species followed by determination as inorganic tin by photometric methods.	2 µg as Sn	240
$(C_6H_5)_nSnX_{4-n}(n = 1-3)$	Solvent extraction, followed by conversion to hydride with LiAlH₄ and separation and determination by GC–EC (electron capture detection)	0.015 µg dm⁻³ on a 200 cm³ sample	235
$(C_4H_9)_3SnX$	Solvent extraction followed by direct photometric determination with dithizone	1 mg dm⁻³	236

Table 3.9 (cont)

Compound investigated	Method	Approximate detection limit	Reference
$((C_4H_9)_3Sn)_2O$	Solvent extraction and direct photometric determination on the haematein complex	0.1 mg dm^{-3}	237
'Organotins'	Solvent extraction, wet-ash followed by photometric determination with phenyfluorone	1 mg dm^{-3} on a 200 cm^3 sample	241
'Organotins'	Solvent extraction, wet-ash followed by photometric determination with phenyfluorone	0.01 mg dm^{-3} on a 50 cm^3 sample	243
$(C_4H_9)_3SnX$	Initial separation by steam distillation followed by extraction and determination by inverse voltammetry as inorganic tin	0.01 mg dm^{-3}	194
$(C_6H_5)_3SnX$	Concentration by evaporation, extraction and determination by inverse voltammetry as above	0.01 mg dm^{-3}	194
$(C_6H_5)_3SnX$	Solvent extraction and direct spectrofluorimetric determination with 3-hydroxyflavone	0.004 mg dm^{-3} on a 50 cm^3 sample	238

3.8.2.4 Soils and sediments

It is well known that organotin compounds adsorb strongly on to soil and sediments (sect. 3.4), and although good recoveries are possible with 'spiked' samples most investigators recognize that it is virtually impossible to extract quantitatively these compounds from 'real life' aged samples of soils and sediments and, thus no reliable methods have as yet been developed.

3.8.2.5 Air

In 1968, a threshold value of 0.1 mg m^{-3} was recommended[248] in the USA for organotins in air and appears to have been generally adopted as a Threshold Limit Value (TLV) for these compounds. Jeltes[249] developed a method for the determination of bis(tributyltin) oxide in air at this level by using a high volume air sampler and assuming that the tributyltin derivative was present as an aerosol and so could be collected on a filter. The air sampler is fitted with two glass fibre filter discs of 16 cm diameter and a sample of 25 m^3 is passed through at a rate of 30 to 40 m^3 h^{-1}. The filters are either extracted with methyl isobutyl ketone in a Soxhlet apparatus and the extract examined by AA using a nitrous oxide flame, or they are extracted with toluene and subjected to GC after *in situ* pyrolysis (at 390 °C) of the tributyltin compound, the resulting 1-butene peak being used for measurement of the trialkyltin species.

Zimmerli and Zimmerman[250] determined tetra- tri- and dibutyltin compounds in air by absorbing the compounds on Chromosorb 102 and extracting the absorbed compounds with diethyl ether containing 0.3 per cent hydrochloric acid. After evaporation to about 10 cm^3, the solution is treated with methylmagnesium choride in tetrahydrofuran. The resulting methylbutyltin derivatives are subjected to GC separation and the tin content is determined using a tin-specific FPD; the authors claim that 0.05 μg m^{-3} of butyltin compounds can be measured on as little as 1 m^3 of sample.

Tricyclohexyltin compounds have, at present, a limited application (sect. 3.3.1) and, therefore, are of perhaps less interest environmentally than other commercially available organotins. However, Bunyatyan and Oganesyan[251] determined tri-, di- and monocyclohexyltin compounds in air and water: 30 litres of air are passed through a filter at a rate of 3 litre min^{-3}, the filter is extracted with hexane, the solvent evaporated to low volume and subjected to TLC separation. The separated spots are extracted and the organotin species converted to the inorganic tin form and determined photometrically with dithiol; a sensitivity of about 0.02 mg m^{-3} is claimed.

3.8.2.6 Molluscs and other marine organisms

Although a relatively large number of studies on the effect of organotins particularly bis(tributyltin) oxide, on marine life (especially oysters) have been made over the past three to four years, only a few have involved the actual analysis of tissue. The majority of the studies are, either with ^{14}C-labelled bis(tributyltin) oxide, or, more usually, exposure of the organism to water of a nominal tributyltin concentration (obtained by serial dilution of a stock solution of the compound). Assessment of the effects of trialkyltin species in water, at these concentrations, is by visual examination of the aquatic organism under study.

Ward et al.[57] analysed whole fish or tissues of muscle, liver, embryo and viscera by dispersing them in concentrated hydrochloric acid, extracting the dispersion with hexane, adding tropolone and separating the tributyl- and dibutyltin species using an alkali wash,[252] the tin content being determined by GFAA. Waldock and Miller[124] also used this method, together with the GC–MS technique of Meinema et al.,[145] after extraction of the organotin compound into hexane from the dispersed tissue in acid. They carried out an extensive survey of the organotin content of oyster tissue and of the local seawater and found that, generally, the two methods gave good agreement.

Tugrul et al.[160] analysed fish, macro-algae and limpets for methyltin compounds by reacting the homogenized material with sodium borohydride and determining the tin compounds as the hydrides in a method similar to that of Hodge et al.[120]

3.8.2.7   Biological material

Investigations into the metabolism of organotin compounds have been of interest for a number of years but the lack of sensitive methods for the separation and determination of tin compounds in biological material led to the use of radiometric methods for these earlier studies.[226]

The development of a spectrofluorimetric method using 3-hydroxyflavone for determining organotin compounds led Aldridge et al.[36,253] to determine trimethyltin compounds in brain tissue. However, most of the recent work on the determination of organotin compounds in biological material appears to have been by the Japanese workers at Tokyo and Gunma Universities, who have published a series of joint papers on the determination of trialkyltin homologues in biological material,[254] tetraalkyltin compounds,[255] separation and determination of butyltin compounds[56] and triphenyltin and its metabolites.[256] The methods are based essentially on homogenization and extraction with organic solvent followed by GC separation (except in the case of the phenyltin compounds where an alumina column is used) and determination by GFAA where required. This work is summarized in the review by Wada et al.[11]

More recently, this group of workers developed a fluorimetric method for the determination of dialkyltin compounds using morin as the complexing reagent[239] and they have extended this to include a prior separation using HPLC, to determine dialkyltin homologues in various biological tissues.[257]

3.8.2.8   Analysis for solubility of organotin compounds in water

The measurement of the solubility of organotin compounds is of interest when investigating the use of organotin compounds in antifouling paints and polymers. From the figures for solubility given in Table 3.8 (sect. 3.5) it can be seen that widely different solubilities have been found for the same compound. At least three different groups of workers[141,142,144] used the same principle of having an excess of the compound in water and depended on either filtration and/or centrifuging to separate completely the solute from the aqueous phase, before determination of the solvated compound. The difficulties inherent in this method are obvious and may, perhaps, account for some of the widely differing results. Soderquist and Crosby[146] reported the solubility of triphenyltin hydroxide in water by 'lining' the surface of a one- or a nine-litre glass container with a film of the compound, and filling the container with water. After eight days of slow stirring, an aliquot of the water is removed and the tin content determined. Consistent values for the solubility of triphenyltin hydroxide were obtained using this method.

An alternative approach is that of Brooker and Ellison,[258] based on the addition of the compound dissolved in a water-miscible solvent, e.g. ethanol or acetone, to a fixed volume of water (50 cm^3), in microlitre amounts, until a turbidity is produced. This method overcomes the problems associated with the separation of undissolved solute from the aqueous phase and has been used[143] to determine the solubility of a number of organotin compounds in both distilled and artificial seawater.

3.9   Summary

Although there has already been much interest in the environmental chemistry of organotin compounds it is apparent that there is still scope for further research, especially in the aqueous chemistry of these compounds and in the further refinement of analytical techniques, particularly with respect to the methyltin species.

In general, most commercially used organotins are characterized by relatively low mobilities in environmental media, having low aqueous solubilities, low vapour pressures and high affinities for soil and organic sediments. While it is true to say that some commercial organotins, particularly the biocidal triorganotin species, have acute effects on plant and animal life, the exposures are mainly localized, and result from direct applications or accidental spillage. In certain specialized areas, e.g. harbours, due to continuous release of organotins from antifouling paints, coupled with their low mobility, short-term elevated concentrations of organotins could occur. However, it has been shown that organotins will degrade under environmental conditions and so serious long-term pollution hazards should not occur.

It is of interest to note that, in 1978, an association of world organotin manufacturers – The Organotin Environmental Programme Association

(ORTEPA) — was formed to promote and foster the dissemination of scientific and technical information on the environmental effects of organotin compounds. This action by the organotin producers will help ensure that organotin compounds continue to be used safely and effectively.

3.10   Acknowledgements

The International Tin Research Council, London, is thanked for permission to publish this review. Mrs L. Hobbs (ITRI) and Dr P.J. Smith (ITRI) are gratefully acknowledged for assisting with literature searching and valuable comments on the manuscript respectively.

References

1. Zuckerman J J, Reisdorf R P, Ellis H V, Wilkinson R R 1978 In *Organometals and Organometalloids: Occurrence and Fate in the Environment* (eds Brinckman F E, Bellama J M) 1978 *ACS Symp Ser* No. 82, pp 388–424
2. Anon 1980 *Environmental Health Criteria, 15: Tin and Organotin Compounds — A Preliminary Review*. World Health Organisation, Geneva
3. Plum H 1981 *Inf Chim* **220**: 135–9
4. Bock R 1981 *Residue Reviews*, vol 79 (ed Gunther F A). Springer Verlag, New York
5. Craig P J 1982 In *Comprehensive Organometallic Chemistry*, Vol 2 (eds Abel E W, Stone F G A, Wilkinson G). Pergamon Press, Oxford, pp 979–1020
6. Brinckman F E 1981 *J Organomet Chem Library* **12**: 343–76
7. Jones P A, Millson M F 1982 Organotins in the Canadian Environment A Synopsis *Environ Can Econ Tech Rev Rep EPS 3-EC-82-1*. Environ Impact Control Directorate, Quebec
8. Chau Y K, Wong P T S 1981 In NBS Spec Pub No 618 (eds Brinckman F E, Fish R H). US Dept Commerce, Washington D C, p 65
9. Blunden S J, Hobbs L A, Smith P J 1984. In *Environmental Chemistry* (ed Bowen H J M) *RSC Spec Per Rep* **3**: 49–77
10. Bennet R F 1983 *Ind Chem Bull* **2**: 171–6
11. Wada O, Manabe S, Iwai H, Arakawa Y 1982 *Jpn J Ind Health* **24**: 24–54
12. van der Kerk G J M, Luijten J G A 1954 *J Appl Chem* **4**: 314–9
13. Blunden S J, Smith P J, Sugavanam B 1984 *Pestic Sci* **15**: 253–7
14. Barnes J M, Magos L 1968 *Organomet Chem Rev* **3**: 137–50
15. Ascher K R S, Nissim S 1964 *World Rev Pest Control* **3**: 188–211
16. Smith P J 1978 *Toxicological Data on Organotin Compounds*. ITRI Publication No 538

17. Sijpesteijn A K, Luijten J G A, van der Kerk G J M 1969 *Fungicides: An Advanced Treatise* vol 2 (ed Torgeson D C). Academic Press, New York, p 331
18. Polster M, Halacka K 1974 *Tag Ber Akad Landwertsch-Wiss DDR Berlin* **126**: 117–22
19. Linden E, Bengtsson B E, Svanberg O, Sundstrom G 1979 *Chemosphere* **8**: 834–51
20. Tooby T E, Hursey P A, Alabaster J J 1975 *Chem Ind*: 523–6
21. Anon 1979 Plictran Miticide *Tech Inf Bull*. Dow Chemical Company, Midland, Michigan
22. Hammann I, Büchel K L, Bungarz K, Born L 1978 *Pflanzenschultz-Nachr Bayer Engl Ed* **31**: 61–83
23. Sylph A 1984 *Bibliography on the Toxicity of Organotins to Aquatic Animals*. ITRI Bibliography No 11
24. Cremer J E 1958 *Biochem J* **68**: 685–92
25. Kimmel E C, Fish R H, Casida J E 1977 *J Agric Food Chem* **25**: 1–9
26. Selwyn M J 1976 In *Organotin Compounds: New Chemistry and Applications* (ed Zuckerman J J) *ACS Adv Chem Ser* **157**: 204–26
27. Aldridge W N 1976 *Organotin Compounds: New Chemistry and Applications* (ed Zuckerman J J) *ACS Adv Chem Ser* **157**: 186–96
28. Aldridge W N 1978 Proc 2nd Int Conf Si Ge Sn and Pb Compds (eds Gielen M, Harrison P G) *Rev Si Ge Sn Pb Compds Special Issue*. Freund Publishing House, Tel Aviv, p 9
29. Davies A G, Smith P J 1980 *Adv Inorg Chem Radiochem* **23**: 1–77
30. Elliot B M, Aldridge W N, Bridges J W 1979 *Biochem J* **177**: 461–70
31. Cain K, Griffiths D E 1977 *Biochem J* **162**: 575–80
32. Taketa F, Siebenlist K, Kasten-Jolly J, Palosaari N 1980 *Arch Biochem Biophys* **203**: 466–72
33. Luijten J G A, Klimmer O R 1978 *Toxicological Assessment of Organotin Compounds*. ITRI Publication No 538, pp 11–20
34. Brown A W, Aldridge W N, Street B W 1979 *Neuropath Appl Neurobiol* **5**: 83
35. Ross W D 1981 *Am J Psychiatry* **138**: 1092–5
36. Aldridge W N, Brown A W, Brierley J B, Verschoyle R D, Street B W 1981 *Lancet*: 692–3
37. Seinen W 1979/80 *Vet Sci Commun* **3**: 279–87
38. Seinen W 1981 *Immunol Consid Toxicol* **1**: 103–19
39. Innes J R M, Ulland B M, Valerio M G, Petrucelli L, Fishbein L, Hart E R, Palotta A J, Bates R R, Falk H L, Gart J J, Klein M, Mitchell I, Peters J 1969 *J Natl Cancer Inst* **42**: 1101–14
40. Anon 1978 *US Natl Cancer Inst Carcinogen Tech Rep Ser* No 139
41. Sheldon A W 1975 *J Paint Technol* **47**: 54–8
42. Anon 1971 *Evaluations of Some Pesticide Residues in Food 1971*. FAO/WHO Rome, pp 521–42
43. Anon 1979 *US Natl Cancer Inst Carcinogen Tech Rep Ser* No 183

44. Anon 1977 *Evaluation of Some Pesticide Residues in Food 1977.* FAO/WHO Rome, pp 229–61
45. Mott K E, 1984 WHO, Geneva (personal communication)
46. Bridges J W, Davies D S, Williams R T 1967 *Biochem J* **105**: 1261–7
47. Arakawa Y, Wada O, Yu T H 1981 *Toxicol Appl Pharmacol* **60**: 1–7
48. Iwai H, Wada O 1981 *Indust Health* **19**: 247–253
49. Iwai H, Wada O, Arakawa Y, Ono T 1982 *J Toxicol Environ Health* **9**: 41–9
50. Freitag K D, Bock R 1974 *Pestic Sci* **5**: 731–9
51. Kimmel E C, Casida J E, Fish R H 1980 *J Agric Food Chem* **28**: 117–22
52. Brown R A, Nazario C M, de Tirado R S, Castrilton J, Agard E T 1977 *Environ Res* **13**: 56–61
53. Evans W H, Cardarelli N F, Smith D J 1979 *J Toxicol Environ Health* **5**: 871–7
54. Anger J P 1975 PhD Thesis, Université Rene Descartes de Paris
55. Blair E H 1975 *Environ Qual Safety Suppl* **3**: 406–9
56. Iwai H, Wada O, Arakawa Y 1981 *J Anal Toxicol* **5**: 300–6
57. Ward G S, Cram G C, Parrish P R, Trachman H, Slesinger A 1981 *Aquat Toxicol Haz Asses: 4th Conf ASTM STP 737* (eds Branson D R, Dickson K L). Am Soc Testing and Materials, p 183
58. Fish R H, Kimmel E C, Casida J E 1976. In *Organotin Compounds: New Chemistry and Applications* (ed Zuckerman J J) *ACS Adv Chem Ser* **157**: 197–203
59. Prough R A, Stalmach M A, Wiebkin P, Bridges J W 1981 *Biochem J* **196**: 763–70
60. Evans C J 1974 *Tin Its Uses* **100**: 3–6
61. Evans C J 1981 *Speciality Chemicals* **1**: 25–30
62. Evans C J 1974 *Tin Its Uses* **101**: 12–5
63. Evans C J 1981 *Speciality Chemicals* **1**: 12–5
64. Wehner W, Wirth H O 1980. In *22 Deutscher Zinntag: Organozinn-chemie* (eds Neuman W P, Gielen M) *Rev Si Ge Sn Pb Compds Special Issue*. Freund Publishing House, Tel Aviv, p 7
65. Bokranz A, Plum H 1975 *Industrial Manufacture and Use of Organotin Compounds*. Schering A.G., Bergkamen
66. Gitlitz M H, Moran M K 1983 Encyclopedia of Chemical Technology, vol 23, 3rd edn. John Wiley, New York, p 42
67. Kumar Das V G 1975 *Planter (Kuala Lumpur)* **51**: 355–64
68. Sugavanam B 1980 *Tin Its Uses* **126**: 4–6
69. Ascher K R S 1979 *Phytoparasitica* **7**: 117–37
70. Evans C J 1980 *Tin Its Uses* **86**: 7–9
71. Evans C J 1976 *Tin Its Uses* **110**: 6–7
72. Ascher K R S, Moscowitz J 1969 *Int Pest Control* 11 (Jan/Feb), 17–20
73. King S 1980 *Tin Its Uses* **124**: 13–6
74. Crowe A J, Hill R, Smith P J 1978 *Tributyltin Wood Preservatives*. ITRI Publication No 559

75. Crowe A J, Hill R, Smith P J, Cox T R G 1979 *Int J Wood Pres* **1**: 119–24
76. Richardson B A 1973 *Stone Ind* **8**: 2–6
77. Richardson B A, Cox T R G 1974 *Tin Its Uses* **102**: 6–10
78. Anon 1979 *Wood Preservatives Approved by the Swedish Wood Preservation Institute*. Svendska Traskyddinstitutet, Stockholm
79. Blunden S J, Chapman A H, Crowe A J, Smith P J 1978 *Int Pest Control* **20**: 5–8, 12
80. Blunden S J, Hill R 1984 *Inorg Chim Acta* **87**: 83–5
81. Anon 1984 *Water Soluble Organotin Biocides*. ITRI Publication No. DS4
82. Hill R, Smith P J, Ruddick J N R, Sweatman K W 1983 *Int Res Gp Wood Pres* Doc. No IRG/WP/3229. ITRI Publication No 629
83. Hill R 1984 ITRI Unpublished work
84. Evans C J, Hill R 1981 *J Oil Col Chem Assoc* **64**: 215–23
85. Nösler G H 1970 *Gesund Disinfekt* **62**: 10–3, 65–9, 175–6
86. Hudson P B, Sanger G, Sproul E E 1959 *J Am Med Assoc* **169**: 89
87. Evans C J, Hill R 1983 *Rev Si Ge Sn Pb Compds* **7**: 55–125
88. Davies A G, Smith P J 1982. In *Comprehensive Organometallic Chemistry*, Vol 2 (eds Abel E W, Stone F G A, Wilkinson G). Pergamon Press, Oxford, p 519
89. Castelli V J, Andersen D M, Becka A M 1983 *Abst 4th Int Conf Organomet Coord Chem Ge Sn Pb*. Montreal, Quebec
90. Brecker L R 1981 *Pure Appl Chem* **53**: 577–82
91. Klimsch P, Kühnert P 1969 *Plaste Kautschuk* **16**: 242–51
92. Niemann H 1973 *Inf Chim* **119**: 119–36
93. Fuller M J 1975 *Tin Its Uses* **103**: 3–7
94. Knoll W 1968 *Chemistry and Technology of Silicones*. Academic Press, New York
95. Karpel S 1980 *Tin Its Uses* **125**: 1–6
96. Anon 1977 Fascat 4101 *Monobutyltin Esterification Catalyst*. Tech Data Sheet M and T Chemicals Inc
97. Greenberg R R, Gordon G E, Zoller W H, Jacko R B, Neundorf D W, Yost K J 1978 *Environ Sci Technol* **12**: 1329–32
98. Senich G A, 1982 *Polymer* **23**: 1385–7
99. Dietz G R, Banzer J D, Miller E M 1977. In *Safety and Health with Plastics Society of Plastics Engineers*. Denver, Colorado, p 25
100. Boettner E A, Ball G L, Hollingsworth Z, Aquino R 1981 *EPA Rep No EPA-600/1-81-062*. EPA Health Effects Res Lab Cincinatti, Ohio
101. Figge K, Freitag W, Beiber W D 1979 *Dtsch Lebensm-Rundsch* **75**, 333–45
102. Klimmer O R, Nebel I U 1960 *Arzneim-Forsch* **10**: 44–8
103. Heller J, Brauman S K, 1977 *Nat Inst Health Rep NI/DTB-79/01* No. PB-299744, Bethesda, Md
104. Guess W L, Stetson T B, 1968 *J Am Med Assoc* **204**: 580–4

105. Thust U, 1979 *Tin Its Uses* **122**: 3–5
106. Anon. 1981 *Specifications and Technical Data: Organotin Compounds.* Schering Industrial Chemicals, Bergkamen
107. Anon 1979 *Tributyltin Oxide: Notes on Handling and Toxicity*, Tech Service Note, Albright and Wilson, Oldbury
108. Maguire R J, Carey J H, Hale E J 1983 *J Agric Food Chem* **31**: 1060–5
109. Laye P G 1982 University of Leeds (unpublished work)
110. Keiser D, Kanaan A S 1969 *J Phys Chem* **73**: 4264–9
111. Barnes R D, Bull A T, Poller R C 1973 *Pestic Sci* **4**: 305–17
112. Suess A, Eben Ch 1973 *Z-Pflanzenkrankh Pflanzenshutz* **80**: 288–94
113. Leeuwangh P, Nijman W, Wisser H, Kolar Z, de Goby J J M 1976 *Med Fac Landbouww Rijkuniv Ghent* **41**: 1483–90
114. Harris L R, Andrews C, Burch D, Hampton D, Maegerlein S 1979 *Rep No DTNSRDC/SME-78/2A* D W Taylor Naval Ship R and D Centre, Bethesda, Md
115. Barug D, Vonk J W 1980 *Pestic Sci* **11**: 77–82
116. Slesinger A E 1977 *Proc 17th Ann Marine Coatings Conf*, Biloxi, Miss
117. Katsumura T J.P. 21 360 165/1975
118. Bollen W B, Tu C M 1972 *Tin Its Uses* **94**: 13–5
119. Omar M, Bowen H J M 1982 *J Radioanal Chem* **74**: 273–82
120. Hodge V F, Seidel S L, Goldberg E D 1979 *Anal Chem* **51**: 1256–9
121. Jackson J-A A, Blair W R, Brinckman F E, Iverson W P 1982 *Environ Sci Technol* **16**: 110–9
122. Maguire R J, Chau Y K, Bengert G A, Hale E J, Wong P T S, Kramar O 1982 *Environ Sci Technol* **16**: 698–702
123. Andreae M O, Byrd J T, Froelich P N 1983 *Environ Sci Technol* **17**: 731–7
124. Waldock M J, Miller D 1983 *Min Agric Fish Food Rept* CM 1983/E: 12. Int Council Expl Sea, Mar Environ Qual Commit
125. Alzieu C, Thibaud Y, Héral M, Boutier B 1980 *Rev Trav Inst Pêches Marit* **44**: 301–48
126. Alzieu C, Héral M, Thibaud Y, Dardignac M J, Feuillet M 1981 *Rev Trav Inst Pêches Marit* **45**: 101–16
127. Soudan F 1982 *Méd et Nut* **18**: 391–3
128. Thain J E 1983 *Min Agric Fish Food Rept* CM 1983/E:13. Int Council Expl Sea, Mar Environ Qual Commit
129. Waldock M J, Thain J, Miller D 1983 *Min Agric Fish Food Rept* CM 1983/E: 52. Int Council Expl Sea, Mar Environ Qual Commit
130. Waldock M J, Thain J E 1983 *Mar Poll Bull* **14**: 411–5
131. Cardarelli N F 1976 *Controlled Release Pesticide Formulations.* Chemical Rubber Co, Cleveland, Ohio
132. Cardarelli N F 1975 In *Molluscicides in Schistosomiasis Control* (ed Cheng T C). Academic Press, New York, p 177
133. Duncan J 1980 *Pharmacol Ther* **10**: 407–29

134. Tobias R S 1978. In *Organometals and Organometalloids: Occurrence and Fate in the Environment* (eds Brinckman F E, Bellama J M). *ACS Symp Ser* **82**: 130–48

135. Brinckman F E, Parris G E, Blair W R, Jewett K L, Iverson W P, Bellama J M 1977 *Environ Health Perspect* **19**: 11–24

136. McGrady M M, Tobias R S 1964 *Inorg Chem* **3**: 1157–63

137. McGrady M M, Tobias R S 1965 *J Am Chem Soc* **87**: 1909–16

138. Blunden S J, Smith P J, Gillies D G 1982 *Inorg Chim Acta* **60**: 105–9

139. Heron P N, Sproule J S G 1958 *Indian Pulp Paper* **12**: 510–17

140. Vind H P, Hochman H 1962 *Proc Am Wood Pres Assoc* **58**: 170–8

141. Beiter C, Engelhart J E, Freiman A, Sheldon A W 1974 *Proc Am Chem Soc Symp Mar Freshwater Pestic.* Atlantic City, New Jersey

142. Ozcan M, Good M L 1980 *Proc Am Chem Soc Div Environ Chem.* Houston, Texas.

143. Chapman A H 1981 ITRI (unpublished work)

144. Chromy L, Uhacz K 1978 *J Oil Col Chem Assoc* **61**: 39–42

145. Mienema H A, Berger-Wiersma T, Verslius-de Haan G, Gevers E Ch 1978 *Environ Sci Technol* **12**: 288–93

146. Soderquist C J, Crosby D G 1980 *J Agric Food Chem* **28**: 111–7

147. Meyling A H, Pitchford R J 1966 *Bull World Health Org* **34**: 141–6

148. Anon 1973 *Vendex Miticide.* Tech Data Bull, Shell Chemical Co, San Ramon, California

149. Davies A G, Goddard J P, Hursthouse M B, Walker N P C 1983 *J Chem Soc Chem Commun*: 597–8

150. Guard H E, Coleman W M, Cobet A B 1982 *Abstr 185th Nat Meet Div Environ Chem Am Chem Soc.* Las Vegas, Nevada

151. Pettine M, Millero F J, Macchi G 1981 *Anal Chem* **53**: 1039–43

152 Nazarenko V A, Antonovich V P, Nevshaya E M 1971 *Russ J Inorg Chem* **16**: 980–2

153. Challenger F, Higginbottom C 1935 *Biochem J* **29**: 1757–78

154. Challenger F 1955 *Quart Rev Chem Soc* **9**: 255–86

155. Challenger F 1959 *Aspects of the Organic Chemistry of Sulphur.* Butterworths, London

156. Challenger F 1978 In *Organometals and Organometalloids: Occurrence and Fate in the Environment* (eds Brinckman F E, Bellama J M). *ACS Symp Ser* **82**: 1–22

157. Braman R S, Tompkins M A 1979 *Anal Chem* **51**: 12–19

158. Chau Y K, Wong P T S, Bengert G A 1982 *Anal Chem* **54**: 246–9

159. Byrd J T, Andreae M O 1982 *Science* **218**: 565–9

160. Tugrul S, Balkas T I, Goldberg E D 1983 *Mar Poll Bull* **14**: 297–303

161. Huey C, Brinckman F E, Grim S, Iverson W P 1974 *Proc Int Conf Transp Persist Chem Aquat Ecos*, Ottawa, p 73

162. Guard H E, Cobet A B, Coleman W M 1981 *Science* **213**: 770–1

163. Hallas L E, Means J C, Cooney J J 1982 *Science* **215**: 1505–7

164. Chau Y K, Wong P T S, Kramar O, Bengert G A 1981. In *Proc Internat Conf Heavy Metals in the Environment (Amsterdam)* (ed Ernst W H O). CEP Consultants, Edinburgh
165. Craig P J, Rapsomanikis S 1982 *J Chem Soc Chem Commun*: 114
166. Anon 1973 *Evaluations of Some Pesticide Residues in Food.* Monographs 1974 *WHO Pesticide Residue Series* No 3 WHO, Geneva, p 448
167. Craig P J, Rapsomanikis S 1984 In *Proc DoE/NBS Workshop Environ Spec Monitor Needs* (eds Brinckman F E, Fish R H) *Nat. Bur Stand (US) Spec Publ No 618.* US Dept Commerce, Washington D C, p 54
168. Dizikes L J, Ridley W P, Wood J M 1978 *J Amer Chem Soc* **100**: 1010−2
169. Fanchiang Y T, Wood J M 1981 *J Amer Chem Soc* **103**: 5100−3
170. Thayer J S 1978. In *Organometals and Organometalloids: Occurrence and Fate in the Environment* (eds Brinckman F E, Bellama J M). *ACS Symp Ser* **82**: 188−204
171. Thayer J S Aug 1981 *Abst 10th Internat Conf Organomet Chem*, Toronto.
172. Thayer J S 1983 Paper presented at *4th Int Conf Organomet Coord Chem Ge Sn Pb* Montreal, Canada
173. Thayer J S 1979 *Inorg Chem* **18**: 1171−2
174. Manders W F, Olsen G J, Brinckman F E, Bellama J M 1984 *J Chem Soc Chem Commun*: 538−40
175. Craig P J 1980 *Environ Technol Letts* **1**: 225−34
176. Bruggemann J, Barth K, Niesar K H 1964 *Zentralbl Veterinarmedzin* **11**: 4−19
177. Bruggermann J, Klimmer O R, Niesar K H 1964 *Zentralbl Veterinarmedzin* **11**: 40−8
178. Hardon H J, Besemer A F H, Brunick H 1962 *Dtsch Lebensm-Rundsch* **58**: 349−52
179. Dunn P, Oldfield D 1973 *Austral Def Sci Ser* Tech Note 298
180. Watkins D A M 1974 *Chem Ind*: 185−93
181. Skinner H A 1964 *Adv Organometal Chem* **2**: 39−114
182. Poller R C 1978 *Rev Si Ge Sn Pb Compds* **3**: 243−77
183. Davies A G 1976. In *Organotin Compounds: New Chemistry and Applications* (ed J J Zuckerman) *ACS Adv Chem Ser* **157**: 26−40
184. Davies A G 1981 *J Organomet Library* **12**: 181−91
185. Cenci P, Cremonini B 1969 *Ind Sacc Ital* **62**: 313−6
186. Getzendaner M E, Corbin H B 1972 *J Agric Food Chem* **20**: 881−5
187. Akagi J, Sakagami Y 1971 *Bull Inst Publ Health Japan* **20**: 1−4
188. Massaux F 1971 *Café Cacao Thé* **15**: 221−34
189. Chapman A H, Price J W 1972 *Int Pest Control* **14**: 11−12
190. Chapman A H, Price J W 1973 ITRI Unpublished work
191. Klötzer D, Thust U 1976 *Chem Tech (Leipzig)* **28**: 614−16
192. Klötzer D 1977 *Zentralinst Kernforsch Rossendorf Dresden Zfk*: 84−7

193. Komora V F, Popl M 1978 *Holztech* **19**: 145–7
194. Woggon H, Jehle D 1975 *Nahrung* **19**: 271–5
195. Mazaev V T, Golovonov P V, Igumnov A S, Tsay V H 1976 *Gig Sanit*: 17–20
196. Odeyemi O, Ajulo E 1982 *Water Sci Tech* **14**: 133–42
197. Sherman L R, Yazdi M, Hoang H 1983 *Abstr 4th Int Conf Organomet Chem Ge Sn Pb* Montreal, Quebec
198. Blunden S J 1983 *J Organometal Chem* **248**: 149–60
199. Barug D 1981 *Chemosphere* **10**: 1145–54
200. Blair W R, Olsen G J, Brinckman F E, Iverson W P 1982 *Microb Ecol* **8**: 241–51
201. Sheldon A W 1978 *Proc 18th Annual Marine Coatings Conf*, Monterey, California
202. Stein V T, Küster K 1982 *Wasser Abwasser Forsch* **15**: 178–80
203. Henshaw B G, Laidlaw R A, Orsler R J, Carey J K, Savory J G 1978 *Record of the 1978 Annual Convention of the BWPA* Cambridge: 19–29
204. Plum H, Landseidel H 1980 *Holz Roh Werkstoff* **38**: 461–5
205. Edlund M L, Hintze W, Jermer J, Ohlsson S V 1982 Swedish Wood Preservation Institute, Stockholm, Rept No 143
206. Ohlsson S V, Hintze W W 1983 *J High Res Chromatogr Chromatogr Commun* **6**: 89–94
207. Jermer, L, Edlund M L, Henningsson B, Hintze W, Ohlsson S 1983 *Int Res Gp Wood Pres* Doc No IRG/WP/3219
208. Beiter C B, Arsenault R D 1982 *Proc Amer Wood Pres Assoc* **77**: 58–63
209. Hintze W, Ohlsson S 1983 *Int Res Gp Wood Pres* Doc No IRG/WP/3250
210. Orsler R J, Holland G E 1984 *Int Res Gp Wood Pres* Doc No IRG/WP/3287
211. Jensen B, Landseidel H A B, Plum H 1984 *Int Res Gp Wood Pres* Doc No IRG/WP/3275
212 Hill R, Chapman A H, Samuel A, Manners K, Morton G 1984 *Int Res Gp Wood Pres* Doc No IRG/WP/3311
213 Dudley-Brendell T E, Dickinson D J 1982 *Int Res Gp Wood Pres* Doc No IRG/WP/1156
214. Carey J K 1980 PhD Thesis. Univ of London, UK
215. Sutter H-P 1980 MSc Thesis. Portsmouth Polytechnic, UK
216. Orsler R J, Holland G E 1982 *Internat Biodet Bull* **18**: 95–8
217. Wise L E 1952 In *Wood Chemistry*, Vol 1 (eds Wise L E, Jahn E C). Reinhold, New York
218. Neumann W P 1970 *The Organic Chemistry of Tin*. Wiley, New York
219. Poller R C 1970 *The Chemistry of Organotin Compounds*. Academic Press, New York
220. Blunden S J, Chapman A H 1982 *Environ Technol Lett* **3**: 267–72
221. Blunden S J, Chapman A H 1983 *A Review of the Degradation of*

Organotin Compounds in the Environment and their Determination in Water. ITRI Publication No 626

222. Monaghan C P, Kulkarni V I, Ozcan M, Good M L 1980 *US Govt Report Office of Naval Research*. Tech Rep No 2 AD-AP87374
223. Jewett K L, Brinckman F E 1981 *J Chromatogr Sci* **19**: 583–93
224. Smith P J, Crowe A J, Allen D W, Brooks J S, Formstone R 1977 *Chem Ind*: 874–5
225. Blunden S J, Hill R 1984 *J Organometal Chem* **267**: C5–8
226. Price J W, Smith R 1978 *Handbook of Analytical Chemistry*, Part 3, Vol 4 $\alpha\gamma$ Tin (ed Fresenius W). Springer-Verlag, Heidelberg
227. Brinckman F E, Blair W R, Jewett K L, Iverson W P 1977 *J Chromatogr Sci* **15**: 493–503
228. Parks E J, Brinckman F E, Blair W R 1979 *J Chromatogr* **185**: 563–72
229. Vickrey T M, Howell H E, Harrison G V, Ramelow G J 1980 *Anal Chem* **52**: 1743–6
230. Schatzberg P, Adema C M, Jackson D F 1983 *Abst 4th Int Conf Organomet Coord Chem Ge Sn Pb* Montreal, Canada
231. Camail M, Loiseau B, Margaillan A, Vermet J L 1983 *Analusis* **11**: 358–9
232. Maguire R J, Huneault H 1981 *J Chromatogr* **209**: 458–62
233. Wright B W, Lee M L, Booth G M 1979 *J High Res Chromatogr, Chromatogr Commun* **2**: 189–90
234. Simon N, Welenic A J, Aldridge M E 1978 *US Dept of Commerce Rept* AD/A-058566
235. Soderquist C J, Crosby D G 1978 *Anal Chem* **50**: 1435–9
236. Chromy L, Uhacz K 1968 *J Oil Col Chem Assoc* **51**: 494–8
237. Pettis R W, Philip A T, Woodford J M D 1972 *Aust Def Sci Service* Rept no 516
238. Blunden S J, Chapman A H 1978 *Analyst* **103**: 1266–9
239. Arakawa Y, Wada O, Manaba M 1983 *Anal Chem* **55**: 1901–4
240. Freitag K D, Bock R 1974 *Z Anal Chem* **270**: 337–46
241. Chernorukova Z G, Zabotin K P 1978 *Fiz-Khim Metody Anal* **3**: 95–7
242. Sherman L R, Carlson T L 1980 *J Anal Tox* **4**: 31–3
243. Mor E, Beccaria A M, Poggi G 1973 *Ann Chim* **63**: 173–80
244. Battais A, Bensimon Y, Besson J, Durand G, Pietrasankta Y 1982 *Analusis* **10**: 426–32
245. Andreae M O, Byrd J T 1984 *Anal Chim Acta* **156**: 147–57
246. Andreae M O 1984 Florida State University Tallahasee, Florida (personal communication)
247. Schaefer C H, Muira T, Dupas Jr E F, Wilder W H 1981 *J Econ Entomol* **74**: 597–600
248. Anon 1968 *Ann Meet Am Conf Gov Ind Hyg*, St Louis, MO
249. Jeltes R 1969 *Ann Occup Hyg* **12**: 203–7
250. Zimmerli B, Zimmerman H 1980 *Fresenius Z Anal Chem* **304**: 23–7
251. Bunyatyan Yu A, Oganesyan G O 1983 *Gig Sanit* **3**: 55–7

252. Anon M & T Chemical Inc Standard Test Method AA-33
253. Aldridge W N, Fleet B W 1981 *Analyst* **106**: 60–8
254. Arakawa Y, Wada O, Yu TH, Iwai H 1981 *J Chromatogr* **216**: 209–17
255. Arakawa Y, Wada O, Yu TH, Iwai H 1981 *J Chromatogr* **207**: 237–44
256. Manabe M, Wada O, Iwai H, Matsui H, Manabe S, Ono T 1981 *Jap J Ind Health* **23**: 312–3
257. Yu TH, Arakawa Y, 1983 *J Chromatogr* **258**: 189–97
258. Brooker P J, Ellison M 1974 *Chem Ind*: 785–7

Chapter 4

Organolead compounds in the environment

4.1 Introduction

The first reported synthesis of a compound containing a lead–carbon bond was that of Löwig,[1,2] who in 1853 reacted a sodium–lead alloy with ethyl iodide to produce impure hexaethyldilead. At about the same time Cahours[3] independently reported reacting lead metal with ethyl iodide to produce an organolead compound. From these tentative beginnings organolead chemistry has developed into one of the largest areas of organometallic chemistry, largely prompted by the discovery by Midgley and Boyd[4] in 1922 that tetraethyllead can act as an efficient antiknock agent in petrol engines.

The alkyl and aryl group compounds of the type R_4Pb, R_3PbX, R_2PbX_2 and $(R_2Pb)_n$(R being an alkyl or aryl group and X a negative ion) were the first to be investigated. During the period 1915–25 these compounds were systematically produced and classified.[5] The discovery of the organolead metal derivatives (e.g. triphenylplumbyllithium, $(C_6H_5)_3PbLi$) by Gilman and co-workers,[6] the organolead hydrides, amides and phosphides and other groups of organolead compounds (e.g. aryllead triacylates[7]) followed quickly, so by 1965 more than 1200 organolead compounds were known. Several excellent reviews of organolead chemistry are available,[5,8–11] and the subject is reviewed annually in the Royal Society of Chemistry (London) Specialist Periodical Report, *Organometallic Chemistry*, and in the annual *Organolead Chemistry* published in the Journal of Organometallic Chemistry Library.

Despite the huge diversity of organolead compounds the environmental chemistry of organic lead is dominated by a small number of tetraalkyllead compounds, their salts, and their decomposition products. The reason for this is two-fold; firstly, anthropogenic emissions of organic lead are almost solely due to the manufacture and use of tetramethyllead and tetraethyllead and the three mixed tetraalkyl intermediate compounds as gasoline additives, and,

Table 4.1 Nomenclature of environmentally occurring organolead compounds

Compound	Abbreviation	Name
$(CH_3)_4Pb$	TML	Tetramethyllead
$(CH_3)_3(C_2H_5)Pb$	TMEL	Trimethylethyllead
$(CH_3)_2(C_2H_5)_2Pb$	DMDEL	Dimethyldiethyllead
$CH_3(C_2H_5)_3Pb$	METL	Methyltriethyllead
$(C_2H_5)_4Pb$	TEL	Tetraethyllead
$(CH_3)_3PbX$	TriML	Trimethyllead salt†
$(CH_3)_2PbX_2$	DiML	Dimethyllead di-salt†
$(C_2H_5)_3PbX$	TriEL	Triethyllead salt†
$(C_2H_5)_2PbX_2$	DiEL	Diethyllead di-salt†

† X = counter-ion (possibly halide)

secondly, environmental alkylation of lead, if it occurs, almost certainly involves the addition of methyl groups to lead.

The scope of this chapter, therefore, is limited to those compounds containing a lead−carbon bond which occur in the environment. These compounds and their abbreviations are listed in Table 4.1.

The organolead halide salts (e.g. $(CH_3)_3PbCl$, $(CH_3)_2PbCl_2$) are high melting point, highly crystalline ionic compounds, soluble in water and relatively stable in solution. Although the trialkyllead salts have high melting points, some volatility is perceptible as indicated by their sternutatory properties.[12] The five tetraalkyllead compounds are dense, moderately volatile liquids, sensitive to both heat and light, and have tetrahedral geometries in the vapour phase. Important physical properties of tetramethyllead and tetraethyllead are shown in Table 4.2.

In 1977 about 10 per cent of refined lead consumption in the western world was used for the manufacture of organolead gasoline additives, reduced to about 5 per cent in 1983[13] with a UK consumption of tetraalkyllead of 12 000 tonnes per year in 1973.[14] About 1 per cent of this organolead is emitted

Table 4.2 Physical properties of TML and TEL

Compound	Density (g cm⁻³) at 20 °C	BP (°C) at 760 mmHg	Vap. pressure (mmHg) at 20 °C	MP (°C)	Refractive index, η_D^{20}
TML	1.995	110	23.7	− 30.2	1.5120
TEL	1.650	200*	0.26	− 130.2	1.5198

* Decomposes below this temperature.

unchanged[15] from engines via the exhaust. Further emissions to the atmosphere, caused by evaporative losses of fuel from fuel tanks, carburettors and spillage during the manufacture and transfer of antiknock compounds, imply an annual release rate of tetraalkyllead of perhaps 250 tonnes yr^{-1} in 1973 in the UK (see below), although this figure may now be reduced by the lowering of the maximum permissible lead content of petrol.

The significant anthropogenic inputs of tetraalkyllead to the environment outlined above may be compounded by the possibility of a natural source of organic lead.[16] There has been much debate in recent years on this subject,[17] and it is, at the time of writing, still unresolved. However, even if the conversion rates involved are very low the continuous net production of organolead compounds in the environment may still constitute an important part of the global geochemical cycle of the element, hitherto not fully recognized.

4.2   Sources and uses of tetraalkyllead (TAL)

4.2.1   Use in gasoline

Lead is added to gasoline in varying proportions of the five tetraalkyllead compounds in order to prevent the spontaneous premature combustion of the fuel mixture (knocking). Because of the complex nature of gasoline, which is a blend of varying amounts of hydrocarbons distilling in the range 30−210 °C, a blend of TAL compounds is used as the fuel additive. The effectiveness of TAL as an antiknocking agent relies on its ability to be easily oxidized to lead oxide. This advances as a fine mist in front of the flame front in the combustion cylinder, scavenging the peroxy radicals which are responsible for the pre-ignition reactions. The presence of sulphur compounds in the fuel has a deleterious effect on the effectiveness of the TAL.

About 16 per cent of total US lead consumption in 1974 was used for the manufacture of lead additives[18] and on a global basis substantially less than 10 per cent of total lead consumption is used in this way. Table 4.3 shows consumption of refined lead for manufacture of gasoline additives. Although the use of these compounds is falling in many countries due to legislative action, in other countries consumption may continue to rise. The present maximum permissible level in gasoline in the UK is 0.4 g dm^{-3} and has fallen to 0.15 g dm^{-3} in 1985. However recommendations have been made to phase out the use of lead additives altogether.[19] In the US lead-free gasoline only must be used, by law, in all cars of model year 1975 or later and is required to prevent poisoning of exhaust system catalysts which are fitted to limit emissions of oxides of nitrogen, carbon monoxide and hydrocarbons. The maximum permitted lead content of gasoline in different countries is shown in Table 4.4 and progress made in the reduction of lead in gasoline in member states of the European Community is shown in Table 4.5.

About 75 per cent of the organolead additive in fuel is emitted from the

Table 4.3 Consumption of refined lead for the manufacture of gasoline additives (tonnes × 10³)§

	1976	1977	1978	1979	1980	1981	1982	1983*
United States	217.5	211.3	178.3	186.9	127.9	111.4	119.3	(85.0)
Europe	87.9	89.6	96.7	93.1	97.5	86.9	82.3	(83.0)
United Kingdom	56.8	55.4	60.8	58.9	63.3	58.1	54.3	(55.0)
Germany FR	7.3	9.3	9.4	9.6	9.9	8.4	8.0	(8.0)
France	14.0	12.3	14.7	15.1	14.5	12.4	12.0	(12.0)
Italy†	9.8	12.6	11.8	9.5	9.8	8.0	8.0	(8.0)
Mexico	9.6	8.7	6.6	6.6	6.4	8.1	8.0	(8.0)
Others (Canada and Greece)	(25.0)	(26.0)	(28.0)	(28.0)	(28.0)	(28.0)	(28.0)	(28.0)
Total	340.0	335.6	309.6	314.6	259.8	234.4	237.6	(204.0)
Total world‡ consumption of refined lead (all uses)	3838	4108	4145	4192	3949	3884	3796	(3750)
Lead in gasoline as % of total	8.9	8.2	7.5	7.5	6.6	6.0	6.3	(5.4)

§ Data collated by the International Lead and Zinc Study Group and reproduced by permission of the Lead Development Association, London.

The table shows the consumption of lead for the manufacture of additives and does not take into account changes in stocks. The table does not indicate the consumption of lead in gasoline by country – most of the United Kingdom, German and Greek, and much of the United States production of additives is exported.

* Estimated.

† Figures for 1979–81 have been revised from those quoted in previous reports to take account of the closure in 1979 of a plant which manufactured additives principally for the export market.

‡ Excluding socialist countries.

Table 4.4   Maximum permissible lead content of petrol by country, at February 1983*

	Lead content (g dm⁻³)		Lead content (g dm⁻³)
Belgium	0.4	Sweden	0.15
Denmark	0.4 premium 0.15 regular	Switzerland	0.15
		Yugoslavia	0.6
France	0.4	Australia	0.8
FR Germany	0.15	*except:*	
Greece	0.4	(i) Sydney, Newcastle and	
Ireland	0.4	Wollongong	
Italy	0.4	districts of NSW (ii) Victoria	0.4 0.3
Luxembourg	0.4	(iii) Tasmania	0.45
Netherlands	0.4	Canada	0.77
UK	0.4	Japan	0.31 premium 0.02 regular
Austria	0.4 premium 0.15 regular		unleaded
Finland	0.4	New Zealand	0.84 (0.45 from
Norway	0.4 premium 0.15 regular	South Africa	1984) 0.84
Portugal	0.635	USA	0.29
Spain	0.65 98 RON 0.6 96 RON 0.48 90 RON		0.013 unleaded

* After Ref. 19.

exhaust as a complex mixture of inorganic lead salts, formed by reaction in the combustion chamber with organic halide scavenger additives (1,2-dichloroethane and 1,2-dibromoethane, known as EDC and EDB). The major components formed are lead bromochloride, $PbBrCl$, and the mixed salts $PbBrCl.2NH_4Cl$ and α-$2PbBrCl.NH_4Cl$.[20,21] These double salts are formed by the reaction of $PbBrCl$ with ammonium chloride, itself formed by the reaction of HCl (produced by the scavengers) with ammonia, which is a minor natural component of the ambient air. A number of other compounds are also emitted, and many others are formed by reaction of these primary compounds in the atmosphere,[22,23] mainly with neutral and acid sulphates.

The amount of organolead additive emitted unchanged in the exhaust has been found to vary considerably with type of vehicle and with driving conditions. Estimates have varied from 0.11 per cent of the total lead in the exhaust, corresponding to 0.023 per cent of the input lead,[24] to about 2 per cent of the exhausted lead.[25] City driving will produce more unchanged TAL in the

Table 4.5   Reductions in maximum permissible lead content of petrol in EEC Member States*

	Lead content (g dm⁻³)	Effective date		Lead content (g dm⁻³)	Effective date
Belgium	0.84		Ireland	0.64	
	0.55	Jan. 1978		0.4	Sept. 1982
	0.45	Oct. 1978	Italy	0.635	
	0.4	July 1981		(Prem)	Apr. 1964
Denmark	0.7			0.635	
	0.4	Jan. 1978		(Reg)	Dec. 1967
	0.15			0.4	July 1981
	(Reg)	July 1982	Luxembourg	0.4	
	0.15		Netherlands	0.84	
	(Prem)	July 1984		0.4	Jan. 1978
France	0.64	Dec. 1966		0.15	1986
	0.55	Apr. 1976			
	0.5	Jan. 1979	United		
	0.4	Jan. 1981	Kingdom	0.64	Jan. 1973
FR Germany	0.4	Jan. 1972		0.55	Nov. 1974
	0.15	Jan. 1976		0.5	Dec. 1976
				0.45	Jan. 1978
Greece	0.84	Apr. 1966		0.4	Jan. 1981
	0.63	March 1979		0.15	Jan. 1986
	0.5	Feb. 1980			
	0.4	Jan. 1981			

* After Ref. 19

exhaust than constant speed highway driving.[24] In older cars a significant proportion of fuel was lost by evaporation from the fuel tank and carburettor which will also contribute tetraalkyllead compounds to the atmosphere. Crankcase blow-by gases may be a significant source of TAL, although the use of positive crankcase ventilation devices, mandatory in the UK since 1972 and in many other countries, will have much reduced this source. In older cars about 45 per cent of the total hydrocarbon emissions resulted from evaporative and blow-by gas losses,[26] which corresponds with similar findings for the proportional loss of tetraalkyllead by these routes.[27]

There is considerable loss of unburned fuel from two-stroke engines (mopeds, motorcycles, lawn mowers, outboard engines, etc.). Up to 50 per cent loss has been estimated[28] and this may represent a further significant source of TAL to the atmosphere. Evaporative losses of TAL undoubtedly occur during fuel handling operations, both during transport and sale, as indicated by the elevated atmospheric concentrations of TAL found in the vicinity of petrol filling stations.[29] Such losses were estimated as 40 tonnes of lead in the US in 1968,[30] but will be now much reduced.

It is very difficult to arrive at a reliable estimate of the total release of organic lead to the atmosphere through its use as a fuel additive. Overall, perhaps 2 per cent of the TAL in fuel will reach the atmosphere unchanged, although some workers have estimated up to 10 per cent loss, but without obvious justification.[31] It has also been estimated that a total emission of about 7000 tonnes of organolead compounds occurred in the western world in 1975[32] which implies a total emission of about 500 tonnes per annum for the UK in the same period. However, on the basis of 2 per cent loss of TAL to the atmosphere a *maximum* total emission of about 250 tonnes per annum would be expected for the UK.

4.2.2   Other uses of tetraalkyllead (TAL)

The use of TAL for industrial and commercial applications other than as a fuel additive has been severely restricted by environmental and public health considerations. Organolead compounds have been used as alkylating agents in the manufacture of mercurial fungicides and have been proposed for use as wood and cotton preservatives, pesticides, lubricant oil additives, antifouling agents and polyurethane foam catalysts.[33] Triphenyllead has been used on a limited scale as a marine paint antifouling additive.

4.2.3   Manufacture of lead alkyls

The production of lead alkyls is primarily of tetramethyl- and tetraethyllead (TML and TEL), the three mixed alkyl compounds TMEL, DMDEL, and MTEL being prepared subsequently by reaction of TML and TEL in the presence of a Lewis acid. The most important manufacturing process by far is the alkylation of a sodium−lead alloy, although the electrolysis of an alkyl Grignard reagent may be used.[34] Figure 4.1 shows a flow diagram of the processes involved in TEL manufacture. TML is made by a similar process, methyl chloride being used in place of ethyl chloride, and with the addition of a catalyst such as aluminium chloride and toluene as a diluent. Manufacture is treated in some detail here in view of the large scale of the process.

Molten lead and molten sodium are combined in a ratio of 9 : 1 by weight and the resultant alloy is solidified and flaked. The reaction takes place over a period of one hour or more and in the presence of an acetone catalyst at a temperature of $70-75$ °C and a pressure of $350-420$ kPa (equation 4.1)

$$4NaPb + 4C_2H_5Cl \rightarrow (C_2H_5)_4Pb + 4NaCl + 3Pb \qquad [4.1]$$

After releasing the autoclave pressure the contents are transferred to steam-stills. Unreacted ethyl chloride is removed by distillation and the TEL is steam distilled. The resultant TEL/water mixture is separated and the TEL purified by air blowing and/or washing with dilute oxidizing agents.

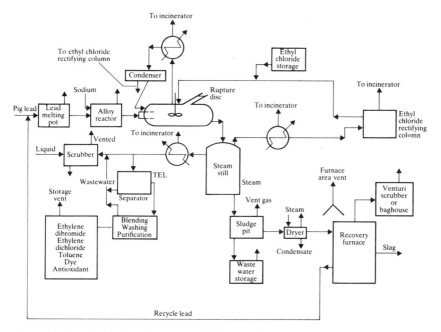

Fig. 4.1 Simplified flow diagram of TEL manufacturing process

The steam-still residue is sluiced to a sludge pit to allow recovery of the unreacted lead and the atmosphere above the sludge pit is extracted producing an alkyllead-rich effluent stream. The sludges are leached with water, the pit sediments are removed, dried and fed to a reverberatory furnace for recovery of inorganic lead. This is then recycled directly to the start of the process. An alkyllead yield of about 85 per cent may be obtained by this method.

Emissions of gaseous lead alkyls arise from the process vents at a rate of about 2 g (Pb) kg^{-1} of product in the manufacture of TEL, and about 75 g (Pb) kg^{-1} of product in the manufacture of the more volatile TML. Emissions from the sludge pit are estimated at 0.6 g (Pb) kg^{-1} of product for both TML and TEL manufacture.[35] The failure of rupture discs fitted to the autoclaves can cause fugitive emissions of lead alkyls, but these are contained and not vented to the atmosphere.

Contaminated gaseous effluent streams may be cleaned of TML by the use of activated carbon adsorption plant followed by recovery of adsorbed lead by steam stripping.[36] TEL may be removed by scrubbing the gaseous effluent stream with an involatile hydrocarbon.

The liquid effluent streams from a tetraalkyllead manufacturing plant may contain 1–10 mg (Pb) dm^{-3} of alkyllead compounds and several methods have been proposed to reduce this. One solution[37] is to pass the effluent through Amberlite 200 ion-exchange resin in the hydrogen form. The resin is then eluted with sodium hydroxide and the organic lead dealkylated by chlorination, but this has not been used in practice.

The Ethyl Corporation has proposed the use of sodium borohydride, $NaBH_4$, a strong reducing agent, in the presence of iron(II) ions supplied by iron(II) sulphate, $FeSO_4.7H_2O$. The preferred pH range is $9-11$ and addition of a water-soluble high molecular weight anionic polyelectrolyte aids precipitation.[38]

The Associated Octel Company has patented a method involving reduction/fixation.[39] The effluent at pH $7.5-9$ is passed down a column of clean acid-washed zinc particles. Occasional backwashing, sufficient to fluidize the zinc, is necessary to maintain extraction efficiences of $80-90$ per cent over long periods ($120-160$ days).

Contact with activated carbon for $12-30$ minutes has been shown to reduce organic lead concentrations from $10-20$ mg (Pb) dm^{-3} to $0.1-0.8$ mg (Pb) dm^{-3}. Carbon is steadily removed from the bed and replaced with activated carbon.[40]

4.2.4   Industrial exposure and control

In the UK certain types of work have been identified as posing a risk of substantial alkyllead exposure to workers;[41] other types of work have been identified as not likely to pose an appreciable risk of exposure. These are summarized in Table 4.6.

Table 4.6   Work and alkyllead exposure in the UK.*

Types of work where there is liable to be significant exposure to alkyllead unless adequate controls are provided	Examples of industries and processes where such work could be carried out
1. Production of concentrated lead alkyls	Lead alkyl manufacture
2. Blending of lead alkyls into gasoline	Blending processes at oil refineries
3. Entry into fixed or mobile plant and vessels, e.g. road tankers, rail tankers or sea tankers carrying leaded gasoline, e.g. for inspection, cleaning and maintenance purposes	Plant and vessel entry at oil refineries, at transport terminals and any place where such work is carried out
Types of work where there is *not* liable to be significant exposure to alkyllead	
1. Any exposure to lead alkyl vapours from leaded gasoline where the lead content is limited under the Motor Fuel (Lead Content of Petrol) Regulations	2. Work with such leaded gasolines including, e.g. the filling of petrol driven vehicles on garage forecourts

* See Ref. 41.

The Threshold Limit Values (TLV) in the UK for a time-weighted average (TWA) over an eight-hour shift are as follows:

Inorganic lead dust and fume in air	0.15 mg (Pb) m^{-3}
Tetraethyllead in air	0.10 mg (Pb) m^{-3}
Tetramethyllead in air	0.15 mg (Pb) m^{-3}

The maximum concentrations to which workers may be exposed for up to 15 minutes continuously, with at least 60 minutes between exposure periods (given that the TLV–TWA is not exceeded) are given by the TLV short-term exposure limits (STEL). These are as follows:

Inorganic lead dust and fume in air	0.45 mg (Pb) m^{-3}
Tetraethyl lead in air	0.30 mg (Pb) m^{-3}
Tetramethyl lead in air	0.45 mg (Pb) m^{-3}

For workers in the UK exposed to concentrated lead alkyls a measure of lead in urine is made at six-weekly or more frequent intervals and blood leads are measured annually. Any worker with a urinary lead value in excess of 0.8 μmol dm^{-3} is suspended from lead work until considered medically fit to return.[42]

4.2.5 Environmental production of tetraalkyllead

Evidence for the environmental formation of tetraalkyllead is available from several sources to date but is mainly circumstantial in nature. That the problem is still unresolved is due, not least, to the non-reproducibility of many of the experimental data in this field.

4.2.5.1 Experiments with environmental media

The first evidence that sediment systems can convert certain organic and inorganic lead compounds into tetramethyllead (TML) was presented by Wong *et al.*[43] It was observed that certain lake sediments produced TML without the addition of lead compounds and that there was no direct relationship between lead concentrations in the sediment and the amount of TML produced. It was found that addition of trimethyllead acetate always enhanced TML production, and that addition of lead(II) nitrate only occasionally did so. Control experiments suggest a microbiological origin for the TML.

Using a micro-organism culture obtained from an aquarium Schmidt and Huber[44] reported production of TML from lead(II) acetate. When di- and trimethyllead salts in aqueous solution were incubated in this way their concentrations decreased more rapidly ($\times$ 5–10) than that expected by chemical disproportionation routes, and so a biological methylation process was postulated.

Triethyllead salts added to sediments produced tetraethyllead (TEL) only, not TML, so suggesting a sulphide–mediated disproportionation route, as in equation 4.2[45]

$$2 R_3Pb^+ + S^{2-} \rightarrow (R_3Pb)_2S \qquad\qquad [4.2]$$

$$(R_3Pb)_2S \rightarrow R_4Pb + R_2PbS$$

The possibility of TML production by both chemical disproportionation and methylation has been investigated[46–48] but widely differing estimates of the relative importance of the two routes were made. The conversion of lead(II) salts to TML in seeded water and sediment samples has been reported.[46,49] Two out of three sediments produced TML when lead(II) nitrate was added[46] but no control experiments were performed and analysis was by gas chromatography (GC) retention times. Experiments with British Columbia sediments (including controls) with analysis by gas chromatography–mass spectroscopy (GC–MS), suggested a 0.03 per cent conversion of added lead nitrate to TML.[47]

Notwithstanding the diverse reports of lead methylation in environmental media under laboratory conditions reviewed above some groups have been unable to detect such a process,[50,51] and have suggested that TML produced by sediments is due to sulphide-mediated formation of $((CH_3)_3Pb)_2S$. Experiments reported by Ahmad *et al.*[52] in which lead(II) salts reacted with alkyl iodides in aqueous solution to produce TML were not repeatable by other workers.[53,54] The latter suggest that aluminium foil used to wrap stoppers in these experiments contacted lead(II) acetate solution forming metallic lead which then reacted with methyl iodide to produce TML, and this explanation is now widely accepted. Craig[55] incubated trimethyllead acetate solution with lake sediments, both sterilized and unsterilized, and in all cases similar amounts of TML (4 per cent) were evolved. He suggested that biological methylation is not a necessary mechanism in this process, and that a relatively simple chemical disproportionation can account for the methylation. In a sophisticated series of experiments using labelled carbon and lead tracers biomethylation products from cultures were not detected but did confirm that sulphide-induced chemical conversion of organic lead salts to tetraalkyllead is possible.[56] In recent work, however, alkylation of labelled lead in intertidal sediments has been demonstrated.[57] Labelled lead(II) nitrate containing ^{210}Pb was added to intertidal sediments in the laboratory. Vapour phase organic lead evolved by the sediments was collected in iodine monochloride solution, selectively extracted into dithizone in carbon tetrachloride and back-extracted into acid. The ^{210}Pb activity of the acid was determined by gamma spectroscopy. Production of vapour phase organic lead containing ^{210}Pb was observed from some, but not all, sediment samples.

4.2.5.2   Experiments with chemical systems

The direct methylation of lead(II) salts by the methyl donor methylcobalamin (CH_3CoB_{12}) has not been detected,[45,58–61] while iodomethane or methyl iodide (CH_3I) has been both reported to react with lead(II) salts in aqueous solution to produce TML[62] and not to do so.[63] The methylation of dialkyllead salts (lead(IV)) by CH_3CoB_{12} to form TML has been reported.[59] This would suggest that if monomethyl lead species could be converted to the dimethyl form then cobalamin-mediated methylation to TML could occur. However, the monomethyl species are very unstable and unless they can be stabilized (possibly by coordination to natural ligands), are unlikely to exist for sufficient time for net methylation to occur.

The methylation of a lead(II) salt, $Pb(NO_3)_2$, in aqueous solution by a dimethylcobalt(III) macrocyclic complex[64] has been recently demonstrated (see Ch. 8). The reactivity of lead(II) in water was tested towards CH_3CoB_{12}, methyl iodide, three models for S-adenosylmethionine and a *trans*-dimethylcobalt(III) complex. Only the latter was found to produce tetramethyllead by reaction with lead(II). Analysis was by GC with an electron capture detector (ECD). As the ECD is not element-specific two different columns were used with different retention times for TML to verify the compounds found.

Although this dimethylcobalt(III) complex is a better carbanion donor towards lead(II) than the naturally occurring methylcobalamin it is possible that environmental conditions may prevail which enhance the carbanion (CH_3^-) donor capacity of CH_3CoB_{12}, so making direct methylation of lead(II) possible. A fuller discussion of this area is given in Chapter 8.

4.2.5.3 Environmental evidence for methylation

Indirect evidence for the natural methylation of lead is provided by the reported presence of TAL in fish tissue.[65-67] However, detection of TAL in the environment is not, in itself, evidence of biomethylation, as selective concentration of these species from pollution sources may be taking place.

Further circumstantial evidence arises from measurements of tetraalkyllead and total lead in atmospheric samples from rural areas.[68,16] On the assumption that particulate lead and tetraalkyllead will be lost from the atmosphere at similar rates (with mean lifetimes typically of several days), it was predicted that rural air samples would have TAL to total lead ratios similar to those found in the urban source areas (1–7 per cent). However, a significant departure from expected values was found, with unusually high TAL to total lead ratios being found in maritime air masses, as identified by backward air mass trajectories.

In order to identify the likely size of this source area further measurements of the TAL to total lead ratios in air samples were obtained from the same sites.[69] Similar ratios were found in concurrent samples taken from a coastal and an inland site, indicating that the source is not local, nor is it a point source, but comes from a large area, and its effect is not diminished by air travelling 30 km inland. Box model calculations based on measured TAL in air concentrations also suggest that the source area is very large.

Measurements made in the Outer Hebrides, north-west Scotland, of TAL and particulate lead concentrations in a north-easterly air stream have confirmed that high organic lead to total lead ratios may be found in maritime air masses which have not recently received anthropogenic inputs of lead.[70]

Recent measurements of lead concentrations in snow and ice samples from Antarctica, including a prehistoric ice block, have been made using ultraclean sampling and analytical techniques. These data suggest that prehistoric snow and ice contain about 1.7 pg (Pb) g^{-1}, of which about 0.7 pg g^{-1} can be accounted for as being derived from silicate dust. Some of the excess lead will originate from volcanic emissions or from sea spray, but it is calculated that about 0.65 pg g^{-1} must be due to an unknown natural source of prehistoric lead. These other natural contributions could be from plant leaves, from direct volatilization from rocks, from alkylation processes or some other unknown natural source.[71] Of these, alkylation of lead to gaseous TAL and subsequent decomposition to particulate lead is probably the only process capable of supplying the approximately 10 000 tonnes yr^{-1} to the atmosphere required to account for this excess lead in snow.[57]

4.3 Analytical techniques for the determination of organolead compounds in environmental samples

Analysis of TAL and the organic lead salts in environmental samples is more complex than the determination of inorganic lead. Although many reagents and techniques are available for the collection and analysis of these compounds a considerable number are unreliable as they are also sensitive to inorganic lead. The importance of using a species-specific method, free from interference from inorganic lead and other organometallic and organic compounds (e.g. iodomethane, methyl iodide) cannot be overemphasized. Obtaining sufficient sensitivity is also, of course, another consideration when choosing an appropriate technique as environmental samples usually contain extremely low levels of these compounds.

4.3.1   Air samples

Two differing methodologies are available for the determination of organolead compounds in air. The first is based upon collection of the compounds in iodine monochloride solution.[72-74] and the second upon various sample trapping techniques followed by desorption into and separation by GC with an element-specific detection system.[75-82] Several comprehensive reviews of techniques and measurements of organic lead in air are available.[83,84]

The recommended chemical method of air analysis is that of Hancock and Slater,[72] as modified for 24–48 hour sampling.[73] The method involves collection of tetraalkyllead in iodine monochloride solution, where it is converted into dialkyllead ions. These are then selectively extracted by dithizone in carbon tetrachloride while inorganic lead remains complexed in the aqueous phase by the addition of ethenediamine tetraacetic acid (EDTA). After back-extraction of the organic lead into dilute nitric acid/hydrogen peroxide the amount of lead present is determined by flameless atomic absorption spectroscopy (AA). The sensitivity of the method is of the order of 0.25 ng (Pb) m^{-3} for a 48-hour sample. The method does, however, suffer from two faults: (a) it does not differentiate between the five tetraalkyllead species; and (b) it is liable to interference from vapour-phase tri- and dialkyllead, or from particulate tri- and dialkyllead should these pass the cellulose membrane prefilter. It should, therefore, be considered to give a measure of the total vapour-phase organic lead concentration.

The alternative and preferred method of analysis of TAL in air employs a sample collection and preconcentration step followed by GC separation and quantification with an element-specific detector, such as an AA spectrometer. Two sample collection methods have been used, cryogenic trapping of the analytes at -80 °C or lower, or collection on a porous polymer. Detection systems have varied from a modified graphite furnace AA[78] to the use of flame AA with the GC eluate being introduced into a silica or recrystallized alumina tube held within the acetylene flame.[77,79–81] In these systems hydrogen is also introduced into the tube by mixing with the eluate. It burns with a small diffusion flame which facilitates atomization of the lead molecules. A recent review of these various interfacing techniques is available.[81] Other detection systems have been used, e.g. a microwave plasma detector.[85]

Problems common to methods employing an adsorption medium are those of sample decomposition and non-quantitative recovery. The use of isotope dilution mass spectrometry allows the possibility of correcting for these losses by adding known amounts of deuterium-labelled $^2_1$H$_{12}$-TML and $^2_1$H$_{20}$-TEL to the sampling columns in advance. The detection limit for TML by one such method[82] is 20 pg m^{-3}. The use of a chemical prefilter (e.g. FeSO$_4$.7H$_2$O) to remove atmospheric ozone has been successfully used to reduce decomposition of TAL in the sampling tube.[81]

The separate determination of the tri- and dialkyllead compounds in the atmosphere has not, to date, been achieved. However, it has been suggested that discrepancies between measurements made by a total organic lead method (e.g. iodine monochloride) and a TAL-specific method (e.g. GC–AA) in parallel samples may be due to the presence of vapour-phase di- and trialkyllead compounds.[81,86] As these salts have an appreciable vapour pressure at ambient temperature their presence in the vapour phase in the atmosphere may be expected.

4.3.2   Water samples

In aqueous solution degradation of TAL takes place to yield various products. It is necessary, therefore, to be able to determine R$_4$Pb, R$_3$Pb$^+$, R$_2$Pb^{2+} and lead(II) at low concentrations and in the presence of each other. There are three distinct methods of analysis which provide the required speciation and sensitivity at the present time:

1. Hexane extraction of TAL with analysis by GC−AA or GC−MS, followed by differential pulse anodic stripping voltammetric (ASV) determination of R_3Pb^+ and R_2Pb^{2+}.[87] Detection limits of about 0.2 μg dm^{-3} (TAL as Pb) and 0.02 μg dm^{-3} (R_3Pb^+ and R_2Pb^{2+} as Pb) may be expected.
2. Extraction and purification of R_3Pb^+ and determination by graphite-furnace AA,[88] with a detection limit of 0.02 μg dm^{-3} (R_3Pb^+ as Pb).
3. Extraction and butylation of all the lead species present in the sample by a Grignard reagent and determination by GC−AA or by GC−microwave plasma emission detection[89,90] with detection limits of about 0.1 μg dm^{-3} (as Pb).

    Of these methods direct extraction by hexane and GC−AA (or preferably GC−MS) determination for TAL and ASV determination of trialkyllead and dialkyllead offer the greatest sensitivity. However, the butylation technique allows all the lead species present in a sample to be simultaneously determined by one analytical method. The quoted detection limit can be substantially improved by the processing of larger samples and by changes in the volumes of the extraction media.[89]

4.3.3   Sediments and biological samples

For the determination of alkyllead in sediments and biological samples basically similar methods are used as for water samples. However, rather more exacting extraction procedures are required. These include the following methods:

1. The ionic alkyllead species may be extracted in the presence of a mixture of salts into toluene, back-extracted into dilute nitric acid and determined by ASV.[91] Inorganic lead is masked by EDTA and a detection limit of 0.01 μg g^{-1} obtained.
2. TAL can be extracted into hexane in the presence of EDTA and separation and analysis carried out by GC−AA.[66,92] Detection limits of 0.01 μg g^{-1} for sediment (5 g) and 0.025 μg g^{-1} for fish (2 g) were reported. Less sophisticated analysis of the extract by flameless AA has also been used.[65]
3. TAL has also been extracted from sediments by steam distillation, collected in hexane and analysed by GC−MS with a detection limit of less than 0.03 ng g^{-1},[93] and by thermal desorption with collection on GC packing material and GC−AA analysis, with a detection limit of about 0.5 ng g^{-1}.[67]

A method has also been devised specifically for the determination of R_3Pb^+,

R_2Pb^{2+} and lead(II) in urine.[94] After sample pretreatment, hydride generation and flameless AA determination allowed detection limits of 0.25 ng cm^{-3} for $(C_2H_5)_3Pb^+$ and $(C_2H_5)_2Pb^{2+}$ (as Pb) and 5 ng cm^{-3} for inorganic lead(II).

As with air and water samples the preferred method of analysis of the alkyllead species in sediments and biological tissue is one which allows the specific and simultaneous determination of all species by one method. It is conceivable that the polar ions could be extracted into benzene after chelation and a butylation technique[89] applied. At present hexane extraction and GC−AA or GC−MS determination for TAL and ASV analysis for di- and trialkyllead offers the best sensitivity.

For water, sediment and biological samples both sample processing and storage conditions will significantly affect the determined levels of organic lead. Increased storage time, higher temperatures and the incidence of light will all promote the degradation of TAL. The effect of storage containers is of great importance as surface adsorption of TAL from water may effectively remove it from solution. Collection and storage vessels should always themselves be extracted as part of the analytical technique, and care should be taken to avoid contamination during storage, and cross-contamination (particularly with standard solutions) during analysis. Blank extractions and determinations should be carried out as a matter of course.

4.4 Environmental sink processes for organolead compounds

4.4.1 Tetraalkyllead in the atmosphere

Both homogeneous and heterogeneous mechanisms for the removal of TAL compounds from the atmosphere have been investigated.[95,96] Homogeneous gas phase reactions with the hydroxyl radical (HO), triplet atomic oxygen (O^{3}P) and ozone (O$_3$) together with photolytic decomposition were found to be the most important breakdown mechanisms, as is typical for hydrocarbons in the atmosphere. The rates of these various removal processes are highly dependent upon sunlight intensity, reactive species concentrations and other variables and it can be seen from Table 4.7, which summarizes the estimated removal rates, that the dominant variable is the concentration of the hydroxyl radical. Tropospheric hydroxyl concentrations have recently been reviewed,[97] and estimates of atmospheric TAL half-lives, $\tau_{\frac{1}{2}}$, derived from equation 4.3 below

$$\tau_{\frac{1}{2}} = \frac{\ln 2}{\ln \left(1 - \dfrac{y}{100} \right)} \qquad [4.3]$$

where y = decay rate per cent h^{-1} and using the reviewed [HO] values, are shown in Table 4.8.

Table 4.7   Estimated upper limit rates of tetramethyllead (TML) and tetraethyllead (TEL) decay in the middle of the day in a moderately polluted irradiated atmosphere‡

| Decay path | Concentration of reactive species | | Decay rate (% h⁻¹) | | | |
| | s* | w* | TML | | TEL | |
			s	w	s	w
HO attack	$(1-3) \times 10^{-7}$ ppm	$(1-2) \times 10^{-8}$ ppm	8–21	1–1.5	51–88	7–13
Photolysis	$z \sim 40°$†	$z \sim 75°$†	8	2	26	7
O₃ attack	100–200 ppb	40 ppb	1–2	0.5	9–17	4
O(^{3}P) attack	10^{-8} ppm	10^{-9} ppm	<0.1	≪0.1	<0.1	<0.1
Particulates	$200\ \mu g\ m^{-3}$	$200\ \mu g\ m^{-3}$	—	—	0.03	0.03
Total	—	—	16–29	3–4	67–93	17–23

* s, summer; w, winter.
† z, solar zenith angle.
‡ After Ref. 95.

Table 4.8   Decay rates and estimated atmospheric half-lives for TML and TEL

Decay path	Season	TML	TEL
Rate constant R_4Pb –	(Ref. 95)	7.8×10^5	70.2×10^5
HO p.p.m.$^{-1}$h^{-1}	(Ref. 96)	5.6×10^5	10.3×10^5
	geometric mean	6.6×10^5	27.0×10^5
HO concentrations	Summer	8×10^{-8} ppm	
(Ref. 97)	Winter	2.4×10^{-8} ppm	
Decay rate due to	Summer	5	22
HO attack % h^{-1}*	Winter	2	6
Decay rate due to	Summer	1.7	8.0
photolysis and O_3 attack % h^{-1}	Winter	0.4	2.6
Total atmospheric	Summer	7	30
decomposition rate % h^{-1}	Winter	2	8
$\tau_{\frac{1}{2}}$ (hours)	Summer	10	2
	Winter	34	8

* Using geometric mean of rate constants.

It can be seen that atmospheric half-lives of about 10 hours (TML) and 2 hours (TEL) may be estimated for the summer months, and about 34 hours (TML) and 8 hours (TEL) for the winter. Thus, while TML and TEL are relatively stable in the dark in a reaction vessel filled with purified air (with half-lives of 320 and 100 hours respectively for TML and TEL),[95] under normal tropospheric conditions rather short half-lives may be expected, precluding transport over large distances. TAL from anthropogenic sources is unable, therefore, to cause significant elevation of atmospheric concentrations in truly remote locations.

Heterogeneous reactions involving adsorption of TAL on to particles have been investigated[95] by exposing atmospheric particulate material to gaseous TAL compounds. Neither direct physical adsorption nor surface reaction in the presence of NO_2 in dry air was found to be significant.

Loss of TAL from the atmosphere by washout, rainout, dry deposition and other mechanisms has not been studied. Considering the insolubility of TAL in water and its occurrence in the vapour phase it is unlikely that the two former processes will be significant, although they could be for the soluble trialkyllead salts. Dry deposition at the earth's surface could possibly be significant given the reactivity of TAL and the probable occurrence of solid-phase salts in the atmosphere.

The degradation products from atmospheric reactions of TAL have not been characterized. It is possible that trialkyllead salts provide a relatively stable intermediate in the inevitable transformation of TAL to lead(II).

4.4.2   Tetraalkyllead in the hydrosphere

TML and TEL are sparingly soluble in water (e.g. TEL: 0.2–0.3 mg dm^{-3} at 0–38 °C),[98] their apparent greater solubility in seawater being due to adsorption by suspended material.[99] Concentrations of TAL taken up into flowing water are much lower than the saturation values.[100] In seawater it has been suggested that TAL would lie on the sea bed in a separate liquid phase, slowly dissolving into the seawater. Some would evaporate to the atmosphere, but most would decompose *in situ*.[101]

In aqueous solution degradation of tetraalkyllead compounds takes place by the stages shown in equation 4.4[102]

$$R_4Pb \rightarrow R_3Pb^+ \rightarrow R_2Pb^{2+} \rightarrow Pb^{2+} \qquad [4.4]$$

TEL in distilled water in the dark is relatively stable with only 2 per cent decomposition to $(C_2H_5)_3Pb^+$ after 77 days.[103] When exposed to sunlight TEL decomposes to $(C_2H_5)_3Pb^+$, only 1 per cent TEL remaining after 15 days. Aqueous solutions of TML also react to give $(CH_3)_3Pb^+$, 41 and 74 per cent of the TML remaining after 22 days in the light and dark respectively.

The trimethyl and triethyllead ions are very stable in aqueous solution, and cations such as copper(II) and iron(II) which promote the decomposition of the tetraalkyl compounds have no effect on the trialkyl salts. Both TML and TEL are totally adsorbed from aqueous solution on to silica, as are the trialkyl salts, and the decomposition of all these compounds is accelerated in the presence of silica.[103] It is suggested, therefore, that although TAL in the dark in a pure water system would persist for a considerable period, in natural waters in the presence of sediment, light and a variety of ions, half-lives of a few days may be expected.[100,103]

4.5   Environmental concentrations of organic lead compounds

4.5.1   Organic lead in air

Tetraalkyllead may be present in the atmosphere in two forms, in the vapour phase and thus separable from inorganic lead by filtration, or in association with atmospheric particles. As discussed above the various analytical techniques available are rather susceptible to interferences from inorganic lead and other organic compounds, and the wide variation in reported concentrations for tetraalkyllead vapour in air has been attributed to the use of non-specific analytical methods.[83] The TAL concentrations are best considered as a percentage of the total lead in air, as this allows comparison of data from different sites without the need to correct for topography or meteorological conditions. A recent comprehensive review of this topic is available.[84]

4.5.1.1 Urban concentrations

At most sites the typical contribution of organic lead to total lead lies within the range 1–10 per cent, with absolute concentrations in the range ~ 10–200 ng (Pb) m⁻³. Concentrations of organic lead in air obtained by reliable techniques are summarized in Table 4.9. The composition of the atmospheric TAL is strongly correlated with the local gasoline composition. In Antwerp, for example, 35 per cent of the lead in petrol was found to be TML, 16 per cent TMEL, 7 per cent DMDEL, 1.7 per cent MTEL and 41 per cent TEL with a mean petrol lead content of 0.44 g dm⁻³.[86] TML was found to be relatively enriched in the atmosphere, presumably due to its higher vapour pressure and chemical stability. However, all five TAL species were found close to traffic. The same study found that TAL concentrations inside residential buildings were similar to those outside.

At roadside sites the percentage of TAL to total lead is determined by the predominant driving mode, so, for example, alongside motorways where vehicles drive at constant high speeds less than 1 per cent of the total lead will be found as TAL. At urban sites with interrupted traffic flows TAL may account for 5–10 per cent of the total lead.

Measurements in a central London street showed a 30–35 per cent decrease in TAL concentrations between 5 and 14 m above the street, with a similar decline for particulate lead.[104] In the vicinity of petrol stations high percentages of TAL may be observed, due to evaporative losses from spillage and the displacement of TAL vapour from fuel tanks. Up to 400 ng (Pb) m⁻³ TAL was found close to a filling station. During the filling of large fuel reservoirs at filling stations surprisingly high concentrations of TAL (2 μg (Pb) m⁻³) were found 2 m away.[86]

High percentages of TAL have also been found in enclosed car parks where vehicles are started from cold, with consequent inefficient fuel combustion and low engine temperatures. In a tunnel with fast, freely-moving traffic low TAL levels were found (~ 15 ng (Pb) m⁻³) confirming that warmed-up vehicles travelling at a constant speed are minor contributors of atmospheric organolead.[86] Once traffic flow was impeded airflow through the tunnel (and hence dilution) was reduced and vehicle residence time increased and a sharp increase in the airborne TAL level was observed (~ 150 ng (Pb) m⁻³).

Although there have been relatively few measurements of tetraalkyllead by entirely specific techniques, it appears that typically it is present at rather low concentrations in urban air, usually representing about 1–5 per cent of total lead.[83]

The second form of organic lead in air is that associated with atmospheric

Table 4.9 Measurements of organic lead in air

Location		No. samples	Averaging time	AL conc(ng(Pb)m⁻³)†		Ratio AL/total Pb (%)†		Ref.
				range	mean	range	mean	
Averaging time-days								
Los Angeles	300 yds from major highway	6	4–15 d	47–110	78	1.5–4.0	2.4	147
Los Angeles	Urban air 3rd floor	1	days		100		5	148
Lancaster (UK)	Roadside, accelerating traffic	3	10 d	75–260	160	4.5–5.7	5.0	95
Shap (UK)	Roadside, remote section of motorway	4	1 d	0.8–2	1.4	0.22–0.46	0.33	73
Lancaster (UK)	Rural sites	33	1–2 d	0.5–230	19	1.5–33	9.5	73
London	City Centre Street at (1) 4.9m, (2) 14 m heights	7	1 d	24–190, 16–130	94, 65	3.2–9.8, 3.6–13	6.2, 6.7	73, 73
Outer Hebrides	Rural	20	1 d	1.0–8.2	3.0	4–48	22	70
Averaging time < 1 day								
Baltimore	20 m downwind major highway	6	2 h	26–75	53	1.5–5.2	3.2	25
Stockholm	City Centre (a) crossing (b) queue of traffic	5	3–8 h	120–1300	510			149

	From rural-city centre							
Frankfurt		104	2 h	1–170		0.3–24		150
London	Roadside, on kerb	6	30 min	30–110	68	0.9–4.1	2.1	83
	Tunnel	9	2 h	57–130	92	0.37–0.83	0.55	25
	Tunnel	1	30 min		20		0.1	83
	Petrol station	2	30 min	240–590	420	3.9–9.7	6.8	83
	Multi-storey car park	2	30 min	1500–5400	4200	10–13	11	151
	Basement car park	4	2 h	560–1000	850		30	150

Measurements made with species-specific TAL methods

Stockholm	Urban	12	10 h	11–77 TML	39	not reported		82
Copenhagen	Urban	2	2 h	185–195 TML, TEL	195	not reported		82
	Residential	6	2–26 h	5.3–60 TML, TEL	34			
	Rural		16–26 h	0.5–2.5 TML	1.5			
Antwerp	Urban	9	1 h	49–109 TML, TEL	83	4.6–12	8	86*
	Residential	7		3.2–14 TML, TMEL, DMEL	7	0.6–3.4	2	
	Highway	10		14–44 TML, TEL	24	0.8–3.0	1.7	
	Tunnel	15		12–162	39	0.2–1.4	0.5	
	Car-repair shop	6		100–290	205	5.9–14	10	
	Petrol station	21		17–410	149	2.7–35	12	
	Rural	6		0.3–3.9 TML, TMEL	2	0.1–0.7	0.5	

*Ratio expressed as AL/Particulate Pb (%)

†AL = alkyllead

particles. This has been found to account for 0.2−1.2 per cent of the particulate lead at an urban site, i.e. less than 10 per cent of the total organic lead present.[105]

No specific measurements have been reported to date of the TAL derivatives, tri- and dialkyllead salts, in the atmosphere. However, the difference found between the total organolead concentrations (obtained by a chemical method) and the tetraalkyllead concentrations (obtained by a species-specific GC−AA method) in concurrent air samples suggests that there are significant quantities of vapour phase tri- and dialkyllead compounds in the urban atmosphere.[86] Details for organic lead in air are given in Table 4.9.

4.5.1.2   Rural concentrations

To date very few published data are available for TAL concentrations in rural air, and none are available for truly remote locations. A typical urban 'background' concentration for TAL of 10−20 ng (Pb) m^{-3} may be expected,[84] and in the absence of a significant source of TAL other than from fuel additive usage, air advected to rural areas with this composition may be expected to be diluted to give a TAL concentration of a few ng (Pb) m^{-3}, but with a similar ratio of TAL to total lead as in the urban source air.

Measurements made at rural sites in the north-west of England showed AL concentrations in the range 0.3−230 ng (Pb) m^{-3}, or 1.5−33 per cent of the total lead.[16] Time-weighted pollution roses for these data are shown in Figs 4.2−4.4 and, as discussed above, a natural source of tetraalkyllead can account for these high concentrations and ratios. Confirmation of the existence of high ratios of organic lead to total lead in other maritime air masses has been obtained. A mean organic lead to total lead concentration of 22 per cent was found in fourteen air samples collected in the Outer Hebrides, north-west Scotland.[70]

TAL concentrations were found to be below the limit of detection (0.1 ng (Pb) m^{-3}) in air samples taken from two rural locations near Beijing, China.[106] In rural Denmark two samples gave TML concentrations of 0.5 and 0.7 ng (Pb) m^{-3} and two others 2.2 and 2.5 ng (Pb) m^{-3}.[82]

4.5.2   Organic lead in water

Few measurements have been made to date of organic lead compounds in natural waters. Apart from direct emission into waters, from fuel spillage or accident during transport for example, the tri- and dialkyllead decomposition productions of atmospheric TAL are quite soluble in water and so may be washed out by rain into surface waters.

A survey of waters in Birmingham, UK, revealed detectable amounts (>2 μg (Pb) dm^{-3}) of TAL only in a few road drainage grids.[107] The TAL was found in association with sediments and the predominance of TEL over TML was attributed to the greater volatility of TML at ambient temperatures.[108]

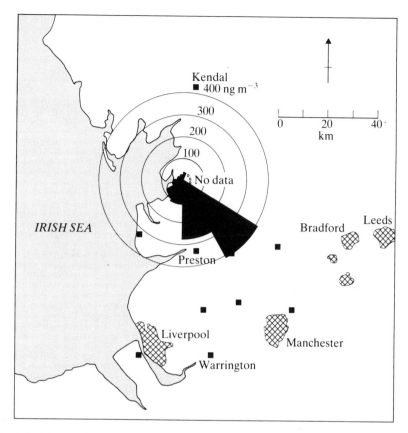

Fig. 4.2   Time-weighted pollution rose for particulate lead concentrations measured at three rural sites around Lancaster, UK

In contrast, analysis of run-off waters from the M6 motorway, in rural north-west England, showed levels of TAL compounds in the range 0–36 μg (Pb) dm^{-3}. Analysis was by GC–MS and TEL was always found to be <2 μg (Pb) dm^{-3}.[109] Simultaneous analysis by ASV of the same samples for their organolead salt contents showed insignificant levels of R$_3$Pb$^+$ and R$_2$Pb^{2+}. Storage of the water samples was accompanied by a slow conversion of R$_4$Pb to R$_3$Pb$^+$, which remained relatively stable in the dark.

TAL was not detected (with detection limits of 0.2 and 0.4 μg (Pb) dm^{-3} for TML and TEL respectively) in four samples of water taken from the River Clyde estuary or in four samples from the River Lune estuary, north-west England.[110]

Rainwater samples simultaneously obtained at six locations in or near Antwerp, Belgium, were found to contain 28–330 ng (Pb) dm^{-3} trialkyllead with an apparent correlation with local traffic density.[111] The organolead in these samples accounted for about 0.1 per cent of the total precipitated lead.

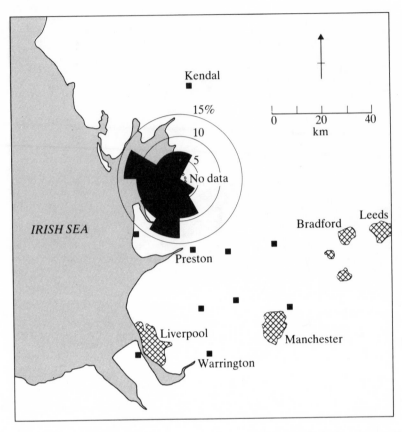

Fig. 4.3   Time-weighted pollution rose for vapour-phase organic lead concentrations measured at three rural sites around Lancaster, UK

Samples were also taken from surface waters and the trialkyllead concentrations found to be less than 20 ng (Pb) dm^{-3}, except for road drainage waters from a highway (70, 140 and 130 ng (Pb) dm^{-3}).

Recent measurements of the ionic alkyllead species in the Mersey estuary and the Manchester Ship Canal below the point where liquid effluents from a tetraalkyllead manufacturing plant are discharged showed a maximum concentration of 97 μg (Pb) dm^{-3}. However, alkyllead concentrations rapidly diminished with distance from the point of discharge, being ~1 μg (Pb) dm^{-3} at 30 km.[112]

4.5.3   Organic lead in biological samples

The presence of TAL compounds has been reported in fish from Halifax, Canada, at levels in the range 10 ng (Pb) g^{-1} to 4.8 μg (Pb) g^{-1} corresponding

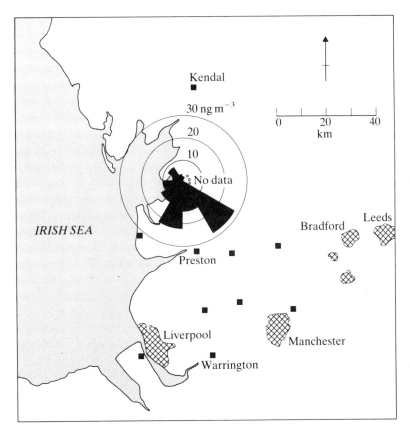

Fig. 4.4 Time-weighted pollution rose for the percentage ratio organic lead/total lead in the same sites (Reprinted by permission from *Nature*, **275**, (5682) pp. 738–40, copyright © 1978 Macmillan Journals Ltd – Figs 4.2–4.4)

to 10–90 per cent of the total lead content.[65] There must, however, be some doubt as to the specificity of the analytical technique used. Mussels collected in the vicinity of the wreck of the *SS Cavtat* in the Adriatic Sea, where a cargo of about 200 tonnes of antiknock compound was lost, were found to contain about 19 μg (Pb) g^{-1} of 'volatile organic lead', or 12 per cent of the total lead.[113] However, samples taken from an apparently unpolluted site had 'volatile organic lead' contents of about 13 μg (Pb) g^{-1}, or 16 per cent of the total lead.

In an extensive survey of fish from various lakes and rivers in Ontario, Canada, TAL compounds were found in 17 (of 107) samples at levels of 1–10 ng (Pb) g^{-1} wet weight, or less than 10 per cent of the total lead.[92] All five TAL species were found. Samples of water, vegetation, algae, weeds and sediments taken in the same survey (total of 126 samples) did not contain TAL above the limits of detection (0.1–0.5 ng (Pb) g^{-1}).

Samples of macoma from estuarine locations in north Wales and north-west England had TAL contents of less than 0.02 μg (Pb) g^{-1}, and R$_2$Pb^{2+} contents of less than 0.01 μg (Pb) g^{-1} and at two sites 0.03 and 0.05 μg (Pb) g^{-1}. Mean total lead contents of 1.1, 1.3 and 1.8 μg (Pb) g^{-1} were found at the three sites.[91] Samples of whiting and cod caught in the English Channel off Torbay were found to have TAL, R$_3$Rb$^+$ and R$_2$Pb^{2+} all below the detection limit (0.01 μg (Pb) g^{-1} wet weight).[91]

High concentrations of trialkyllead were found in birds found dead or dying on the Mersey estuary, UK, in 1979. Nearly all the analysed birds (mainly dunlin) had a total lead concentration in the liver in excess of 10 μg (Pb) g^{-1} (wet weight) with 30—70 per cent of the lead present in the trialkyl form.[114,115]

Analysis of the birds' invertebrate prey, the bivalve baltic tellin and polychaete worms, showed elevated concentrations of trialkyllead in samples taken from the Mersey estuary (0.2 μg (Pb) g^{-1} in polychaetes and 1.0 μg (Pb) g^{-1} in the baltic tellin) compared with those taken from the nearby Dee estuary (less than 0.1 μg (Pb) g^{-1} for both species). It is not known whether this incident of poisoning was due to an undetected release of organic lead compounds into the estuary, but monitoring elsewhere has confirmed that this was a very localized problem.

In one study the amount of trialkyllead has been measured in the brains of individuals with no occupational exposure to organic lead.[116] Twenty-two individuals of age 12—54 years had a median concentration of 0.014 μg (Pb) g^{-1} R$_3$Pb$^+$, or about 20 per cent of the total lead content of the brain. There was significantly more R$_3$Pb$^+$ in the brains of persons living on the lower floors of the houses (in central Copenhagen) than in those of persons living on the upper floors or in the suburbs or villages. Median total lead contents did not follow this trend, however; those living on lower floors in the city having the lowest median total lead content (0.067 μg (Pb) g^{-1}). Analysis for R$_3$Pb$^+$ was by a selective extraction and AA determination. This does not, unfortunately, unequivocally identify the lead species measured as trialkyllead, and doubt has been cast upon the reported levels.[117]

4.5.4 Organic lead in dust and sediments

Levels of total organic lead (TAL, tri- and dialkyllead) were measured in 46 samples of street dust from Lancaster, UK.[118] Concentrations in the range 0.4—7.4 μg (Pb) g^{-1} were found, being <1 per cent of the total lead.

One study failed to detect TAL in sediments from lakes and rivers where TAL had been detected in some fish.[92] Trialkyllead was not investigated.

Measurements of TAL and tri- and dialkyllead have not, to date, been reported in soils, crops and food.

4.6 Human metabolism and biological effects of organic lead

4.6.1 Uptake of lead

Tetraethyllead (TEL) is readily absorbed through the skin,[119] absorption increasing with the area of skin involved. Rabbits absorb lethal amounts of TEL in about one hour when the pure liquid is applied to their naked skin, the absorbed lead being distributed throughout the entire organism.[120] However, the possibility of human contact with liquid TAL, other than through occupational exposure or persistent contact with leaded gasoline (e.g. for washing and degreasing purposes), is slight.

The uptake of inhaled TAL vapour has been studied by volunteer inhalation of ^{203}Pb-labelled TEL and TML.[117] The labelled compounds were prepared by a Grignard reaction and the vapour passed into a perspex box, giving an airborne concentration of about 1 μg (Pb) m^{-3}. Air was inhaled through a particulate filter and exhaled through bubblers containing petroleum ether to collect exhaled TAL.

The amount of lead deposited in the subject was calculated from the TAL concentration in the box, the volume of air inhaled and the amount trapped after exhalation. Whole body counting of the 279 keV gamma rays from the ^{203}Pb immediately after exposure confirmed the quantity of labelled lead retained in the body. Further measurements of the radiation emitted from the head, liver, thighs, blood and excreta were made at intervals. After about 12 days the radioactive decay of ^{203}Pb (which has a half-life of 52 hours) reduced activities below measured levels.

The uptake of TAL vapour after exposure for 1–4 minutes was ~50 per cent for TML and ~40 per cent for TEL (three subjects each). The uptake of inhaled TAL vapour is regulated by a reversible transfer from the air in the lung to the blood as it circulates through the lung:

$$\text{TAL in air} \overset{\text{lung}}{\rightleftharpoons} \text{TAL in blood}$$

and these values probably approximate to the equilibrium uptake, due to the rapid loss of TAL from the blood stream. TML was found to have a mean residence time in the blood stream of only 13 seconds. About 40 per cent of the TML initially deposited was lost by exhalation in 48 hours.

Labelled lead in the blood (red cells and plasma) steadily decreased after inhalation ceased. After about 20 hours (TEL) and 8 hours (TML) ^{203}Pb began to reappear in the red cells. The fall in blood-^{203}Pb was seen as a response to exhalation and to conversion of the TAL to the water-soluble tri- and dialkyl forms in the liver.

It has recently been demonstrated that TML and TEL added to soil are quickly converted to water-soluble lead compounds (presumably the trialkyllead salts) and that these have a high toxicity to spring wheat, producing a yield depression at high concentrations. Similar amounts of inorganic lead added to the soil resulted in considerably less uptake of lead by the plants at all stages of growth.[121]

4.6.2   Metabolism of alkyllead

After acute and lethal exposure to TEL the highest concentrations of triethyllead are found in the liver with smaller amounts in the other organs (see Table 4.10).[122-4] The total lead content of the blood is not as high as in inorganic lead poisoning, although more blood lead will be found after TML exposure compared with TEL exposure.

Table 4.10   Distribution of lead in three TEL poisoning victims (in $\mu g(Pb)g^{-1}$ wet weight)

Organ	Reference 123	122	124
Liver	23.5	34.2	41.0
Kidney	7.9	19.0	12.0
Pancreas	—	12.8	—
Brain	7.4	11.0	10.0
Muscle	—	—	8.0
Heart	—	9.4	8.0
Spleen	2.9	—	6.5

In experiments on rabbits the highest concentration of triethyllead were found in the liver,[125] and similarly in mice and rats.[126] In experiments on young Mallard ducks dosed with 1 per cent aqueous solution of trimethyllead chloride at levels ranging from 0 to 65 mg (Pb) kg^{-1} body weight the highest concentrations were again found in the liver and kidneys, followed by brain, muscle and body fat.[127,128] Tissue analysis showed that trimethyllead chloride was absorbed unchanged and there was no evidence for the formation of dimethyllead or inorganic lead during the 21-day study period.

The experimental evidence available suggests, therefore, that tetraalkyllead is degraded to the trialkyllead form, either by metabolic activity in the mammalian liver,[126] or by physico-chemical processes in plants,[129] and this is confirmed by the presence of trialkyllead compounds in the liver of TEL intoxication victims.[130-133]

The biological residence time of organolead compounds is not clear. For ducks the half-life of trimethyllead is thought to be short,[127] but in humans subjected to [203]Pb[TAL] exposure the TAL had a mean retention time of 300–500 days.[116] In patients recovering from TEL poisoning sequential urinary lead measurements suggest mean lifetimes of less than 150 days.[134]

The toxicokinetics, toxic effects and toxic and metabolic mechanisms of alkyllead have been thoroughly reviewed elsewhere.[135]

4.6.3 Toxicology

TEL has no toxic action at all on the phytoflagellate *Poterioochromonas malhamensis* in the dark, even at high concentrations. However, upon illumination similar cultures are severely inhibited, suggesting that the apparent toxicity of TEL is due to the photolytically produced triethyllead derivative.[136,137] An example of this effect is shown in Fig. 4.5. Confirmation of this was obtained by exposing *P.malhamensis* and *Euglena gracilis* to micromolar quantities of trimethyllead. The same inhibition of growth was observed and giant spherical cells developed in the cultures, each containing up to twelve nuclei. Triethyllead was found to be about twice as toxic to the algae as trimethyllead, and *P.malhamensis* was more sensitive than *E.gracilis*.[138] TML also greatly inhibited the growth of algae in the light, and lead ions were found to penetrate the cell and deposit within concretion

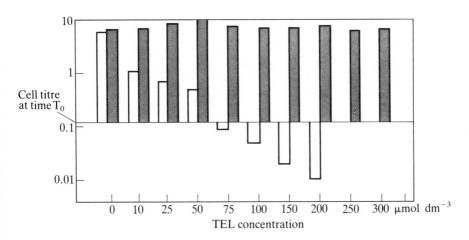

Fig. 4.5 Inhibition of cellular growth of *Poterioochromonas malhamensis* after treatment with TEL under light (open bars) and dark (closed bars) conditions (after Ref. 136)

bodies.[139] Higher aquatic species show a different response than do algae. The water-soluble trialkyl compounds do not pass through the gill membrane, whereas the hydrophobic TAL compounds are readily absorbed and metabolized[140] and the trialkyllead compounds are thus apparently less toxic than tetraalkyllead.[141] TML is less toxic to fish than TEL.[142]

The LD_{50} (i.e. the quantity of substance per unit body weight given as a single dose required to kill half the exposed population within 14 days) of trimethyllead chloride to the Mallard duck was found to be 30 mg (Pb) kg^{-1}.[126] For rats the LD_{50} for the same compound is < 36 mg (Pb) kg^{-1}, for TML 80 mg (Pb) kg^{-1}, for TEL 15 mg (Pb) kg^{-1}, and for triethyllead chloride 20 mg (Pb) kg^{-1}.[32] The greater toxicity of TEL compared with TML is due to its faster dealkylation *in vivo*.[143]

The symptoms of alkyllead poisoning in man differ from those due to inorganic lead but are equally non-specific: sleep disturbance, nausea, anorexia and vomiting being the most commonly presented.[143] The lethal dose for man is not known but can be estimated from the known LD_{50} for rats. On this basis the actual oral dose for an adult is about 0.25 g TEL and more than 1 g TML. At the present time the effects of long-term low level exposure to organolead compounds are unquantified, but no effects due to occupational exposure of this type have been reported.

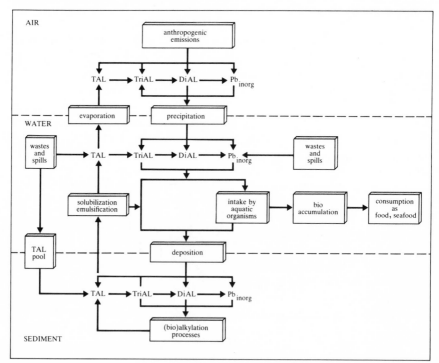

Fig. 4.6 Simplified (bio)geochemical cycle for organic lead compounds in the environment (after Ref 146) Reproduced with permission

4.6.2 Summary of human health effects

The toxic action of organolead compounds was recognized early in the commercial use of TEL as an antiknock compound. About 100 fatal cases of TEL poisoning have been reported, but none involving TML.[32] The main occupational hazard from these compounds now arises from the cleaning of large tanks in which leaded gasoline has been stored, sufficiently stringent precautions now being taken to eliminate exposure during TAL manufacture.[10] Other potentially significantly exposed occupational groups include filling-station and enclosed car-park staff. Non-occupational exposure to TAL occurs through the use of leaded gasoline as a cleaning agent, and through the deliberate sniffing of fuel, particularly by children.[32] No effective treatment for TAL intoxication is known. Chelating agents (e.g. EDTA) which are used in the management of inorganic lead poisoning have no effect.

No assessment has been made to date of the health effects, if any, of non-occupational exposure to organic lead through its use as an antiknock agent. However, it may be pointed out that in urban air the TAL concentration is typically about 10^{-3} times its Threshold Limit Value (cf. inorganic lead concentrations typically about 10^{-2} times the TLV).

4.7 The biogeochemical cycle of organic lead

Much consideration has been given to the biogeochemical cycle of inorganic lead and other heavy metals and several global models have been developed with values assigned to the magnitude of the various metal reservoirs and the flux rates between them.[144] Although these models obviously take into account the very substantial contribution made by organolead fuel additives to the input of inorganic lead to the atmosphere, none, to date, has fully considered the role played by organic lead itself in the overall cycle. Emissions of vapour-phase organolead compounds from vehicle use, spillage of liquid compounds, environmental production of organolead, and inputs from other sources have not been incorporated into the overall biogeochemical cycle of the metal, although their importance has recently been suggested.[57]

Two environmental cycles for TAL and its derivatives have been proposed,[137,145,146] both summarizing the main routes of these compounds through the environment. One such cycle is shown in Fig. 4.6.

It can be seen from Fig. 4.6 that a crucial role is played in the biogeochemical cycle of organic lead by environmental alkylation processes. In the absence of such natural source processes the environmental existence and concentrations of these compounds would depend entirely upon anthropogenic emissions (i.e. through their use as gasoline additives). However, if natural alkylation does occur (as the weight of evidence now suggests)[57] then this environmental cycle will still take place and inorganic lead deposited in sediments (previously thought to be an ultimate sink for lead) may be remobilized and again enter the cycle as organic lead.

References

1. Löwig C 1853 *Anal Chem* **88**: 318
2. Löwig C 1853 *J Prakt Chem* **60**: 304
3. Cahours A 1853 *Compt Rend* **36**: 1001
4. Midgley T Jr, Boyd T A 1922 *Ind Eng Chem* **14**: 894
5. Krause E, Von Grosse A 1937 *Die Chemie der Metall-Organischen Verbindungen*. Borntraeger, Berlin, pp 372–429
6. Leeper R W, Summers L, Gilman H 1954 *Chem Rev* **54**: 101–67
7. Panov E M, Kocheshkov K A 1952 *Dokl Akad Nauk SSSR* **85**: 1037
8. Willemsens L C, 1964 *Organolead chemistry*. Int Lead Zinc Res Org, New York
9. Willemsens L C, Van der Kerk G S M 1965 *Investigations in the Field of Organolead Chemistry*. Int Lead Zinc Res Org, New York
10. Shapiro H, Frey F W 1968 *The Organic Compounds of Lead*. Wiley, New York
11. Harrison P G 1982 In *Comprehensive Organometallic Chemistry* (eds Abel E W, Stone F G A, Wilkinson G). Pergamon, London
12. Saunders B C, 1950 *J Chem Soc*: 684–7
13. World Bureau of Metal Statistics 1979 *World Metal Statistics* vol 32. World Bureau of Metal Statistics, London pp 72–8
14. Central Unit of Environmental Pollution, Department of the Environment 1974 *Lead in the Environment and its Significance to Man*. HMSO, London
15. Harrison R M, Laxen D P H 1981 *Lead Pollution, Causes and Control*. Chapman and Hall, London, p 10
16. Harrison R M, Laxen D P H 1978 *Nature (London)* **275**: 238–40
17. Craig P J 1982 In *Comprehensive Organometallic Chemistry* (eds Abel E W, Stone F G A, Wilkinson G). Pergamon, London
18. Robinson I M 1978 In *The Biogeochemistry of Lead in the Environment* (ed Nriagu J O). Elsevier, Amsterdam
19. Royal Commission on Environmental Pollution 1983 9th Report *Lead in the Environment*. HMSO, London
20. Hirschler D A, Gilbert L F, Lamb F W, Niebylski L M 1957 *Ind Eng Chem* **49**: 1131–42
21. Smith W H 1976 *J Air Pollut Control Assoc* **26**: 531–2
22. Biggins P D E, Harrison R M 1978 *Nature (London)* **272**: 531–2
23. Biggins P D E, Harrison R M 1979 *Environ Sci Technol* **13**: 558–65
24. Hirschler D A, Gilbert L F 1964 *Arch Environ Health* **8**: 297

25. Reamer D C, Zohler W C, O'Haver T C *Anal Chem* **50**: 1449
26. Crawford K C, Linsay R 1970 MOR558F, Shell Int Petrol Co Ltd, London
27. Laveskog A 1971 TPM-BIL-62 (revised 1972), cited in Ref 32
28. Laveskog A 1977 Personal communication, cited in Ref 32
29. De Jonge W R A, Chakraborti D, Adams F C 1981 *Environ Sci Technol* **15**: 1217–23
30. Engel R E, Hammer D I, Horton R J M, Lane N M, Plumlee L A 1971 *USEPA Report* AP-90, Research Triangle Part NC 34 pp
31. Rickard D T, Nriagu J O 1978 In *The Biogeochemistry of Lead in the Environment* (ed J O Nriagu). Elsevier, Amsterdam
32. Grandjean P, Nielsen T 1979 *Residue Rev.* **72**: 98–148
33. Ref 10 pp 407–26
34. Ref 15 p 80
35. US EPA Report 1977 EPA-450/2-77-012 *Control Techniques for Lead Air Emissions*
36. Health and Safety Commission 1976 Report of HM Alkali Inspectorate *Industrial Air Pollution 1976.* HMSO, London
37. Nozaki M, Hatotano H 1967 *Water Res* **1**: 167–77
38. Ethyl Corporation 1975 *Reduction of dissolved organic lead content in aqueous solution*, Patent Spec 1405080, Sept 1975
39. Associated Octel Co Ltd 1975 *Treatment of dilute solutions of organo-lead ions to reduce the lead content thereof.* Patent Spec 1417078, Dec 1975
40. Otto J M 1978 *Method of removing dissolved organo-lead compounds from water.* US Patent 4070282, Jan 1978
41. Health and Safety Commission 1978 *Control of Lead at Work.* Draft Regs and Draft Approved Code of Practice, HMSO, London
42. Ref 15 pp 99–103
43. Wong P T S, Chau Y K, Luxon P L 1975 *Nature (London)* **253**: 263–4
44. Schmidt U, Huber F 1976 *Nature (London)* **259**: 157–8
45. Jarvie A W P, Markall R N, Potter H R 1975 *Nature (London)* **255**: 217–18
46. Dumas J P, Pazdernik L, Belloncik S, Bouchard D, Vaillancourt G 1977 *Proc 12th Canadian Symp Water Pollut Res* 91–6
47. Thompson J A J, Crerar J A 1980 *Marine Poll Bull* **11**: 251–3
48. Thompson J A J 1981 *Int Conf Heavy Metals in the Environment, Amsterdam.* C E P Ltd, Edinburgh, p 91
49. Huber F, Schmidt U, Kirchmann H 1978 In *Organometals and Organometalloids* (eds Brinckman F E, Bellama J M) A C S Symp Ser No 82, Washington
50. Jarvie A W P, Whitmore A P, Markall R N, Potter H R 1983 *Environ Poll* (B) **6**: 69–79
51. Jarvie A W P, Whitmore A P, Markall R N, Potter H R 1983 *Environ Poll* (B) **6**: 81–94

52. Ahmad I, Chau Y K, Wong P T S, Carty A J, Taylor L 1980 *Nature (London)* **287**: 716–17
53. Jarvie A W P, Whitmore A P 1981 *Environ Technol Lett* **2**: 197–204
54. Snyder W, Bentz J M 1982 *Nature (London)* **296**: 228–9
55. Craig P J 1980 *Environ Technol Lett* **1**: 17–20
56. Reisinger K, Stoeppler M, Nurnberg H W 1981 *Nature (London)* **291**: 228–9
57. Hewitt C N 1985 PhD Thesis, University of Lancaster, UK
58. Taylor R T, Hanna M L 1976 *J Environ Sci Health* **A11**: 201–3
59. Ridley W P, Dizikes L J, Wood J M 1977 *Science* **197**: 329–32
60. Agnes G, Bendle S, Hill H A O, Williams F R, Williams R J P 1971 *Chem Commun*: 850–1
61. Lewis J, Prince R H, Stotter D A 1973 *J Inorg Nucl Chem* **35**: 341–57
62. Whitmore A P 1981 PhD Thesis, University of Aston, UK
63. Harrison R M 1982 Unpublished Report, University of Lancaster, UK
64. Rhode S F, Weber J H 1984 *Environ Technol Lett* **5**: 13–68
65. Sirota G R, Uthe J F 1977 *Anal Chem* **49**: 823–5
66. Chau Y K, Wong P T S, Bengert G A, Kramar O 1979 *Anal Chem* **51**: 186–8
67. Cruz R D, Lovouso C, Georges S, Thomassen Y, Kinrade J D, Butler L R P, Lye J, Van Loon J C 1980 *Spectrochim Acta* **35B**: 775–83
68. Laxen D P H 1978 PhD Thesis, University of Lancaster, UK
69. Birch J 1980 Unpublished Report, University of Lancaster, UK
70. Hewitt C N, de Mora S J, Harrison R M 1984 *Marine Chem* **15**: 189–90
71. Boutron C F, Patterson C C 1983 *Geochim Cosmochim Acta* **47**: 1355–68
72. Hancock S, Slater A 1975 *Analyst* **100**: 422–9
73. Birch J, Harrison R M, Laxen D P H 1980 *Sci Total Environ* **14**: 31–42
74. De Jonghe W R A, Adams F C 1979 *Anal Chim Acta* **108**: 21–30
75. Cantuti V, Cartoni G P 1967 *J Chromatog* **32**: 641–7
76. Coker D T 1976 *Ann Occ Hyg* **21**: 33–8
77. Chau Y K, Wong P T S, Goulden P D 1975 *Anal Chem* **47**: 2279–81
78. De Jonghe W , Chakraborti D, Adams F 1980 *Anal Chim Acta* **115**: 89–101
79. Ebdon L, Ward R W, Leathard D A 1982 *Anal Proc* **19**: 110–14
80. Ebdon L, Ward R W, Leathard D A 1982 *Analyst* **107**: 129–43
81. Hewitt C N, Harrison R M 1984 *Anal Chim Acta* **167**: 277–87
82. Nielsen T, Egsgaard H, Larsen E, Schroll G 1981 *Anal Chim Acta* **124**: 1–13
83. Harrison R M, Perry R 1977 *Atmos Environ* **11**: 847–52
84. De Jonghe W R A, Adams F C 1982 *Talanta* **29**: 1057–67
85. Reamer D C, Zoller W H, O'Haver T C 1978 *Anal Chem* **50**: 1449–53
86. De Jonghe W R A, Chakraborti D, Adams F C 1981 *Environ Sci Technol* **15**: 1217–22
87. Hodges D J, Noden F G 1979 *Proc Int Conf Heavy Metals in the Environment*. CEP Ltd, Edinburgh

88. De Jonghe W R A, Van Mol W E, Adams F C 1983 *Anal Chem* **55**: 1050–4
89. Chau Y K, Wong P T S, Kramar O 1983 *Anal Chim Acta* **146**: 211–17
90. Estes S A, Uden P C, Barnes R M 1982 *Anal Chem* **54**: 2402–5
91. Birnie S E, Hodges D J 1981 *Environ Technol Lett* **2**: 433–42
92. Chau Y K, Wong P T S, Kramar O, Bengert G A, Cruz R A, Kinrade J O, Lye J, Van Loon J C 1980 *Bull Environ Contam Toxicol* **24**: 265–9
93. Slater A, 1983 Associated Octel Co Ltd. (personal communication)
94. Yamauchi H, Arai F, Yamamura Y 1981 *Ind Hyg* **19**: 115–24
95. Harrison R M, Laxen D P H 1978 *Environ Sci Technol* **12**: 1384–92
96. Nielsen O J, Nielsen T, Pagsberg P 1982 *Direct spectrokinetic investigation of the reactivity of OH with tetraalkyllead compounds in the gas phase. Estimate of lifetimes of TAL compounds in ambient air.* Riso Report R-463 Riso National Laboratory Denmark
97. Hewitt C N, Harrison R M 1985 *Atmos Environ* **19**: 545–54
98. Ref 10 p 56
99. Robinson J W, Rhodes I A L 1980 *J Environ Sci Health* **A15**: 201–9
100. Grove J R 1977 *Proc Int Experts Discussion Lead-Occurrence, Fate and Pollution in the Marine Environment.* Rovinj, Yugoslavia
101. Robinson J W, Kiesel E L, Rhodes J A L 1979 *Environ Sci Health* **A14**: 65–68
102. Noden F G 1977 *Proc Int Experts Discussion Lead – Occurrence, Fate and Pollution in the Marine Environment.* Rovinj, Yugoslavia
103. Jarvie A W P, Markall R N, Potter H R 1981 *Environ Res* **25**: 241–9
104. Harrison R M, Laxen D P H, Birch J 1979 *Proc Int Conf Heavy Metals in the Environment, London.* CEP Ltd, Edinburgh
105. Harrison R M, Laxen D P H 1977 *Atmos Environ* **11**: 201–3
106. Jiang S, Ma C, Liu H, Ge J, Li M, Adams F C, Winchester J W, 1982 *Proc 1st Ann Sci Meeting Soc Environ Geochim and Health.* E Carolina University, Greenville, NC, USA
107. Potter H R 1976 PhD Thesis, University of Aston, UK
108. Potter H R, Jarvie A W P, Markall R N 1977 *Water Pollut Control* **53**: 123–8
109. Wilson S J, Harrison R M 1982 Unpublished data, University of Lancaster UK
110. Hewitt C N, Harrison R M 1983 Unpublished data, University of Lancaster UK
111. De Jonghe W R A, Van Mol W E, Adams F C, 1983 *Proc Int Conf Heavy Metals in the Environment, Heidelberg.* CEP Ltd, Edinburgh
112. Riley J P, Towner J V 1984. *Marine Poll Bull* **4**: 153–8
113. Mor E D, Beccaria A M 1977 *Proc Int Experts Discussion Lead – Occurrence, Fate and Pollution in the Marine Environment.* Rovinj, Yugoslavia
114. Head P C, D'Arcy B J, Osbaldeston P J 1980 N W Water Authority Scientific Report No DSS-EST-80-1 Warrington, UK. Cited in Ref 19.

115. Bull K R, Every W J, Freestone P, Hall J R, Osborn D, Cooke A S, Stowe T, 1983 *Environ Pollut* **31**: 239–59
116. Nielsen T, Jensen K A, Grandjean P 1978 *Nature (London)* **274**: 602–3
117. Heard M J, Wells A C, Newton D, Chamberlain A C 1979 *Proc Int Conf Heavy Metals in the Environment, London.* CEP Ltd, Edinburgh, pp 103–8
118. Harrison R M 1976 *J Environ Sci Health* **AII**: 417–23
119. Laug E P, Kunze F M 1948 *J Ind Hyg* **30**: 256
120. Kehoe R A, Thamann F 1931 *Amer J Hyg* **13**: 478
121. Diehl K H, Rosopulo A, Kreuzer W, Judel G K 1983 *Z Pflanzeneraehr Bodenk* **146**: 551–9
122. Bolanowska W, Wisniewska-Knypl J M 1967 *Arch Toxicol* **22**: 278
123. Cassels D A K, Dodds E C 1946 *Brit Med J* **2**: 681
124. Kehoe R A 1976 *Lead Pharmacol Ther Quart* **1**: 161
125. Bolanowska W, Garczynski H 1968 *Med Pract* **19**: 235
126. Cremer J E 1959 *Brit J Ind Med* **16**: 191
127. Whittingham A, Roberts N L 1983 *Proc Int Conf Heavy Metals in the Environment, Heidelberg.* CEP Ltd, Edinburgh
128. Osborn D, Every W J, Bull K R 1983 *Environ Pollut* **31**: 261–75
129. Röderer G, 1982 *J Environ Sci Health* **A17**: 1–10
130. Hayakawa K 1972 *Jap J Hyg* **26**: 526
131. Stevens C D, Feldhake C J, Kehoe R A 1960 *J Pharmocol Exp Ther* **128**: 90–4
132. Bolanowska W 1968 *Brit J Ind Med* **25**: 203–8
133. Bolanowska, Wisniewska – Knypl J M 1971 *Biochem Pharmacol* **20**: 2108–10
134. Yauamura Y, Takakura J, Hirayama F, Yamanchi H, Yoshida M 1975 *Jap J Ind Health* **17**: 223–4. Cited in Ref 116
135. Grandjean P (ed.) 1984 *Biological Effects of Organolead Compounds.* CRC Press, Boca Raton, Fl
136. Röderer G 1980 *Environ Res* **23**: 371–84
137. Röderer G 1981 *Proc Int Conf Heavy Metals in the Environment, Amsterdam.* CEP Ltd, Edinburgh
138. Röderer G 1983 *Proc Int Conf Heavy Metals in the Environment, Heidelberg.* CEP Ltd, Edinburgh
139. Silverberg B A, Wong P T S, Chau Y K 1977 *Arch Environ Contam Toxicol.* **5**: 303–13
140. Wood J M 1977 *Proc Int Experts Discussion Lead – Occurrence, Fate and Pollution in the Marine Environment, Rovinj, Yugoslavia.* Pergamon, London, pp 299–303
141. Maddock B G, Taylor D 1977 *Proc Int Experts Discussion Lead – Occurrence, Fate and Pollution in the Marine Environment, Rovinj, Yugoslavia.* Pergamon, London, pp 233–61
142. Marchetti R 1978 *Mar Pollut Bull* **9**: 206–7

143. Waldron H A, Stöfen D 1974 *Subclinical Lead Poisoning*. Academic Press, London, p 224
144. Nriagu J O 1978 In *The Biogeochemistry of Lead in the Environment* (A) (ed Nriagu J O). Elsevier, Amsterdam
145. De Jonghe W R A, Jiang S, Adams F 1981 *Proc Int Conf Environmental Pollution, Thessaloniki*. p 183
146. De Jonghe W R A 1983 PhD Thesis, University of Antwerp, Belgium.
147. Snyder L J 1967 *Anal Chem* **39**: 591–5
148. Rabinowitz M B, Wetherill G W, Kopple J D 1977 *J Lab Clin Med* **90**: 238–48
149. Allvin B, Berg S 1977 Rep SNV PM 907 Dept Anal Chem Stockholm University
150. Rohbock E, Georgii H-W, Müller J 1980 *Atmos Environ* **14**: 89–98
151. Chamberlain A C, Heard M J, Little P, Newton D, Wells A C, Wiffen R D 1978 Report AERE-R9198, HMSO, London

Chapter 5

Organoarsenic compounds in the environment

5.1   Introduction

The history of organoarsenic chemistry goes back to 1760, when L.C. Cadet de Gassicourt distilled a mixture of arsenic trioxide and potassium acetate in a glass retort luted to a glass receiver. The product, a heavy, fuming liquid inflammable in air and having an intensely disagreeable odour, became known as 'Cadet's fuming arsenical liquid' and was later studied by a number of prominent chemists, including Berzelius, Dumas, Bunsen, and Baeyer. This compound is a mixture whose chief component is $((CH_3)_2As)_2O$. Cadet's work in fact represents the first laboratory synthesis of an organometallic compound. Organoarsenic chemistry received renewed interest in the nineteenth century, when poisoning incidents occurred in Germany and England which were traced to the production of organoarsines by moulds growing on wallpapers with arsenical pigments (Schweinfurt green). The toxic compound was eventually identified as trimethylarsine by Challenger,[1] a discovery which led him to the classical research on the methylation of metals by microorganisms. The studies of Ehrlich on the pharmacological properties of organoarsenic compounds led to the discovery of Salvarsan (arsphenamine, 3,3'-diamino-4,4'-dihydroxyarsenobenzene) (Fig. 5.1), which became a major remedy for syphilis and other infectious diseases.

In this chapter the uses of organoarsenicals in medicine and agriculture, their toxicology, the behaviour of man-made arsenicals in the environment, and the biogeochemical cycle of arsenic in terrestrial and aquatic systems will be examined.

5.2   Uses of organoarsenicals in medicine and agriculture

The first organoarsenical shown to be of therapeutic value was *p*-arsanilic acid (atoxyl).[2] Koch[3] applied this compound to the treatment of sleeping sickness

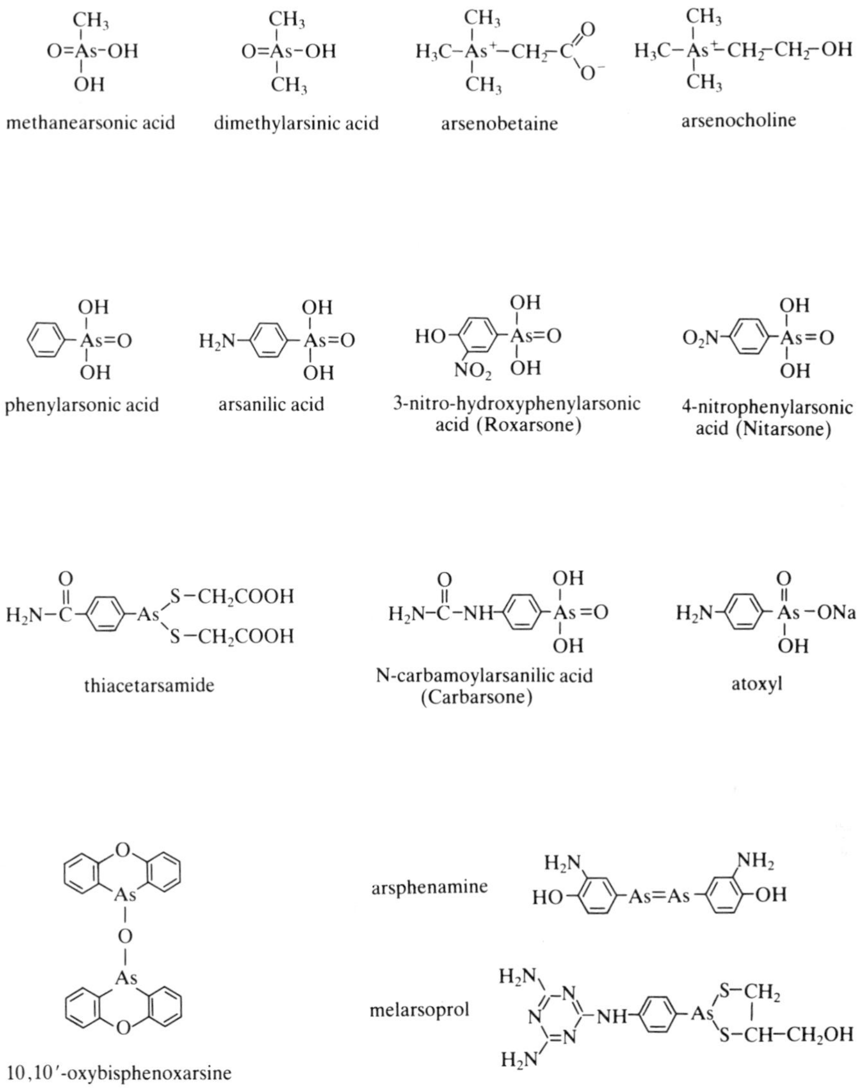

methanearsonic acid

dimethylarsinic acid

arsenobetaine

arsenocholine

phenylarsonic acid

arsanilic acid

3-nitro-hydroxyphenylarsonic acid (Roxarsone)

4-nitrophenylarsonic acid (Nitarsone)

thiacetarsamide

N-carbamoylarsanilic acid (Carbarsone)

atoxyl

10,10′-oxybisphenoxarsine

arsphenamine

melarsoprol

Fig. 5.1 Structures of selected organoarsenic compounds

and his success prompted Ehrlich to initiate studies on the pharmacology of arsanilic acid derivatives, which culminated in the development of arsphenamine and neoarsphenamine, a sulphoxyl derivative of arsphenamine.[4] In the first half of this century, pharmacological manuals usually contained a relatively extensive section on organoarsenical drugs (e.g. Beckman[5]), but due to the availability of more selective and less toxic drugs, the current use of arsenicals in medicine is very limited. The 1983 issue of the *AMA Drug Evaluation Manual* lists only Carbarsone (*N*-carbamoylarsanilic acid) as an obsolete treatment for amoebiasis and Melarsoprol (2-[4-[(4,6-diamino-1,3,5-triazin-2-yl)amino]phenyl]-1,3,2-dithioarsolane-4-methanol) as the drug of choice for treating meningoencephalitis in the late stages of African trypanosomiasis. In veterinary medicine, thiacetarsamide (arsenamide [[(*p*-carbamoylphenyl)arsylene]-dithio]diacetic acid) is still in use for the treatment of heartworm infection in dogs (canine filariasis), especially in the adult stage.[6]

The recent discovery of the carcinostatic (cancer-arresting) action of molecules of the type R—S—As(CH$_3$)$_2$ (R = amino acid, di- or tripeptide, monosaccharides, or lipid) against lymphocytic leukaemia in mice as well as against other carcinoma cell cultures *in vitro* may open novel medical uses of organoarsenicals in the treatment of cancer.[7,8]

Four arsenicals are currently approved by the US Food and Drug Administration for animal feed additives: arsanilic acid and Roxarsone (3-nitro-4-hydroxyphenyl arsonic acid) for increased rate of gain and improved feed efficiency in chickens and swine and for the control of swine dysentery, and Carbarsone and Nitarsone (4-nitrophenyl arsonic acid) as antihistomonads in turkeys.[9] These compounds are rapidly excreted either chemically unchanged or in the form of organoarsenic derivatives; in particular, little or no breakdown to inorganic species occurs during metabolism. These feed additives lead to levels of a few tens of μg g^{-1} in the wastes, which may be released to the environment when such wastes are used as fertilizer. Studies by Morrison[10] and Woolson,[11] however, suggest that arsenicals introduced in this manner do not persist in soils and are rapidly removed by leaching and/or volatilization.

While the veterinary and medical uses of arsenicals represent only a very small fraction of the annual production of arsenic, the application of arsenicals as herbicides and cotton desiccants represented 31 and 15 per cent of the 1981 US market, respectively.[12] Most of this use occurs in cotton production, where the mono- and disodium salts of methanearsonic acid (MSMA and DSMA) are applied as post-emergence grass herbicides at an annual rate of about 10.4 × 10^3 tonnes in the US.[11,13] Dimethylarsinic acid (cacodylic acid) is used as a cotton defoliant prior to harvesting (0.2585 × 10^3 tonnes per year). The major cotton defoliant in current use is inorganic arsenic acid, however, which is used at a current rate of 3.3 × 10^3 tonnes per year.

Additional agricultural uses of arsenic compounds include 40.8 tonnes per year of sodium arsenite for the control of measles and dead arm on table grapes, 136.1 tonnes per year of lead arsenate and 104.3 tonnes per year of calcium arsenate for the control of acidity in grapefruit, and minor amounts of 10,10-oxybisphenoxarsine for fungus control in cotton sailcloth and vinyl films. Calcium arsenate is a component of snail baits and is used for fly control in poultry houses (136.1 tonnes per year) and as a herbicide for the grass *Poa annua* (0.635×10^3 tonnes per year). The use of these compounds has declined drastically from the 1940s, when about 0.77×10^3 tonnes per year were used (all figures for the US from Alden[14]).

Figure 5.1 gives details of some organoarsenic compounds.

5.3 World production of arsenic; other uses

Fitzgerald[12] has estimated the world production of arsenic to be about 31 700–34 500 tonnes in 1980, down from about 70 000 tonnes in 1970. Most of the world production comes from the following countries: Sweden (5400 tonnes), USA (4500 tonnes), France (4535 tonnes), USSR (3600 tonnes), Belgium and the People's Republic of China (about 2700–3600 tonnes each). The US production has also declined, from 13 000 tonnes in 1970 down to 4500 tonnes in 1980. US consumption shows a declining trend as well, from about 27 000 tonnes in the 1960s to about 16 000 tonnes in 1980. The US market for arsenic is divided up into agricultural uses (46 per cent), wood preservatives (36 per cent, mostly copper arsenate),[15] molybdenum ore flotation (8 per cent), glass manufacture (5 per cent, mostly in the form of arsenic acid and arsenic trioxide),[16] and miscellaneous uses (5 per cent), which include semiconductor[17] and drug manufacture.

5.4 Toxicology of arsenic compounds

The toxic activity of arsenic depends very much on its molecular form. This is illustrated in Table 5.1, which lists organic and inorganic arsenic compounds and their oral LD_{50} values for rats. Comprehensive data on the toxicology of arsenic can be found in the Registry of Toxic Effects of Chemical Substances.[18] For comparison, the values for strychnine and aspirin are also shown.[18,19] The large differences in toxicity between the arsenic species reflect both the dependence of the toxic action on molecular structure and the fact that the relatively harmless organoarsenicals are not converted metabolically to the more toxic inorganic species during their passage through animal systems.

The mechanism of toxicity differs with the valence state of arsenic: trivalent arsenic has a high affinity for thiol groups and attaches to thiol groups which are present in the active centres of a number of enzymes.[20,21] This effect is especially strong for enzymes containing two adjacent thiol groups, which makes cross-linking and the formation of arsenic-containing heterocyclic structures possible. Among the enzyme systems inhibited by trivalent arsenic are ketoacid and aldehyde oxidases and succinic, malic, and lactic

Table 5.1   Toxicity of selected arsenic compounds*

Arsine	3
Arsenic trioxide	20
Potassium arsenite	14
Calcium arsenate	20
Methanearsonic acid	700−1800
Dimethylarsinic (cacodylic) acid	700−2600
Phenylarsonic acid	50
Arsanilic acid	216
Atoxyl	75
Arsphenamine	100
Melarsoprol	250
Strychnine†	16
Aspirin†	1000−1600

From Refs 18, 19.

* As LD_{50} in rats, mg kg^{-1}

† For comparison

dehydrogenases. Organic trivalent compounds have especially broad inhibitory activity, e.g. the compound $ClCH{=}CH{-}AsCl_2$ (Lewisite) which was developed in the Second World War as a war gas. The toxic mechanism of pentavalent arsenic is less well understood. It may act directly by a process called arsenolysis, which involves the formation of the unstable arsenate ester of ADP instead of the phosphate ester, ATP. The arsenate ester then hydrolyses non-enzymatically, so that the energy of the ester bond cannot be recovered metabolically, as is the case for ATP. This mechanism uncouples oxidative phosphorylation and leads to the breakdown of energy metabolism. There is a substantial amount of indirect evidence for the formation of arsenate esters with monosaccharides, in which arsenate replaces phosphate (e.g. glucose-6-arsenate instead of glucose-6-phosphate), but the arsenate esters are extremely sensitive toward hydrolysis and have not been positively identified or isolated.[22] It has been suggested that arsenic(V) becomes metabolically reduced to arsenic(III) which then interacts with thiol groups in enzymes as discussed above.

Knowles and Benson[21] have recently proposed a toxic mechanism for arsenic, where arsonous acid $[HAs(OH)_2]$ or its organic analogue

[RAs(OH)$_2$] are the effective toxic species, which bind to the sulphydryl groups of enzymes. Arsenate and arsenite are suggested to be reduced *in vivo* to arsonous acid. However, the evidence for the *in vivo* reduction of arsenic(V) to arsenic(III) in mammals is still rather ambiguous. While some of the earlier workers reported that arsenic(III) was produced in rats and dogs infused with arsenate,[23-25] it is possible that these results were due to inaccuracies in the analytical procedures and to induction of renal failure in the test animals as a result of the high arsenic doses used. Peoples[26] did not observe the formation of arsenic(III) in the blood of dogs infused with arsenate, as long as the doses were kept below the levels of acute toxicity. The *in vivo* oxidation of arsenite to arsenate has been reported to occur in mice.[27] As will be discussed below, biomethylation of inorganic arsenic *in vivo* has been shown in a number of animals as well as man.

As is evident from Table 5.1, most organic arsenic compounds are substantially less toxic than the inorganic forms. This is especially true for the arsenic analogues of choline and betaine (Fig. 5.1), which can be fed at the per cent level in animal diets without causing toxic effects.[28] The non-toxicity of these compounds must be attributed to their inability to bind to sulphydryl groups and to their resistance against metabolic conversion to more toxic forms. When organoarsenic compounds in seafoods, including fish, crustaceans, and seaweeds, are fed to test animals or human subjects, they are rapidly excreted in essentially unchanged form.[29-32]

There is substantial epidemiological evidence that inorganic arsenic may cause lung and skin cancers, especially in connection with occupational exposure (for discussion of this topic see World Health Organization[33] and papers by Furst,[34] Simmon,[35] and Harding-Barlow[36]). There is no laboratory evidence to suggest that organic arsenicals may be carcinogenic. No epidemiological studies on the potential relationship between organoarsenicals and cancer have been conducted.

The presence of a number of inorganic and organic arsenic species in the environment and in foodstuffs and the different toxicities of these compounds makes it extremely difficult to set rational exposure standards.[37] This problem is discussed in several criteria dicuments.[33,38-40]

Details of organoarsenic compounds are given in Fig. 5.1.

5.5 The determination of organoarsenic compounds in environmental samples

The analytical chemistry of arsenic compounds in water has been reviewed recently.[41] In this section a brief overview of the analytical methods suitable for the determination of organoarsenic compounds in natural waters will be given, and then some of the techniques and problems specific to biological matrices will be discussed.

5.5.1   Hydride generation techniques

A number of arsenic species are either volatile, e.g. the methylarsines, or can be easily transformed into volatile species by reduction with sodium borohydride. The volatile species can then be removed from solution by gas stripping, collected on a cold trap, and separated by gas chromatography. They are then detected by an element-specific detector, e.g. a quartz-cuvette atomic absorption spectrophotometer,[42] a d.c.-arc atomic emission detector,[43,44] or a microwave-induced atomic emission spectrometer.[45] It should be noted that the hydride techniques only characterize the species with regard to the number of methyl groups attached to arsenic. It is conceivable that species other than methanearsonic acid, dimethylarsinic acid, and trimethylarsine oxide may also be reduced to mono-, di-, and trimethylarsine, respectively. If this is suspected in a given sample, the borohydride reduction could be preceded by a liquid chromatographic separation using a cation-exchange resin as suggested by Dietz and Perez.[46]

5.5.2   Liquid chromatography techniques

Most organoarsenic species are not volatile and cannot be easily transformed into volatile compounds. Furthermore, they are usually quite polar and highly water-soluble. For these reasons, liquid chromatography with aqueous or polar organic eluents appears highly suitable for the separation of organoarsenic compounds. Two basic approaches have been taken to implement analytical strategies based on liquid chromatography: low pressure column chromatography, which has the advantage that relatively large sample volumes can be processed, and reversed-phase HPLC coupled with element-specific detectors, which provides good separation and high selectivity. Graphite-furnace atomic absorption spectrophotometers with deuterium lamp[47,48] or Zeeman background correction[41] have been used as arsenic-specific detectors. These HPLC–AA systems have been used to separate and determine arsenite, arsenate, methanearsonic acid, dimethylarsinic acid, phenylarsonic acid, arsenobetaine, and arsenocholine, as well as some other organoarsenicals. For environmental analysis, preconcentration may be necessary before this technique can be used, as the sample volume that can be injected is rather small (microlitre range).

5.5.3   Arsenic determination in biological matrices

As will be discussed below, a variety of organoarsenic compounds are found in plants and animals. Some of these compounds have recently been chemically identified, but the structure of several others remains unknown. Under these circumstances, the quantitative determination of the arsenic species in biota remains a very difficult task. Edmonds and co-workers (see references in sect. 5.8) have isolated organoarsenic compounds from marine biota by application of a variety of chromatographic techniques to extracts of large samples (tens of kg) of marine organisms. The compounds were subsequently identified by X-ray crystallography and NMR and IR spectroscopy as well as by elemental analysis. While this approach has resulted in a breakthrough in the structure identification of these compounds, it is not suitable for routine quantitative analysis. Shiomi *et al.*[49] have developed a semi-quantitative method for the determination of arsenobetaine in marine animals based on ion-exchange chromatography and the measurement of arsenic in the eluates by inductively coupled plasma emission spectrometry. Mass spectrometric techniques (Fast Atom Bombardment mass spectrometry (FAB MS) and Field Desorption MS) have also been used to identify arsenobetaine and arsenocholine in extracts of marine animals after chromatographic separation.[50,51] These techniques must also be considered semi-quantitative at best.

Since most of the arsenic in marine organisms is in organic forms (as will be discussed in detail below), and since no techniques are available for the reliable quantitative determination of these compounds, quantitative information is usually limited to operational categories (e.g. 'organic arsenic') which are calculated as the difference between observed inorganic species and total arsenic. In this context the determination of total arsenic is of some importance. Any of a variety of methods can be used to determine total arsenic after digestion (see the reviews by Irgolic *et al.*[41] and Talmi and Bostick[52]). It is, however, important to ensure that all organic arsenic has been digested. The most reliable technique for this purpose appears to be the magnesium nitrate dry-ashing technique.[53] An ashing-aid solution containing magnesium nitrate and magnesium oxide is mixed with the sample and evaporated to dryness. The sample is then ashed at 500 °C for several hours. Several of the wet-ashing techniques described in the literature appear not to be able to digest some of the very stable organoarsenic compounds found in marine organisms.

5.6   Biosynthesis of organoarsenic compounds

The ability to transform inorganic arsenic into organoarsenic compounds is extremely widespread in nature. It has been observed in bacteria, fungi, algae, invertebrates, vertebrates, and man. There appear to be two major pathways, one which involves simple stepwise reduction and methylation of arsenic via methanearsonic and dimethylarsinic acid to the methylated arsines, and another which leads to the production of relatively complex organoarsenic molecules and which is characteristic of aquatic biota. These two modes of organoarsenic biosynthesis will be discussed separately in the following sections.

5.6.1 Reduction and biomethylation of arsenic by non-marine micro-organisms

In the nineteenth century, a number of poisoning incidents took place in Germany and England which were linked to the fact that the victims had been living in rooms which either contained relatively high amounts of arsenic in the plaster used on the walls or where the walls were covered with wallpaper containing arsenical compounds. Gmelin[54] noted a garlic odour in the rooms where symptoms had developed and suspected that the presence of arsenic in the wall coverings led to contamination of the room air with arsenic. Fleck[55] showed experimentally that moulds were able to release a gaseous arsenic compound from the arsenical pigment formulations, which he suspected to be arsine (AsH_3). Gosio[56] identified a number of mould species which were able to produce a volatile arsenic compound from arsenite, among them *Penicillium brevicaule (Scopulariopsis brevicaulis)*. He suggested that the gaseous species produced was diethylarsine, an identification which was subsequently questioned by other investigators. The question was not resolved until 1932, when Challenger identified the gas as trimethylarsine.[1] His work further suggested that methanearsonic acid and dimethylarsinic acid were intermediates in the production of trimethylarsine. He suggested the mechanism for the formation of trimethylarsine shown in Fig. 5.2. He proposed that the methylation occurred by the transfer of carbonium ions from *S*-adenosyl methionine (SAM) to a lone electron pair on trivalent arsenic. This suggestion is supported by the work of Cullen *et al.*[57,58] Based on *in vitro*

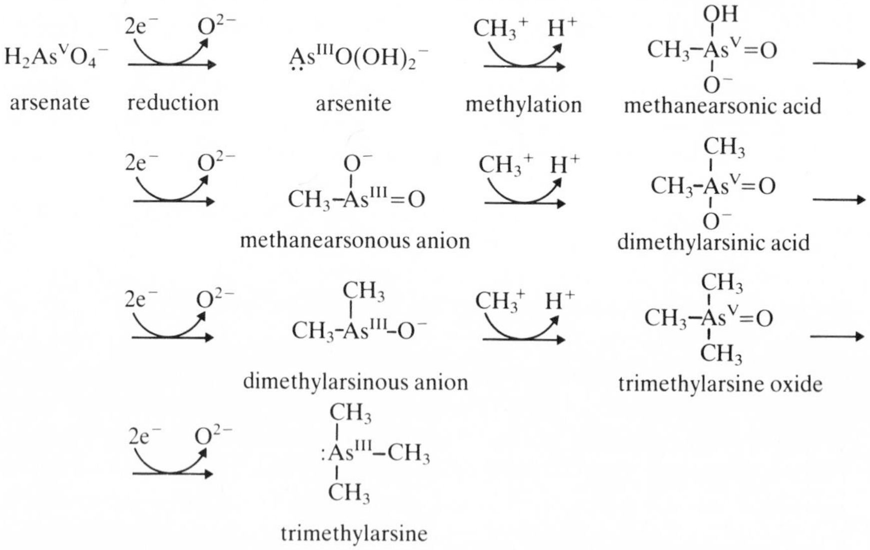

Fig. 5.2 Pathway for the biomethylation of arsenic. Modified after Challenger, 1954 – see Ref. 1

studies which show effective reduction of trimethylarsine oxide to trimethylarsine by a variety of organic thiols (including cysteine, glutathione and lipoic acid), Cullen *et al.*[58] have proposed that organic thiols are involved in the reduction steps via the following mechanism:

$$(CH_3)_3AsO + 2\,RSH \rightarrow (CH_3)_3As + RSSR + H_2O \qquad [5.1]$$

The role of As—S bonds in arsenic biochemistry deserves intense further study.

The work of Challenger on *S. brevicaulis* was later expanded by studies on other fungi. Cox and Alexander[59] showed the production of trimethylarsine from methylated arsenic substrates (methylarsonic and dimethylarsinic acids) by three species of sewage and soil fungi: *Candida humicola*, *Gliocladium roseum* and *Penicillium* sp. Only *C. humicola* was able to methylate inorganic substrates, however. Later work by the same group showed that arsenic biomethylation from inorganic arsenic and monomethylarsonate was inhibited by phosphate, but that dimethylarsinate was transformed to trimethylarsine at an increased rate in the presence of high levels of phosphate.[60]

Chromated copper arsenate (CCA) is widely used as a wood preservative. Since it contains both copper and arsenic, it is related to the pigments which were originally implicated in the formation of trimethylarsine.[54] Work by Cullen *et al.*[61] has shown that CCA can be biomethylated to trimethylarsine by *C. humicola* when this fungus is grown in the presence of CCA either as dilute solution or in the form of treated wood.

Candida humicola, *G. roseum*, and *S. brevicaulis* are able to methylate a variety of organoarsenic compounds with aryl and aliphatic substituents. Benzenearsonic acid and butylarsonic acid were transformed to phenyldimethylarsine and butyldimethylarsine, respectively.[57,62,63] This result is of particular interest since aromatic arsonic acids are in widespread use as animal food additives as discussed above. When these compounds are incorporated into animal wastes and applied to soils, biomethylation and volatilization can take place, as suggested by the observation of garlic-like odours over soils treated with these substances.[64,65]

Biomethylation of arsenic in the presence of atmospheric oxygen is not restricted to fungi. In experiments with lake and river sediments and with pure cultures of the bacteria *Aeronomas* sp., *Flavobacterium* sp., and *Escherichia coli*, Wong *et al.* have shown[66] the production of methanearsonic acid, dimethylarsinic acid and trimethylarsine oxide as well as the volatile arsines dimethyl- and trimethylarsine. Usually these experiments required the addition of substantial amounts of inorganic arsenic for the production of methylarsenic compounds to occur, but in some systems containing high levels of arsenic from a polluted watershed, biomethylation took place even without amending the culture with additional arsenic. The rate of methylation was shown to be pH-dependent, with the highest rates in the pH range from 3.5 to 5.5. This suggests that arsenic mobilization from sediments into the overlying waters is increased by acidification, not only as a consequence of dissolution of solid phases to liberate inorganic arsenic but also by enhanced biomethylation.[67] All experiments were conducted in systems where arsenic levels are at least three orders of magnitude higher than those typical of environments not polluted by arsenic. This leaves some doubt if the biomethylation of arsenic by aerobic bacteria is a process generally occurring in the environment, or if it is induced only by extreme arsenic stress. The same criticism also applies to the work on freshwater algae by Baker *et al.*,[68] which were found to produce methanearsonic acid and dimethylarsinic acid in pure culture at arsenic levels of 5 mg dm^{-3}. Typical arsenic concentrations in freshwaters are around 1 μg dm^{-3}.[69,70] However, the fact that methylarsonic acid and dimethylarsinic acid are commonly detected in freshwaters[69,70] makes it plausible that arsenic biomethylation by freshwater micro-organisms is a common process. Further studies are required to ascertain if the production of these compounds takes place by bacteria in the sediments, or by algae in the water column, as is the case in marine systems (see below).

The mixed microbial communities present in soils are able to produce volatile arsines when soils are treated with methylarsenicals, like methanearsonic acid or dimethylarsinic acid. Braman[71] found dimethyl- and trimethylarsine in the airspace trapped in bell jars over arsenical-treated soils and lawn. This work suggests that methylarsines could be formed and released from soils treated with methyl arsenical herbicide preparations.

Under anaerobic conditions, the biomethylation of arsenic proceeds only to dimethylarsine.[72] The arsine biosynthesis required the presence of H_2, ATP, arsenate, and a methyl donor, originally identified as methylcobalamin (CH_3CoB_{12}). Arsenite and methanearsonate could serve as substrates for dimethylarsine formation in place of arsenate. The reaction would proceed either in the presence of whole cells of the methanogenic bacterium *Methanobacterium* strain MoH or in the presence of cell-free extracts of the same organism, but not with heat-treated cell extracts. The pathway suggested by McBride and Wolfe[72] for the biomethylation of arsenic by methanogenic bacteria differs from the pathway suggested by Challenger in so far as the methylation of arsenic occurs by the transfer of methylcarbanion (CH_3^-) from CH_3CoB_{12} rather than by carbonium ion transfer from SAM. The validity of this mechanism has been more recently questioned by the same group, however.[57] Their more recent work suggests that CH_3CoB_{12} had transferred a methyl group to Coenzyme M, which is the actual methyl donor to arsenite *in vivo*.[73] Coenzyme M was identified by Challenger[74] as 2-mercaptoethane sulphonic acid ($HSCH_2CH_2SO_2OH$). McBride and his co-workers did not find CH_3CoB_{12} to be present in whole cells of *Methanobacterium*, a strong argument against the involvement of CH_3CoB_{12} in arsenic biomethylation.[73]

Under anaerobic conditions, methylation stops at dimethylarsine, which is stable in the absence of oxygen, but which would be oxidized rapidly if produced under oxic conditions. Trimethylarsine, while not altogether stable in the presence of oxygen, is oxidized more slowly than dimethylarsine and provides a better mechanism for anaerobic organisms to remove arsenic from their environment by methylation and volatilization. To test if the arsenic biomethylation by methanobacteria would occur not only in pure cultures but also in ecosystems in which these bacteria play an important role, McBride *et al.*[62] added labelled arsenate to anaerobic sewage digester sludge, to rumen fluid from cattle, and to compost and marine mud. In all experiments, with the exception of the marine mud, they detected the formation of a reactive arsenic compound, which they tentatively identified as dimethylarsine. They found, however, that this compound could not be removed from the system by gas stripping, but became rapidly bound to unknown substances. Sulphydryl groups are likely to be important in this binding process due to the high affinity of trivalent arsenic for sulphur. No formation of arsines was observed in experiments with marine mud, but the authors were not able to ascertain if this mud had been collected from a methanogenic environment.

5.6.2 Arsenic biomethylation by terrestrial higher organisms

The ability to biomethylate arsenic is extremely widespread in higher animals. Braman and Foreback[75] discovered the presence of methanearsonic acid and dimethylarsinic acid in human urine as well as in bird eggshells. Lakso and Peoples[76] investigated the problem of whether the methylarsenicals in animal urine were ingested with their diet, or were biosynthesized in the animal. They showed that cows and dogs were able to synthesize methylarsenicals from either arsenate or arsenite. The ability of dogs to biomethylate arsenic showed that methanogenic rumen bacteria were not essential for arsenic biomethylation by mammals. Marafante *et al.*[77] showed that rats and rabbits formed dimethylarsinic acid as major arsenic metabolite after injection with sodium arsenite. Methanearsonic acid was only found at the level of a few per cent of total As in the urine, while dimethylarsinic acid accounted for 86 per cent of As excreted by rats and 60 per cent by rabbits. These studies had not been able to distinguish if the arsenic biomethylation was linked to the presence of bacteria in the intestine of the animals or if the mammals themselves were able to biomethylate arsenic. Vahter and Gustafsson[78] injected germ-free and conventional mice with arsenate and arsenite. After 48 hours, 90 per cent or more of the injected arsenic had been excreted when doses up to 2 mg kg^{-1} bodyweight were given. When arsenic was injected as arsenic(V) at 0.4 mg kg^{-1}, the germ-free mice excreted 82 per cent of the injected amount as dimethylarsinic acid, about 1 per cent as methanearsonic acid, and the rest as inorganic arsenic. Similar results were obtained when arsenite was injected. The control animals with the normal bacterial infauna excreted arsenic in the same proportion within the experimental limits, demonstrating that intestinal bacteria play little or no role in the methylation of arsenic by mammals, at least in this species.

The metabolism of arsenic in humans following oral ingestion of inorganic arsenic was studied by Crecelius,[31] Tam *et al.*,[79] and Yamauchi and

Yamamura.[80,81] Smith *et al.*[82] investigated the biotransformation of arsenic from airborne exposure to smelter dusts containing arsenic(III) oxide. Typically, these studies found about 50–70 per cent excreted as dimethylarsinic acid, around 20 per cent as methanearsonic acid, and the rest as inorganic arsenic(III) and arsenic(V).

The speciation of arsenic in freshwater fish has not been conclusively studied. Penrose[83] exposed brown trout to inorganic arsenic and found that most of it was converted to a cationic, organic form. This organoarsenic compound was shown to be different from those Penrose had isolated from marine fish. The conversion was much more efficient when the arsenic was given as an oral dose rather than injected, which led Penrose to suggest that the intestinal flora of the fish may be involved in the production of the organic arsenic compound.

Some terrestrial higher plants are also able to synthesize methanearsonic acid and dimethylarsinic acid, as well as possibly methanearsinic acid, from inorganic arsenic. Nissen and Benson[84] showed the production of these compounds by tomato plants (*Lycopersicon esculentum* Mill. cv. Better Boy) under nitrogen- and phosphorus-deficient conditions. Nutrient-sufficient plants did not produce organoarsenic compounds. A number of other higher plant species, including the pines *Pinus halepensis* Mill., *P. pinea* L., and *P. radiata* D. Don., corn (*Zea mays* hybrid NK 177), melon (*Cucumis melo* cv. Honeydew), and pea (*Pisum sativum* cv. Sugarsnap), reduced arsenate to arsenite, but did not produce organoarsenic compounds.

5.7   Metabolism of organoarsenicals

The organoarsenic compounds most commonly found in the environment, methanearsonic acid and dimethylarsinic acid, are very stable with respect to both biological and chemical attack. In fact, the latter will even resist boiling nitric acid for several hours. When these compounds are taken up by plants, they are not degraded to any significant extent.[85] In animals, the ingestion of methanearsonate may result in the further methylation to dimethylarsinate, but demethylation of the latter compound has not been observed in animal studies by Stevens *et al.*[86] The organoarsenic compounds present in seafoods (see below) are transferred through the human body with little or no change. Chapman[87] showed that most of the arsenic in a lobster meal (about 30 mg, a nearly lethal dose had it been in the form of inorganic arsenic) was excreted in unchanged chemical form over the next 48 hours. Similar data were obtained by Coulson *et al.*[88] in experiments with shrimp as the source of the organoarsenic compounds. Crecelius[31] did not find an elevation in the levels of inorganic arsenic or the simple methylarsenicals, methanearsonate, or dimethylarsinate, in human urine following ingestion of crab meat containing 2 mg organic As. Only after hot digestion of the urine with 2 mol dm^{-3}

sodium hydroxide (NaOH) was the arsenic ingested with the crab meat detectable, now in the form of dimethylarsinate. In the urine of persons who consumed large amounts of crustaceans, arsenobetaine was shown to be present, but was not adequately quantified to allow a mass balance of arsenic metabolism.[32,89] Tam *et al.*[30] investigated the excretion of arsenic by fifteen volunteers who had consumed 10 mg As each in the form of the arsenic compounds naturally occurring in witch flounder (*Glyptocephalus cycnoglossus*). While the authors were not able to identify positively the arsenic compound involved, they could exclude the presence of significant amounts of inorganic arsenic, methanearsonic acid, dimethylarsinic acid, or arsenobetaine. Within 8 days, 76 per cent of the ingested As was recovered in unchanged form in the urine, and only 0.33 per cent in the faeces.

At this time, only bacteria have been shown to have the ability to demethylate organoarsenic compounds. This demethylation occurs both in the aquatic environment (discussed below) and in soils. Methanearsonate and dimethylarsinate become rapidly attached to insoluble soil fractions after agricultural applications of these compounds. Iron oxides and clay minerals are particularly effective in adsorbing methanearsenicals.[85] In the soil, the methylarsenic acids can be either oxidized to arsenate and carbon dioxide, or further methylated to the free methylarsines. Evidence for a loss mechanism by volatilization has been presented by Braman[71] who observed the evolution of volatile arsines from arsenic-treated soils (see above) and by Woolson and Kearney,[64] who determined the loss of arsenic added to soils as dimethylarsinic acid, but did not identify the volatile species produced. Under anaerobic conditions, they did not observe any degradation of dimethylarsinic acid to carbon dioxide and inorganic arsenic, but rather the loss of 61 per cent of the applied arsenic as an unknown volatile species over a 24-hour period. Under aerobic conditions, 41 per cent of the applied arsenic was lost by oxidation to carbon dioxide and arsenate, but volatilization of arsenic still accounted for the loss of 35 per cent of the applied dimethylarsinic acid.

Demethylation appears to be the most important loss mechanism for methylarsenicals from aerated soils. Experiments with ^{14}C-labelled methanearsonate applied to soils demonstrate the production of $^{14}CO_2$ as a result of bacterial oxidation. The oxidation rates are typically of the order of a few per cent per month of methanearsonate present.[90,91] The oxidation rates increase with soil organic content and the total soil carbon dioxide evolution rate.

The ability to demethylate arsenic appears to be widespread among soil bacteria. Shariatpanahi *et al.*[92] isolated seven bacterial species from soils and sediments and incubated them with methanearsonate. Five species, *Achromobacter* sp., *Flavobacterium* sp., *Nocardia* sp., *Pseudomonas* sp., and *Alicaligenes* sp., demethylated methanearsonate at rates of 3–5 per cent per 48-hour incubation period. Two other species, *Aeromonas* sp. and *Enterobacter* sp., did not produce $^{14}CO_2$ from methanearsonate, indicating absence of demethylation capability for these species. These two species did, however, reduce and further methylate methanearsonic acid to form mono-, di-, and trimethylarsine.

5.8   The aquatic arsenic cycle

In contrast to the arsenic cycle in terrestrial organisms, which is characterized by the predominance of the simple methylated species (methanearsonic and dimethylarsinic acids, trimethylarsine oxide, and the methylated arsines), the arsenic speciation in aquatic systems is dominated by a number of structurally more complex organoarsenic compounds, only a few of which have been identified at this date. Methanearsonate and dimethylarsinate, while present in seawater and freshwaters, appear to be only degradation products of these organoarsenic compounds.

5.8.1   Biosynthesis of organoarsenic compounds by algae

To satisfy their phosphorus requirement, primary producers have to take up phosphate ion from their environment. This presents algae with a serious problem, since arsenate resembles phosphate rather closely and the two ions are often present at comparable concentrations.[93] In the oligotrophic oceanic gyres, for example, the arsenate concentration is near 20 nmol dm^{-3}, while the phosphate concentration may be anywhere from 100 nmol dm^{-3} to less than 3 nmol dm^{-3}. If the cells were to accumulate arsenate and phosphate at the ratios which are present in seawater and to the high levels characteristic of phosphate in the algal cell fluids, the arsenate concentrations would poison the plants by interaction with sulphur-containing enzymes and by uncoupling oxidative phosphorylation. Two fundamental strategies can be applied to defend the cell against the toxic effects of arsenate: the selectivity of the uptake mechanism can be increased to exclude arsenate, or the arsenate can be rapidly metabolized intracellularly and either transformed into a harmless chemical form which can be retained by the cell or at least into a chemical species which is distinct enough from arsenate that it can be excreted from the cells without interfering with phosphate uptake. Both strategies seem to be used by marine phytoplankton. In the following paragraphs the uptake of arsenate by algae and the subsequent biotransformations will be discussed.

5.8.1.1   Uptake of arsenic by algae

The details of the uptake mechanism for arsenate by marine algae are still poorly understood. There appears to be some inhibitory interaction between arsenate and phosphate,[94–98] but the uptake kinetics suggest non-competitive inhibition.[96] Under certain circumstances, an increase of arsenate uptake with increasing phosphate concentration has been observed.[98] It seems clear from these results that the uptake mechanism of marine algae does discriminate between arsenate and phosphate, and does not show the simple competitive inhibition kinetics observed for the marine yeast *Rhodotorula rubra*.[99] The discrimination factor favouring phosphate over arsenate ranges from one to six.[100]

5.8.1.2 Biosynthesis of organoarsenic compounds

Once arsenate has entered the algal cell, it is rapidly transformed to a variety of organoarsenic compounds. As a consequence, inorganic arsenic usually represents only a relatively small fraction of the total arsenic content of algal materials. In freshwater aquatic plants, Benson and Nissen[100] found that arsenite accounted for less than 10 per cent of total arsenic in all but one of the six species investigated. Marine algae are similarly effective in transforming arsenic to organic forms. The planktonic species *Dunaliella tertiolecta* converts over 95 per cent of its arsenate uptake to arsenolipid compounds.[101] In six samples of marine kelps, organic arsenic represented on average 78 per cent of total arsenic[70] (Table 5.2). This is similar to the percentage found by Sanders[102] for marine macroalgae belonging to the class Phaeophyceae, but higher than found in Chlorophyceae and Rhodophyceae (about 55 per cent organic arsenic). In samples of mixed marine phytoplankton (mostly diatoms), 87–96 per cent of the arsenic was in organic form[70] (Table 5.2). Most authors have not determined the chemical forms of arsenic present, but have only fractionated the cell extracts into water-soluble and lipid-soluble forms.[98,100,103,104] These studies also show that most of the arsenic in algae is in organic forms not soluble in water. The water-soluble fraction also contains a substantial fraction of organic arsenic compounds, so that these studies support the findings that most of the arsenic in algae is in organic form.

The chemical identity of some of the organoarsenic compounds in marine algae has been elucidated in the last few years. The most abundant compounds in terrestrial systems, methanearsonate and dimethylarsinate, are also found in aquatic systems, but they make up only a minor fraction of the organic arsenic in algae.[98,103] They appear to be intermediates in the formation of the more complex organoarsenic compounds. Wrench and Addison[105] showed that they are rapidly formed after addition of radioactive, carrier-free ^{74}As to cultures of *Dunaliella tertiolecta*, which implies that they have a high turnover rate in the algal cells. Isotopic equilibrium was reached in 40 minutes or less for dimethylarsinic acid and a number of arsenolipids. On the other hand, dimethylarsinate and some methanearsonate are the only organoarsenic metabolites found in the culture medium after algae have been growing in it. This suggests that they are the major forms in which organic arsenic is excreted by algae. This could be either the result of the excretion of these compounds immediately after their biosynthesis or the product of the breakdown of larger organoarsenic compounds as suggested by Benson and Nissen.[100]

In addition to methanearsonate and dimethylarsinate, up to twelve soluble

Table 5.2   Speciation of arsenic in some seaweeds from various regions and in phytoplankton from the eastern North Pacific

	As(III)	As(V)	CH$_3$As†	(CH$_3$)$_2$As‡	As$_t$§	% organic¶
Laminaria digitata (France)	—	3.76*	0.073	10.6	57.6	93
Laminaria digitata (Iceland) no. 1	—	23.9*	0.048	0.13	42.0	43
Laminaria digitata (Iceland) no. 2	7.53	14.3	n.d.	0.43	57.6	62
Fucus sp. (Iceland)	2.79	8.9	0.022	0.23	41.4	72
Macrocystis pyrifera (Argentina)	0.07	0.42	0.10	2.0	33.3	99
Ascophyllum nodosum (France)	0.076	0.56	0.015	0.47	36.0	98
Phytoplankton sample no.						
001	0.063	0.030	0.006	0.49	2.01	95
018	0.26	0.094	0.020	1.29	3.52	96
033	0.027	0.43	0.027	0.74	3.40	87

* As(III) + As(V)

† Methanearsonate.

‡ Dimethylarsinate.

§ Total arsenic by MgO/Mg(NO$_3$)$_2$ dry-ashing.

¶ Contains unidentified organic arsenic species.

(Concentrations are given in μg (As) g^{-1} dry weight. The samples were ground with 2 mol dm^{-3} HCl in a tissue grinder and extracted overnight at room temperature.)

organoarsenic compounds have been observed in algae. These compounds can be grouped into water-soluble and lipid-soluble fractions. The arsenolipids were first described by Lunde,[104,106,107] who grew marine algae in the presence of radiolabelled arsenic and characterized several lipid compounds on the basis of their chromatographic behaviour. In a number of subsequent studies the production of these and similar compounds in a wide variety of marine and freshwater organisms was documented, but the chemical identity of the compounds could not be ascertained.[98,103,105] The first major breakthrough in understanding the biochemistry of arsenic in algae was the identification of two dimethylarsenosugars in the brown kelp *Ecklonia radiata*, by Edmonds and Francesconi[108,109] (Fig. 5.3). Benson and Nissen[100] were able to show that

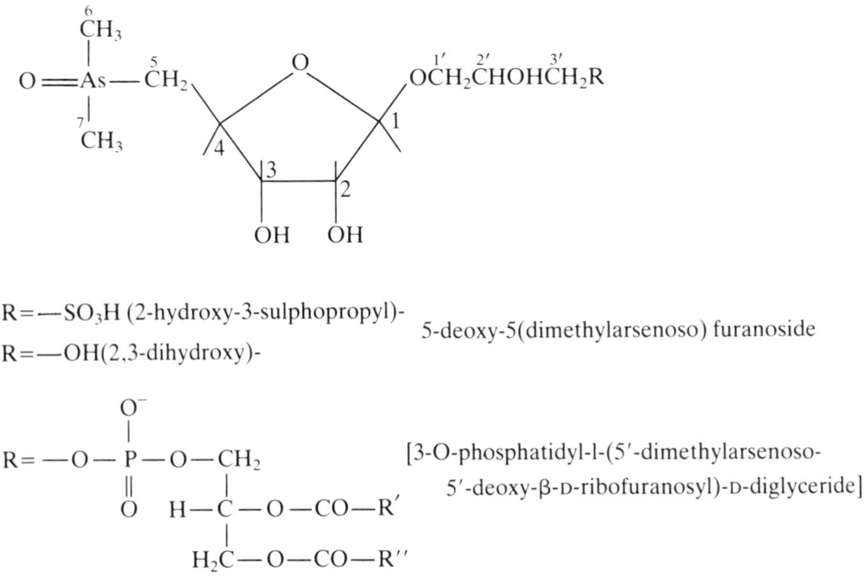

R=—SO$_3$H (2-hydroxy-3-sulphopropyl)-
R=—OH(2,3-dihydroxy)-

5-deoxy-5(dimethylarsenoso) furanoside

$$R= -O-\overset{\overset{\displaystyle O^-}{\displaystyle |}}{\underset{\underset{\displaystyle O}{\displaystyle \|}}{P}}-O-CH_2$$

[3-O-phosphatidyl-l-(5'-dimethylarsenoso-5'-deoxy-β-D-ribofuranosyl)-D-diglyceride]

H—C—O—CO—R'

H$_2$C—O—CO—R''

Fig. 5.3 Structures of the dimethylarsenosoribosides (arsenosugars) found in algae

one of these compounds, 5-dimethylarsenoso-5-deoxy-β-D-ribosyl-glycerol, was identical to the product obtained by enzymatic cleavage of the arsenolipid found in the diatom *Chaetoceros* sp., and suggested the structure in Fig. 5.3 for the arsenolipid. The work of Wrench and Addison[105] suggests that there are additional arsenolipids with substantially different structures in some algae. The three arsenolipids they found in *D. tertiolecta* gave arsenite upon acid hydrolysis, whereas the dimethylarsenosugars from *E. radiata* hydrolyse to dimethylarsinic acid.

5.8.2   Organoarsenic compounds in marine invertebrates and fish

Arsenobetaine and arsenocholine (Fig. 5.1), two substances which have not been found in marine algae, appear to be the dominant organoarsenic compounds in marine invertebrates and fish. Arsenobetaine was first identified to be the major organoarsenical in the Australian rock lobster *Panulirus longipes* cygnus George[110] and later in the American lobster *Homarus americanus*.[111] Subsequent work by the same group demonstrated the presence of arsenobetaine in the dusky shark *Carcharhinus obscurus* Le Sueur and in human urine following ingestion of cooked rock lobster.[32] Norin and Christakopoulos[112] found arsenobetaine to be the only detectable organoarsenical in fish meal, in addition to arsenobetaine in shrimp (*Pandalus borealis*). The existence of arsenocholine in shrimp was subsequently confirmed by Fast Atom Bombardment MS (FAB MS).[51]

Shiomi *et al.*[49] investigated the speciation of arsenic in the flatfish *Limanda herzensteini*, the sea squirt *Halocynthia roretzi*, and the sea cucumber *Stichopus japonicus*. About 90 per cent of arsenic in the flatfish and 60 per cent in the sea cucumber was found to be arsenobetaine, whereas the sea squirt contained two different, as yet unidentified arsenic compounds and no arsenobetaine. Fukui *et al.*[113] found arsenobetaine in the shrimp *Sergestes lucens* and in the American lobster *Homarus americanus* in addition to an unidentified organoarsenical, possibly arsenocholine. Arsenobetaine has also been identified by FAB MS in sole (*Solea solea*), lemon sole (*Microstomus kitt*), flounder (*Platichtys flesus*), dab (*Limanda limanda*), crab (*Cancer cancer*), and shrimp (*Neophrops norvegicus*) by Luten *et al.*[50] These authors suggested that arsenobetaine accounts for essentially all of the arsenic in these organisms.

Edmonds *et al.*[114] have suggested a metabolic pathway for the formation of arsenobetaine and arsenocholine from the arsenosugars found in the kelp *Ecklonia radiata* (Fig. 5.4). This pathway is based upon their observations that arsenosugars similar or identical to those found in *E. radiata* are present in the kidneys of the clam *Tridacna maxima*[115] and that dimethyloxarsylethanol can be formed from these arsenosugars by anaerobic fermentation. It remains yet to be established if the route from dimethyloxarsylethanol to arsenobetaine proceeds via arsenocholine or via dimethyloxarsylacetic acid, and at which level in the food chain these transformations occur.

Most of the organic arsenic in marine animals is found in the water-soluble fraction.[49,103,116–118] None of the lipid-soluble components has yet been identified. However, the lipid-soluble compounds can be hydrolysed to give a water-soluble compound also present in the animal, but different from arsenobetaine and arsenocholine.[98,103] In the marine snails *Littorina littoralis* and *Nucella lapillus*, most of the arsenic is in this unidentified, water-soluble

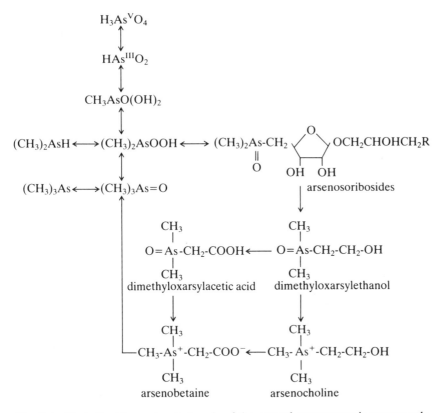

Fig. 5.4   Tentative biogeochemical cycle of the natural organoarsenic compounds

form.[103] These compounds may be related to the arsenosugars found in the clam *Tridacna maxima* by Edmonds *et al.*[115]

The polychaete worm *Tharyx marioni* is unusual in both its arsenic content and speciation. Typical total arsenic concentrations in this worm were found to be about 0.2 per cent (dry weight), with the palp organs containing up to 1.3 per cent As. Most of this arsenic is in the lipid-soluble fraction (61 per cent), while 25 per cent appears to be protein-bound (extractable with cysteine), 4 per cent water-soluble, and the rest remaining in the insoluble residue.[119] The mechanism and function of this remarkable arsenic accumulation remain yet to be elucidated.

5.9   Organoarsenicals in natural waters

As discussed in previous sections, algae are able to transform arsenate into arsenite and a number of organoarsenic compounds. The products excreted by algae and possibly also by aquatic animals are, however, limited to arsenite and the simple methylated species, methanearsonate and dimethylarsinate. Digestion of seawater with methods able to mineralize all organic arsenic, e.g. dry-ashing with a magnesium nitrate ashing aid, did not indicate the presence of significant amounts of arsenic in addition to those represented by inorganic arsenic and the simple methylarsenic acids.[70] The excretion of these compounds by algae, therefore, represents a source of methylated arsenic compounds in the photosynthetically active (euphotic) zone of aquatic systems.

The release of methylarsenicals by anaerobic micro-organisms into sedimentary pore waters and subsequently into the water column proposed by Wood[120] has been looked for by Andreae *et al.* in several likely environments, e.g. the pore waters of the anoxic sediments in the basins off California and in the anoxic waters of the Baltic Sea,[93,121] but no evidence for the existence of such a process could be found. McBride *et al.*[62] looked for the production of methylarsenicals by marine muds in laboratory systems, again without success. The possibility of this process occurring in freshwater anoxic systems, where less sulphydryl groups are present to scavenge trivalent arsenic, remains to be investigated. As discussed above, the only studies pertinent to this question have been conducted in laboratory systems with very high arsenic enrichments.[66] There is some suggestion that dimethylarsinate may be produced in the pore waters of Lake Washington, where Crecelius[122] found concentrations of this species in sedimentary pore waters which were consistently higher than in the overlying waters, but the possibility remains that this is due to the decomposition of deposited algal material rather than production *in situ*.

Due to the high retention of arsenic species in soils, the application of methylarsenical herbicides is not likely to be an important source of methylated arsenic to natural waters. The only known situation where anthropogenic inputs contributed significantly to the methylarsenical concentration in natural waters is the Menominee River, Wisconsin, where effluents from the methylarsenical production by the Ansul Company, Marinette, Wisconsin, had entered the river in large concentrations.[123] The sediments in the Menominee River were found to be a site of demethylation, rather than of the production of methylarsenic compounds.

Consequently, the excretion of the methylarsenic acids by algae is the only important process which leads to the presence of organoarsenicals in natural waters. In a steady-state system, this production has to be balanced by removal processes, in particular the demethylation of the methylarsenic acids and the oxidation of arsenite to arsenate. The demethylation is most likely exclusively due to bacterial action, since methanearsonate and dimethylarsinate are extremely stable with respect to breakage of the arsenic–carbon bond and oxidation by purely chemical means. Seawater samples containing these species have been kept in our laboratory for years under sterile conditions without appreciable change of the methylarsenical concentration. On the other hand, in the presence of bacteria the methylarsenic acids show a substantial decrease over only a few days.[70,102] Sanders[102] has estimated a bacterial removal rate for dimethylarsinate in seawater of about 1 ng dm^{-3} per day, similar to estimates of its production rate.[124]

The actual speciation of arsenic in a given body of water is, therefore, determined by the interplay of the reduction and methylation by marine algae and by bacterial demethylation. In the open ocean, vertical profiles of arsenic species' distribution are typified by the data from a station in the Pacific Ocean off California shown in Fig. 5.5. Arsenate is the quantitatively dominant species throughout the water column, but within the euphotic zone the methylarsenic acids and arsenite are present in appreciable concentrations. The actual concentrations of the methylated species are dependent on the rate of primary production,[93,124] as well as on the distribution of plankton species, since the rate at which methylated species are excreted and the ratio of methanearsonic acid to dimethylarsinic acid varies from one algal species to another.[68,97,98,124] Under eutrophic conditions, the methylated species can account for most of the arsenic in surface waters.[98,121,124] There appears to be a temperature dependence of the production of methylated arsenic species, at least in estuaries. Howard *et al.*[125] have observed that methylated arsenic species are present in the waters of the Beaulieu estuary only when the water temperature is above *c.* 12 °C during the warmer months. On the other hand, we have found methylarsenic acids to be present in seawater samples collected from below the ice in the McMurdo Sound region, Antarctica (Table 5.3). The concentrations here are substantially lower, however, than those found in temperate and tropical waters (Fig. 5.5 and Table 5.2) and would have been below the detection limits of Howard *et al.*[125]

5.10   Arsenic in food chains – biomagnification or biodiminution?

The high concentrations of arsenic in marine algae and animals discussed above have led to the suggestion that arsenic is bioaccumulated (accumulated in biota from seawater) and biomagnified (concentrations increasing in the food chain). This suggestion is misleading for two reasons. First, it involves the comparison between a concentration in water and a concentration in biomass (expressed as a ratio to wet or dry weight) when this concept is applied to the first step in the food chain (water → algae). The meaning of such a comparison is questionable.[126] Secondly, it ignores chemical speciation changes, which are of critical importance for the environmental chemistry and toxicology of arsenic. Since the fundamental problem at the first level in the food chain is the similarity between arsenate and phosphate, it appears more realistic to compare arsenic/phosphorus ratios at different levels in the food chain. Table 5.4 shows that this ratio continuously decreases in the marine food chain, representing a progressive purification of the phosphate pool and elimination of arsenic.[127,128] The biochemical mechanisms discussed above are,

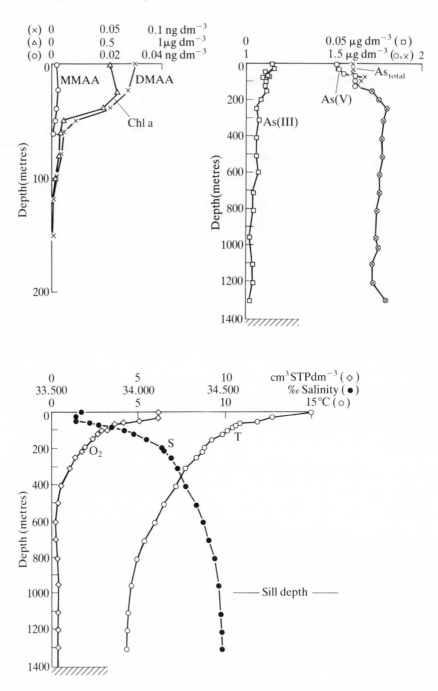

Fig. 5.5 Arsenic species distribution at Station AS3−2 in the Santa Catalina Basin off California (33° 23.9'N, 118° 49.8'W)

Table 5.3 Arsenic speciation in coastal seawater from McMurdo Sound, Antarctica (December 1976; January 1977), the southern California Bight, and the Florida coast

	Depth	As(III)	As(V)	CH₃As	(CH₃)₂As	As$_t$*
McMurdo Sound						
Heald Island	15	<0.05	18.4	<0.09	0.15	18.5
	25	0.15	18.3	0.13	0.11	18.7
	50	0.08	18.4	<0.09	0.07	18.7
	75	0.08	18.3	0.11	0.03	18.5
White Island	15	0.15	18.3	<0.09	0.25	18.8
	50	0.08	17.7	<0.09	0.24	18.1
New Harbor	5	0.12	18.0	<0.09	0.13	18.4
	23	0.12	18.0	<0.09	0.16	18.4
Cape Chocolate	25	0.05	18.8	<0.09	0.17	19.1
	50	0.08	20.7	<0.09	0.15	21.1
S. California Blight						
Tanner Bank	1	—	18.4*	0.24	1.64	20.3
32°40′N, 119°05′W,	35	—	16.5*	0.19	1.43	18.1
23 October 1976)	67	—	18.3*	<0.07	0.15	18.4
St Nicolas Island (24 October 1976)	2	0.28	16.3	0.29	2.13	18.9
Florida coast						
Fort Lauderdale (7 October 1976)	1	10.7	6.51	<0.10	1.73	18.9
Key Largo (7 October 1976)	0.5	7.51	7.29	0.08	2.00	16.9

* As(III) + As(V).

Concentration in nmol dm⁻³. Depth in metres

Table 5.4   Arsenic and phosphorus in seawater and organisms at increasing trophic
levels

	As	P	P/As	Ref.
Surface seawater	0.0015	0.016	11	69
Marine algae	25	4 250	170	70, 127
Crustacea	1.5	9 000	6 000	20, 70
Pisces	1.7	18 000	10 600	128
Mammalia	0.2	43 000	215 000	70

(Concentrations in μg g^{-1} for seawater and mg kg^{-1} dry weight for biological materials)

therefore, quite effective in dealing with the problem of arsenic toxicity both
by depleting this element relative to phosphorus (biodiminution) and by converting it to relatively harmless chemical forms.

5.11   Biological volatilization of arsenic

The experiments of Braman,[71] which showed the formation of volatile arsines
from lawns and soils, and the work of Woolson and Kearney,[64] who observed
the loss of arsenic from soils by volatilization, have led to the suggestion that
arsenic biomethylation may represent a significant source of arsenic to the
atmosphere. However, organic arsenic compounds have not been detected in
the atmosphere away from obvious point sources. Walsh *et al.*[129] observed
only a very minor proportion of atmospheric arsenic to be in the gas form, and
most of this small vapour component appears to be of inorganic character. In
studies on arsenic in aerosols and precipitation by Andreae the presence of
methylated arsenic species at levels above the detection limit was never observed.[130] While it cannot be excluded that methylarsenic species could have been
oxidized to arsenate during their residence time in the atmosphere, it seems
very unlikely that a species as chemically stable as dimethylarsinic acid should
have been completely oxidized under all the various atmospheric conditions
represented by our sample. Volatilization of arsenic from the oceans by
biomethylation can be ruled out as a source of atmospheric arsenic both on
the basis of the work of Sanders and Windom,[97] who were not able to detect
any volatilization of arsenic by laboratory cultures of phytoplankton, and on
the basis of Andreae's data on arsenic in rain,[130] which show very low levels of
arsenic in marine precipitations, all of it in inorganic form. This argument is
further supported by the atmospheric arsenic budget based on the aerosol data
of Walsh *et al.*,[129,131] which precludes the existence of a significant oceanic
arsenic source to the atmosphere.

5.12   Organoarsenic compounds in oil shales and oil shale products

Using a high performance liquid chromatograph coupled to a graphite furnace atomic absorption detector, Fish *et al.*[132,133] have observed the presence of a number of organoarsenic species in extracts from Green River formation oil shale (sample from Anvil Points, Colorado; National Bureau of Standards Reference Material) and in various oil shale process liquids. In methanol extracts of Green River shale they determined methylarsonic and phenylarsonic acids. When the oil shale is heated to 500 °C, pyrolysis of the oil shale kerogen produces shale oil and considerable amounts of process water as a waste product. In such process waters, Fish *et al.*[132] identified substantial amounts of methanearsonic and phenylarsonic acids in addition to arsenate. When shale oils are subjected to size exclusion chromatography, arsenic is found to be associated with molecular weights of about 2000 to 4000.[134] Methanearsonate added to these shale oils becomes associated with the same molecular weight fraction. This observation has led Weiss *et al.*[134] to postulate that organoarsenic compounds in shale oils are associated with iron–humic complexes found in this fraction.

References

1.   Challenger F 1945 *Chem Reviews* **36**: 315–61
2.   Blumenthal F 1902 *Medizinische Woche* p 163
3.   Koch R 1907 *Deutsche Medizinische Wochenschrift* **33**: 1889
4.   Raiziss G W, Gavron J L 1923 *Organic Arsenical Compounds*. Chem Catalog Co, New York
5.   Beckman H 1958 *Drugs: Their Nature, Action and Use*. W B Saunders Co, Philadelphia
6.   Seigmund O H (ed) 1973 *Merck Veterinary Manual* 4th edn. Merck and Co, Rahway, New Jersey
7.   Banks C H, Daniel J R, Zingaro R A 1979 *J Medicinal Chem* **22**: 572–5
8.   Daniel J R, Zingaro R A 1978 *Phosphorus and Sulfur* **4**: 179–85
9.   Calvert C C 1975 In *Arsenical Pesticides, ACS Symp Ser* **7**: 70–80
10. Morrison J L 1969 *J Agr and Food Chem* **17**: 1288
11. Woolson E A 1974 ACS abstract, Pesticide Chem Div, Los Angeles, California
12. Fitzgerald L D 1983 In *Arsenic: Industrial, Biomedical, Environmental Perspectives* (eds Lederer W H, Fensterheim R J). Van Nostrand Reinhold, New York, pp 3–9
13. Abernathy J R 1983 In ibid, pp 57–62
14. Alden J C 1983 In ibid, pp 63–71
15. Baldwin W J 1983 In ibid, pp 99–111

16. Bauer R J 1983 In ibid, pp 45–56
17. Willardson R K 1983 In ibid, pp 72–88
18. Tatken R L, Lewis R J (eds) 1983 *Registry of Toxic Effects of Chemical Substances* 1981–82 edn. US Dept of Health and Human Services, Cincinnati, Ohio
19. Frost D V 1967 *Proc Federation of American Societies for Experimental Biology* **26**: 194–208
20. Schroeder H A, Balassa J J 1966 *J Chron Dis* **19**: 85–106
21. Knowles F C, Benson A A 1983 *Trends in Biochem Sci* **8**: 178–80
22. Zingaro R A 1983 In *Arsenic: Industrial, Biomedical, Environmental Perspectives*, (eds Lederer W H, Fensterheim R J). Van Nostrand Reinhold, New York, pp 327–47
23. Lanz H Jr, Wallace P W, Hamilton G 1950 *Univ of Calif Publ in Pharmacol.* **2**: 263–82
24. Ginsberg J M, Lotspeich W D 1963 *Amer J Physiol* **205**: 707–14
25. Ginsberg J M 1965 *Amer J Physiol* **208**: 832–40
26. Peoples S A 1983 In *Arsenic: Industrial, Biomedical, Environmental Perspectives*, (eds Lederer W H, Fensterheim R J). Van Nostrand Reinhold, New York, pp 125–33
27. Bencko V, Benes B, Cikrt M 1976 *Arch Toxicol* **36**: 159–62
28. Welch A D, Landau R L 1942 *J Biol Chem* **144**: 581–8
29. Westöö G, Rydälv M 1972 *Var föda* **24**: 21–40
30. Tam G K H, Charbonneau S M, Bryce F, Sandi E 1982 *Bull Environm Contam. Toxicol.* **28**: 669–73
31. Crecelius E A 1978 *Environ. Health Perspectives* **19**: 147–50
32. Cannon J R, Edmonds J S, Francesconi K A, Raston C L, Saunders J B, Skelton B W, White A H 1981 *Aust J Chem* **34**: 787–98
33. World Health Organization, Geneva 1981 *Environmental Health Criteria*, vol 18, *Arsenic*
34. Furst A 1983 In *Arsenic: Industrial, Biomedical, Environmental Perspectives* (eds Lederer W H, Fensterheim R J). Van Nostrand Reinhold, New York, pp 151–65
35. Simmon V F 1983 In ibid, pp 166–72
36. Harding-Barlow I 1983 In ibid, pp 203–9
37. Zielhuis R L, Wibowo A A E 1984 In *Changing Metal Cycles and Human Health*, (ed Nriagu J O). Dahlem Konferenzen, Springer-Verlag, Berlin, pp 323–44
38. Pershagen G, Vahter M 1979 *Arsenic: A Toxicological and Epidemiological Appraisal*. SNV PM 1128, The National (Swedish) Environmental Protection Board, Stockholm
39. National Academy of Sciences, Washington DC 1976 *Arsenic*
40. Environmental Protection Agency, Washington DC 1983 *Health Assessment Document for Inorganic Arsenic*. EPA-600/8-83-021F
41. Irgolic K J, Stockton R A, Chakraborti D 1983 In *Arsenic: Industrial, Biomedical, Environmental Perspectives*, (eds Lederer W H,

Fensterheim R J). Van Nostrand Reinhold, New York, pp 282–308 and refs therein

42. Andreae M O 1977 *Anal Chem* **49**: 820–3
43. Braman R S, Johnson D L, Foreback C C, Ammons J M, Bricker J L 1977 *Anal Chem* **49**: 621–5
44. Feldman C 1979 *Anal Chem* **51**: 664–9
45. Talmi Y, Bostick D T 1975 *Anal Chem* **47**: 2145–50
46. Dietz E A Jr, Perez M E 1976 *Anal Chem* **48**: 1088–92
47. Brinckman F E, Jewett K L, Iverson W P, Irgolic K J, Ehrhardt K C, Stockton R A 1980 *J Chromatogr* **191**: 31–46
48. Fish R H, Brinckman F E, Jewett K L 1982 *Environ Sci Technol* **16**: 174–9
49. Shiomi K, Shinagawa A, Azuma M, Yamanaka H, Kikuchi T 1983 *Comp Biochem Physiol* **74**: 393–6
50. Luten J B, Riekwel-Booy G, van den Greef J, ten Noever de Brauw M C 1983 *Chemosphere* **12**: 131–41
51. Norin H, Ryhage R, Christakopoulos A, Sandström M 1983 *Chemosphere* **12**: 299–315
52. Talmi Y, Bostick D T 1975 *J Chromatogr Sci* **13**: 231–7
53. Uthe J E, Freeman H C, Johnston J R, Michalik P 1974 *Journal of the AOAC* **57**: 1363–5
54. Gmelin L 1839 *Karlsruher Zeitung*. November
55. Fleck H 1872 *Zeitschrift für Biologie* **8**: 444–56
56. Gosio B 1897 *Chemische Berichte* 1024–6
57. Cullen W R, Froese C L, Lui A, McBride B C, Patmore D J, Reimer M 1977 *J Organometallic Chem* **139**: 61–9
58. Cullen W R, McBride B C, Reglinski J 1984 *J Inorganic Biochem* **21**: 45–60 and 179–94
59. Cox D P, Alexander M 1973 *Bull Environ Contam Toxicol* **9**: 84–8
60. Cox D P 1975 *Am Chem Soc Symp Ser* **7**: 81–96
61. Cullen W R, McBride B C, Pickett W A, Reglinski J 1984 *Appl and Environ Microbiol* **47**: 443–4
62. McBride B C, Merilees H, Cullen W R, Pickett W 1978 *Am Chem Soc Symp Ser* **82**: 94–115 and refs therein
63. Cullen W R, Erdman A E, McBride B C, Pickett W 1983 *J Microbiol Meth* **1**: 297–303
64. Woolson E A, Kearney P C 1973 *Environ Sci Technol* **7**: 47–50
65. Insensee A R, Kearney P C, Woolson E A, Jones G E, Williams V P 1973 *Environ Sci Technol* **7**: 841–5
66. Wong P T S, Chau Y K, Luxon L, Bengert G A 1977 In *Trace Substances in Environmental Health-XI*, (ed Hemphill D D). University of Missouri, Columbia, pp 100–5
67. Baker M D, Inniss W E, Mayfield C I, Wong P T S, Chau Y K 1983 *Environ Technol Lett* **4**: 89–100
68. Baker M D, Wong P T S, Chau Y K, Mayfield C I, Inniss W E 1983 *Can*

J Fish Aquat Sci **40**: 1254−7
69. Andreae M O 1978 *Deep-Sea Res* **25**: 391−402
70. Andreae M O, Florida State University (unpublished data)
71. Braman R S 1975 *Am Chem Soc Symp Ser* **7**: 108−23
72. McBride B C, Wolfe R S 1971 *Biochemistry* **10**: 4312−17
73. McBride B C, University of British Columbia (personal communication)
74. Challenger F 1978 *Am Chem Soc Symp Ser* **82**: 1−22
75. Braman R S, Foreback C C 1973 *Science* **182**: 1247−9
76. Lakso J U, Peoples S A 1975 *J Agric Food Chem* **23**: 674−6
77. Marafante E, Rade J, Pietra R, Sabbioni E, Bertolero F 1980 In *Proc 3rd Internat Symp on Arsenic and Nickel*, (eds Anke M, Schneider H J, Brucker Chr). Jena, GDR, July 1980, pp 49−55
78. Vahter M, Gustafsson B 1980 In ibid, pp 123−9
79. Tam G K H, Charbonneau S M, Bryce F, Pomroy C, Sandi E 1979 *Toxicol Appl Pharmacol* **50**: 319−22
80. Yamauchi H, Yamamura Y 1979 *Indust Health* **17**: 79−83
81. Yamauchi H, Yamamura Y 1979 *Jap J Indust Health* **21**: 47−54
82. Smith T J, Crecelius E A, Reading J C 1977 *Environ Health Perspectives* **19**: 89−93
83. Penrose W R 1975 *J Fish Res Board Can* **32**: 2385−90
84. Nissen P, Benson A A 1982 *Physiol Plant* **54**: 446−50
85. Hiltbold A E 1975 *Am Chem Soc Symp Ser* **7**: 53−69 and refs therein
86. Stevens J T, Hall L L, Farmer J D, DiPasquale L C, Chernoff N, Durham W F 1977 *Environ Health Perspectives* **19**: 151−7
87. Chapman A 1926 *Analyst* **51**: 548
88. Coulson E J, Remington R E, Lynch K M 1935 *J Nutr* **10**: 255
89. Kurosawa S, Yasuka K, Taguchi M, Yamazaki S, Toda S, Morita M, Uehiro T, Fuwa K 1980 *Agric Biol Chem* **44**: 1993−4
90. Dickens R, Hiltbold A E 1967 *Weeds* **15**: 299−304
91. Von Endt D W, Kearney P C, Kaufman D D 1968 *J Agric Food Chem* **16**: 17−20
92. Shariatpanahi M, Anderson A C, Abdelghani A A 1981 In *Trace Substances in Environmental Health-XV*, (ed Hemphill D D) University of Missouri, Columbia, pp 383−7
93. Andreae M O 1979 *Limnol Oceanogr* **24**: 440−52
94. Planas D, Healey F P 1978 *J Phycol* **14**: 337−41
95. Brunskill G J, Graham B W, Rudd J W M 1980 *Can J Fish Aquat Sci* **37**: 415−23
96. Klumpp D W 1980 *Marine Biol* **58**: 257−64
97. Sanders J G, Windom H L 1980 *Estuarine and Coastal Marine Sci* **10**: 555−67
98. Andreae M O, Klumpp D W 1979 *Environ Sci Technol* **13**: 738−41
99. Button D K, Dunker S S, Morse M L 1973 *J Bacteriol* **113**: 599−611
100. Benson A A, Nissen P 1982 *Developments in Plant Biol* **8**: 121−4

101 Cooney R V 1981 Ph. D. Dissertation, University of California, San Diego

102. Sanders J G 1979 *Estuarine and Coastal Marine Sci* **9**: 95−9

103. Klumpp D W, Peterson P J 1981 *Marine Biol* **62**: 297−305

104. Lunde G 1973 *Acta Chemica Scand* **27**: 1586−94

105. Wrench J J, Addison R F 1981 *Can J Fish Aquat Sci* **38**: 518−23

106. Lunde G 1972 *Acta Chemica Scand* **26**: 2642−4

107. Lunde G 1975 *J Sci Food Agric* **26**: 1257−62

108. Edmonds J S, Francesconi K A 1981 *Nature* **289**: 602−4

109. Edmonds J S, Francesconi K A, 1983 *J Chem Soc Perkin Trans vol I.* pp 2375−82

110. Edmonds J S, Francesconi K A, Cannon J R, Raston C L, Skelton B W, White A H 1977 *Tetrahedron Lett* **18**: 1543−6

111. Edmonds J S, Francesconi K A 1983 *J Chem Soc Perkin Trans vol I*

112. Norin H, Christakopoulos A 1982 *Chemosphere* **11**: 287−98

113. Fukui S, Hirayama T, Nohara M, Sakagami Y 1981 *J Food Hyg Soc Japan* **22**: 513−19

114. Edmonds J S, Francesconi K A, Hansen J A 1982 *Experientia* **38**: 643−4

115. Edmonds J S, Francesconi K A, Healy P C, White A H 1982 *J Chem Soc Perkin Trans* vol I, pp 2989−93

116. Penrose W R, Conacher H B S, Black R, Méranger J C, Miles W, Cunningham H M, Squires W R 1977 *Environ Health Perspectives* **19**: 53−9

117. Benson A A, Summons R E 1981 *Science* **211**: 482−3

118. Lunde G 1977 *Environ Health Perspectives* **19**: 47−52

119. Gibbs P E, Langston W J, Burt G R, Pascoe P L 1983 *J Mar Biol Ass UK* **63**: 313−325

120. Wood J M 1974 *Science* **183**: 1049−52

121. Andreae M O, Froelich P N 1984 *Tellus* **36B**: 101−17

122. Crecelius E A 1975 *Limnol Oceanogr* **20**: 441−51

123. Anderson M A, Armstrong D E, Andren A W, Holm T R, Iverson D G, Stanforth R R, Barta S 1979 *Mass Balance and Speciation of Arsenic in the Menominee River, Wisconsin.* Report to the US Environ Prot Agency, Athens, Georgia

124. Sanders J G 1983 *Can J Fish Aquat Sci* **40**: 192−6

125. Howard A G, Arbab-Zavar M H, Apte S 1982 *Marine Chem* **11**: 493−8

126. Bernhard M, Andreae M O 1984 In *Changing Metal Cycles and Human Health*, (ed. Nriagu J O). Dahlem Konferenzen, Springer-Verlag, Berlin, pp 143−67

127. Bowen H J M 1966 *Trace Elements in Biochemistry*. Academic Press, New York

128. Le Blanc P J, Jackson A L 1973 *Marine Pollut Bull* **4**: 88−90

129. Walsh P R, Duce R A, Fasching J L 1979 *J Geophys Res* **84**: 1710−18

130. Andreae M O 1980 *J Geophys Res* **85**: 5412−18

131. Walsh P R, Duce R A, Fasching J L 1979 *J Geophys Res* **84**: 1719–26
132. Fish R H, Jewett K L, Brinckman F E 1982 *Environ Sci Technol* **16**: 174–9
133. Fish R H, Tannous R S, Walker W, Weiss C S, Brinckman F E 1983 *J Chem Soc Chem Commun* 490–2
134. Weiss C S, Parks E J, Brinckman F E 1983 In *Arsenic: Industrial, Biomedical, Environmental Perspectives*, (eds Lederer W H, Fensterheim, R J). Van Nostrand Reinhold, New York, pp 309–26

Chapter 6

Environmental aspects of organosilicon chemistry and use

6.1 The occurrence of silicon and organosilicon compounds in natural systems by industrial usage or natural formation

Silicon constitutes 16.1 per cent of the earth's crust and is one of the most abundant elements on our planet. It is present in rocks, atmospheric dust, natural water systems and living matter, mainly in its inorganic forms. Many primitive plants and animals such as the diatoms, radiolarians and some forms of flagellates, sponges and gastropods have silicate skeletal structures. Relatively high concentrations are also found in higher plant species belonging to the Cyperaceae, Gramineae and Urticaceae families and most living organisms contain at least trace quantities of silicon. It has been calculated[1] that about 10 000 million tonnes of silica are involved in the biogeochemical cycle.

In biological systems silicon occurs mainly as silica, silicic acid and silicates but it is thought that it may also bind to the oxygen and nitrogen of naturally occurring organic compounds. Engel[2] observed silicon—galactose complexes in rye straw and reported that part of the silicon in plants is bound to the cellulose framework. Schwarz, who carried out extensive studies on the biochemistry and physiology of silicon in plant and animal systems, considered silicon to be specifically associated with collagens, glycosaminoglycans, polyuronides,[3] pectins and alginic acids.[4] He found that the silicon bound to organic compounds was not released on autoclaving, was not dialysable and did not react with ammonium molybdate, which established that these were not simply just mixtures of inorganic silicon and organic compounds. It was also found that silicon-bound materials did not react with dilute acid and base; since strong acid or base was required to hydrolyse silicon–polysaccharide derivatives it was assumed the silicon-bound materials were of this type. Enzymatic hydrolysis did not release the bound silicon but gave lower molecular weight silicon-containing products and it was concluded that the silicon acted as a bridge between oxygens and played some role in the structural organization of glycosaminoglycans and polyuronides.[5] Schwarz[4] and others[6,7] made numerous attempts to extract and identify silicon–organic compounds from tissues but in no instance was a single characterizable product obtained. Holzapul reported that he had isolated and characterized silicon 'esters' of lipids such as cholesterol, lecithin and choline. Holt and Yates[8] demonstrated that such products could be formed *in vitro* by treating the lipids with silicic acid and concluded that the 'ester' products were not present in tissue material but were artefacts formed in the work-up procedure. The only silicon derivative isolated so far,[9] for which there is reasonable evidence that it may have been a genuine tissue component, is the β-thujaplicine (thpl) derivative isolated from Thujaplicata (Fig. 6.1).

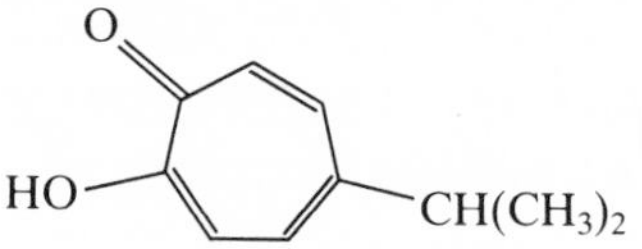

β-thujaplicene (thpl)

Fig. 6.1 β-Thujaplicine

Silicon complex isolated as (thpl)$_3$ SiPF$_6$. Full structure not characterized but may not be an organometallic species (i.e. Si—C bonded)

The tissue extracts were treated with ammonium hexafluorophosphate and the silicon derivative was isolated from the plant material as the β-thujaplicinesiliconhexfluorophosphate ((thpl)$_3$ Si$^+$PF$_6^-$). Optical activity was observed in the extracted and precipitated hexafluorophosphate salt. The optical activity disappeared on hydrolysis at pH 7.5 indicating its connection with the hydrolysable complex. If the complex were formed from thujaplicine and silicic acid in the extraction procedure it would be expected to be racemic. The possibility was considered that a racemic product formed in the extraction process might be resolved by preferential adsorption of an enantiomer on the chiral plant material. Experiments using synthetic reagents did not give any evidence for such a separation and it was concluded that the silicon–thujaplicine complex was derived from the plant tissues.

The evidence that specific organic compounds containing silicon bound to oxygen or nitrogen occur naturally is inconclusive. Evidence that genuine organosilicon compounds having silicon–carbon linkages are formed or occur in natural systems is almost non-existent. Heinen[9] found that bands corresponding to C—O—Si, C—Si, N—Si and Si—H were present in the infrared

(IR) spectra of both the cell wall and fluid material when the bacterium *Proteus mirabilis* was cultured in a silicon-containing medium. This is the only report of silicon–carbon formation in natural systems. Although it is unlikely that silicon–carbon compounds are produced naturally there is no doubt that organosilicon compounds, mainly siloxanes, are dispersed widely throughout our natural system.

The siloxanes are compounds with alternate silicon and oxygen atoms (A)[10]

$$\left[\begin{array}{ccc} & R & R \\ & | & | \\ \text{—Si—O—Si—O—} \\ & | & | \\ & R & R \end{array} \right]_n \qquad \qquad \text{[A]}$$

Oligomeric siloxanes, generally known as silicones, have been produced commercially for the past 30–40 years. The western world's output of silicones in 1982 was 401 780 tonnes and output has generally increased by about 10–20 per cent per annum. There are, therefore, now large quantities of silicones distributed throughout the world. The polydimethylsiloxanes (PDMS where R = CH_3) are by far the largest volume products in the silicone group of materials. The PDMS materials are manufactured in fluid, rubber and resin form, the fluids being the major products and constituting about 60 per cent of total output. Silicones containing vinyl, phenyl and other functional groups are also manufactured for specialized usage, but on a very small scale compared with the polydimethylsilicones.

The silicones as a class show high thermal and chemical stability, low surface tension, water repellancy, lubricity and usually low toxicity. These useful properties have led to their use in a wide range of applications. The liquids are used as dielectric coolants, brake and hydraulic fluids, synthetic lubricants, and also in polishes, paints, antifoams and as an inert base in many domestic, medical and cosmetic preparations. The solids are used in seals, piping, insulation, laminates, coatings and biomedical devices.

The molecular weights and viscosities of the linear polymers increase with chain lengths. The PDMS fluids are classified and marketed by viscosity grade, which ranges from 0.65 centistokes (cSt) for hexamethyldisiloxane to over 60 000 cSt for some silicone greases. It is usually relatively low viscosity linear and cyclic siloxanes which are used in personal care products. Higher molecular weight linear polymers with higher viscosities are used in industrial applications.

The silicones are expensive materials and are generally only employed in situations where their particular properties are required and no cheaper alternative material can be found, and almost always in low volume applications. When used in a bulk material the silicone will usually be present only in trace quantities.

Many silicone-containing products will be disposed of in use, for example the silicone bases in cosmetics will evaporate into the atmosphere or be washed off during bathing. But since they are usually used in such low quantities, the amount of silicone material entering the environment at any one place and at any one time will normally be very low and unlikely to have much impact. However, the ever-increasing production and usage of silicone materials, the possibility of an expansion in their large-scale usage (such as in dielectric coolants and brake fluids) and the knowledge that the silicones are such stable and persistent materials have led to some anxiety that they may build up in the environment with adverse ecological consequences. This has prompted a number of investigations into the toxicological and ecological properties of organosilicon compounds.

Further environmental aspects of the use of solid, elastomeric silicon polymers are discussed in Chapter 9.

6.2   Toxicity and biological properties of organosilicon compounds

This has been a very active area of research for some time now and the subject has been reviewed by a number of authors.[5,11-18]

The PDMS derivatives, which are the species most likely to be encountered in the environment, generally show very low toxicity. Dow Corning, one of the largest manufacturers of silicones, has over the past decade carried out a very extensive programme of research on the health and environmental effects of its products.[19-23] Hobbs and his co-workers at Dow have investigated[19] the effect of some low- to medium-grade (100–300 cSt) Dow Corning fluids on a range of mainly aquatic species, *daphnia*, some benthic organisms, fish and birds. The 48-hour TL_{50} in *daphnia* for a 100 cSt liquid was 44.5 μg g^{-1}. The liquid being lighter than water remained on the surface, many of the daphnia were trapped in the surface material and apparently died of suffocation rather than poisoning. For the fish and benthic species the four-day TL_{50} for a Dow Corning 300 cSt PDMS emulsion was > 10 000 μg g^{-1} and > 1000 μg g^{-1} respectively. The sub-acute oral toxicity of a PDMS 100 cSt fluid to Mallard duck and Bobwhite quail was measured over five days. The LC_{50} in each species was > 5000 μg g^{-1}. Long-term studies conducted on white Leghorn chickens revealed that ingestion of a PDMS 100 cSt fluid over an extended period had no adverse effect on body weight or food consumption. Deaths which did occur over the prolonged period of the study could be attributed to natural physical causes. Reproduction in the birds was normal. Egg production and quality, hatchability and body weight of offspring were similar for control and treated birds. There appeared to be no transfer of PDMS from parent to offspring, since organosilicon levels were below detection limits in both eggs and chickens.

Silicones of the type shown below (Fig. 6.2) are commonly used as inert bases in medical and cosmetic preparations.

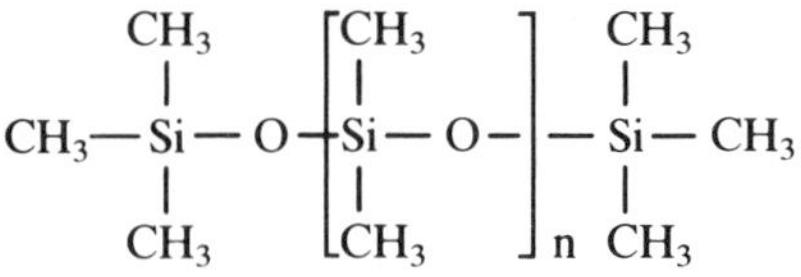

polydimethylsiloxane
'Dimethicone'

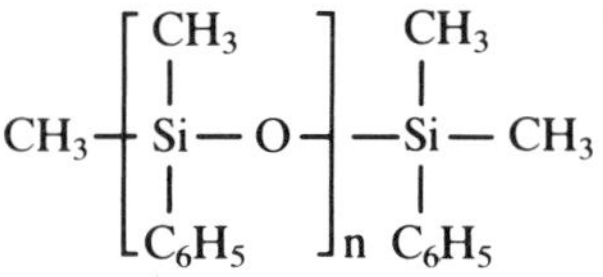

polymethylphenylsiloxane
'Phenyldimethicone'

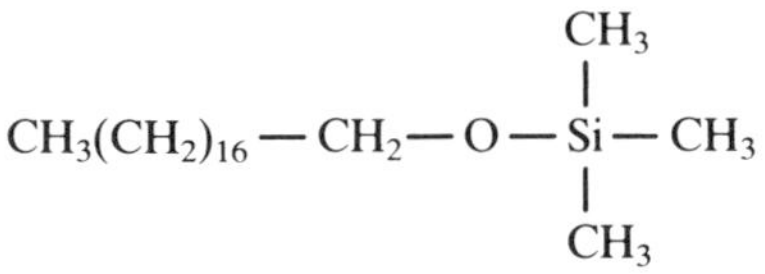

Stearoxytrimethylsilane

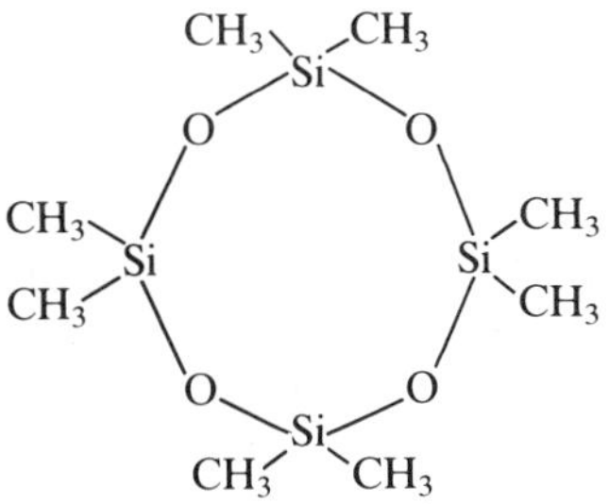

octamethylcyclotetrasiloxane
'Cyclomethicone' (tetramer)

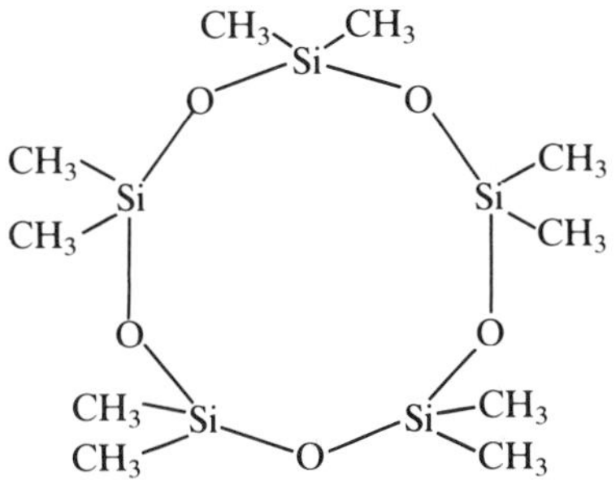

decamethylcyclopentasiloxane
'Cyclomethicone' (pentamer)

Fig. 6.2 Some types of silicone materials used in personal care products

Since personal care products may come into contact with eyes or skin, or be inhaled, both manufacturing and regulatory bodies require that the ingredients in such products be absolutely safe and have no troublesome effects even when used over a prolonged period of time. Therefore, this class of silicones have undergone even more elaborate testing than other silicones in both animal and human subjects.[20] As with other silicones they generally show very low toxicity. Oral LD_{50} values for dimethicone, phenyldimethicone and cyclomethicone (Fig. 6.2) were 38 g kg^{-1} and for stearoxy trimethylsilane 15.0 g kg^{-1}. Dietary studies on mice, rabbits, rats and dogs carried out over extended periods of time, in some cases up to two years, established that these materials lacked any significant oral toxicity, carcinogenicity or mutagenicity. Tests on human subjects revealed that they were neither skin irritants nor sensitizers, nor were they absorbed through the skin since organosilicon derivatives were not detectable in either the urine or breath of treated animals. Neat silicone material can produce a mild temporary eye irritation but the diluted product is without effect. Indeed some silicone-based shampoos have been found to produce less eye discomfort than simple soap solutions. There is no apparent inhalation toxicity; rabbits and rats exposed to highly exaggerated atmospheric concentrations showed no ill effects. Monkeys were used to evaluate the long-term effects on primate species of cyclomethicone, the common base for anti-perspirant formulations. Over 90 days they were subjected twice daily for 20-minute periods to concentrations of about 2 mg dm^{-3} of cyclomethicone. Electrocardiac, neurophysiological, respiratory, haemotological and clinical tests revealed no unfavourable results. Dietary studies using ^{14}C-labelled dimethicone polymers in animals and non-labelled fluids in human subjects established that only relatively low molecular weight silicones are absorbed from the gastrointestinal tract, and that these are eliminated via expiration and urination. The materials eliminated in respiration were unaltered low molecular weight cyclic tetramer and pentamer. The materials found in the urine were siloxanols formed by hydrolysis of Si—O—Si bonds; these products were probably formed in the stomach since cleavage of the Si—O—Si linkage can be catalysed by both acids and bases.[10] The high molecular weight compounds were not absorbed and were excreted in the faeces.

The concentrations used in all of these studies on personal care products were highly exaggerated compared with those which would be received by a human subject using silicone-based products in either aerosol or any other form. The gastric absorption of the low molecular weight materials is not a serious cause for concern since silicon-based products (cosmetics, ointments, etc.) are generally applied topically. They may be inhaled but are unlikely to be eaten except perhaps when used in lipsticks and lipsalves, and then not in great quantity. However, even if they are eaten their toxicity is so low that they are unlikely to produce any ill effects. Indeed there are reports of PDMS preparations being used to alleviate digestive disorders.[24]

The work of the Dow group on the health and environmental effects of silicone materials has been at the same time both detailed and wide ranging; no other group has carried out such exhaustive studies. Other work, although of a more fragmented nature, has generally served to confirm the Dow findings. Magg and Alzieu[25] found that seawater saturated with PDMS fluid had no toxic activity towards phytoplankton, molluscs, crustaceans and fish. Mann and his co-workers,[26] in their study on the effect of silicone oils on rainbow trout, observed that fish given dry food soaked with silicone oil (190 mg per day) for 28 days showed no histopathological changes or any difference in behaviour or growth over the controls. Silicone oils were included in a toxicological assessment[27] of heat transfer fluids proposed for use in solar energy applications. The oils were evaluated for acute and oral toxicity in the rat, dermal and ocular irritation in the rabbit and mutagenic potential in the Ames test. The oral 24-hour LD$_{50}$ for silicone oils was found to be 24 g kg^{-1}. None of the fluids was mutagenic as measured by the Ames test, or more than mildly irritating to rabbit skin. Fluids heated to high temperature to simulate working conditions showed no significant mutagenic activity at doses ranging from 1 to 10 000 mg per culture plate. Among 300 materials evaluated for the toxicity of their pyrolysis gases,[28] the silicone materials which included two rubbers, an elastomer, a foam, a transformer liquid and six resins, were the least toxic.

There are only a few reports of PDMS materials producing troublesome effects in living systems. Silicone oils have been found to hasten leaf senescence;[29] this may be a consequence of the anti-transpirant properties of silicones.[30] In a study carried out in South-West Africa, silicone oil pollution[31] was mooted as the possible cause of the falling-off in algal growth in sewage ponds. Since the algal material was used as fertilizer, efforts were made to determine the cause of the decline in algal growth. It was found that the silicone oil hydrolysed under ambient conditions to give siloxanols which were considered to be responsible for the deleterious effects. It is unlikely that the siloxanols themselves were the toxic agents since the toxicity of the methylsiloxanols is no greater than that of other PDMS species. It is, however, possible that in this very complex system the siloxanols reacted with some other reagents to give the toxic products.

Although the polydimethylsiloxanes, in general, and, in particular, the silicones used in personal care products have proved to be essentially inert, this is not absolutely true for all siloxanes. There is the very occasional siloxane which does have biological activity. Palazzolo and his co-workers[32] observed that an equilibrated copolymer of mixed cyclosiloxanes ($[((C_6H_5)(CH_3)SiO)_x((CH_3)_2SiO)_y]$ where $x + y = 3$ to 8), administered orally, produces androgen depressant activity in rats, rabbits and monkeys. This material also reduced fertility in female rats and rabbits. Similar effects were noted when the copolymer was applied topically to rabbits but not to monkeys and man.[33] The material is absorbed dermally by rabbits but not by primates. Subsequently the components of the polymer mixture were separated and identified and their bioactivity evaluated.[34,35] Activity was found only among arylmethyl substituted linear di- and trisiloxanes, cyclotri- and tetrasiloxanes. The cyclosiloxanes were in general more active than the linear derivatives and the cyclotetrasiloxanes more active than the cyclotrisiloxanes. Activity is also highly dependent upon stereochemistry, the 2,6-*cis*-diphenylhexamethyl-cyclotetrasiloxane (I) is about 100 times more active than the *trans* isomer (II) shown in Fig. 6.3.

The methylphenylsiloxanes are not the only organosilicon compounds with biological activity. In many instances silylated derivatives of compounds known to have drug activity have been prepared.[5,11–18] The silylated drugs can show activity comparable with that of the parent drug but are much more commonly less active than the parent compound, which is not surprising since the original compound had already been selected for use because it was the most active member of its class. In silylated drugs, the silyl group may replace an active hydrogen on oxygen or nitrogen or one of the carbons. When attached

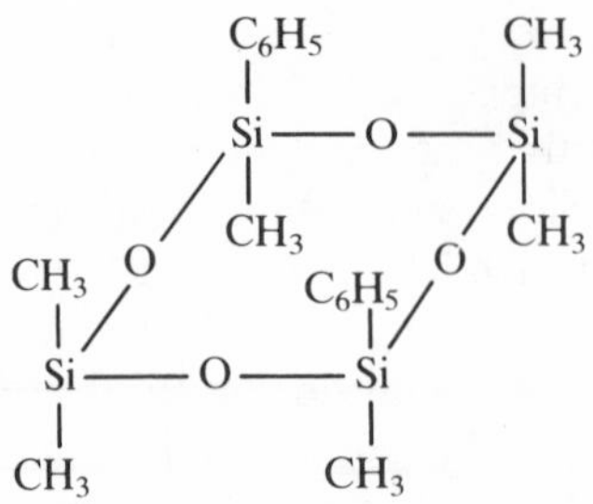

Fig. 6.3 2,6-Diphenylhexamethylcyclotetrasiloxane isomers

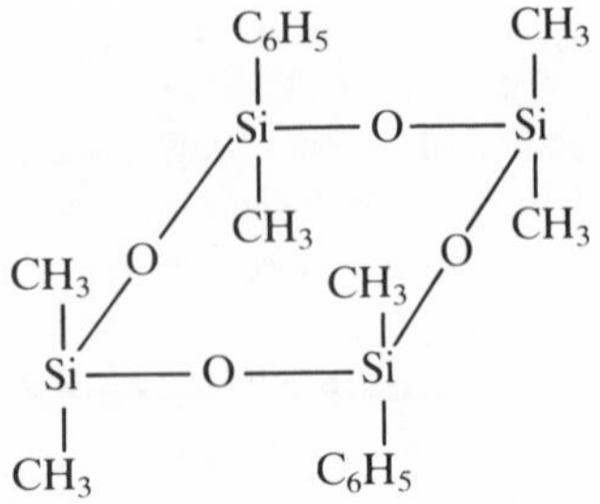

to oxygen or nitrogen the silyl group is readily hydrolysed and blocking of polar centres in a drug by silylation is a potential route to pro-drugs (compounds converted *in vivo* to active materials). Because polar centres are protected, the silylated pro-drug can cross lipophilic membranes and then release the less lipophilic parent drug on hydrolysis in the body fluids. Silylation of known bioactive compounds has produced derivatives with marginally different and occasionally useful properties but it has not yet yielded any really new exciting drugs.

Numerous organosilicon derivatives have been used to control plant growth. Most are chloroalkyl silicon derivatives and examples are given in Ref. 36. Their mode of operation is hydrolysis to generate the plant hormone ethylene. This is similar to the mechanism of plant growth regulation shown by the related organophosphorus compounds (Ch. 10). These materials are characterized by low toxicity, both acute and longer term, and rapid degradation.

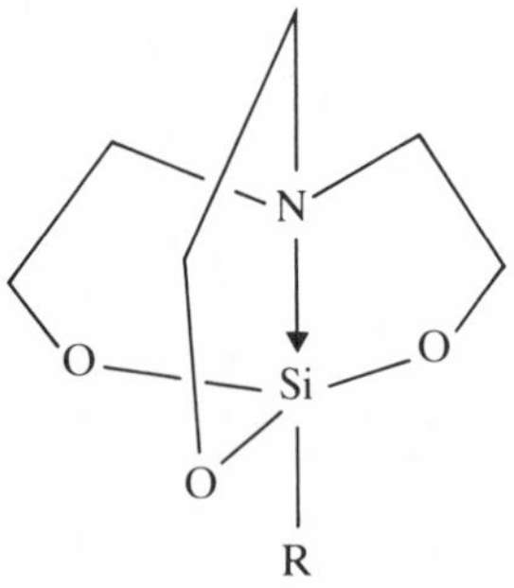

Fig. 6.4 I-substituted silatrane

From the bioactivity point of view the most interesting and exciting organic compounds are the silatranes which have no known carbon analogues. In 1964 Voronkov and his co-workers[37] reported that compounds of the general formula $RSi(OCH_2CH_2)_3N$ displayed very high biological activity.[38-39] These com-

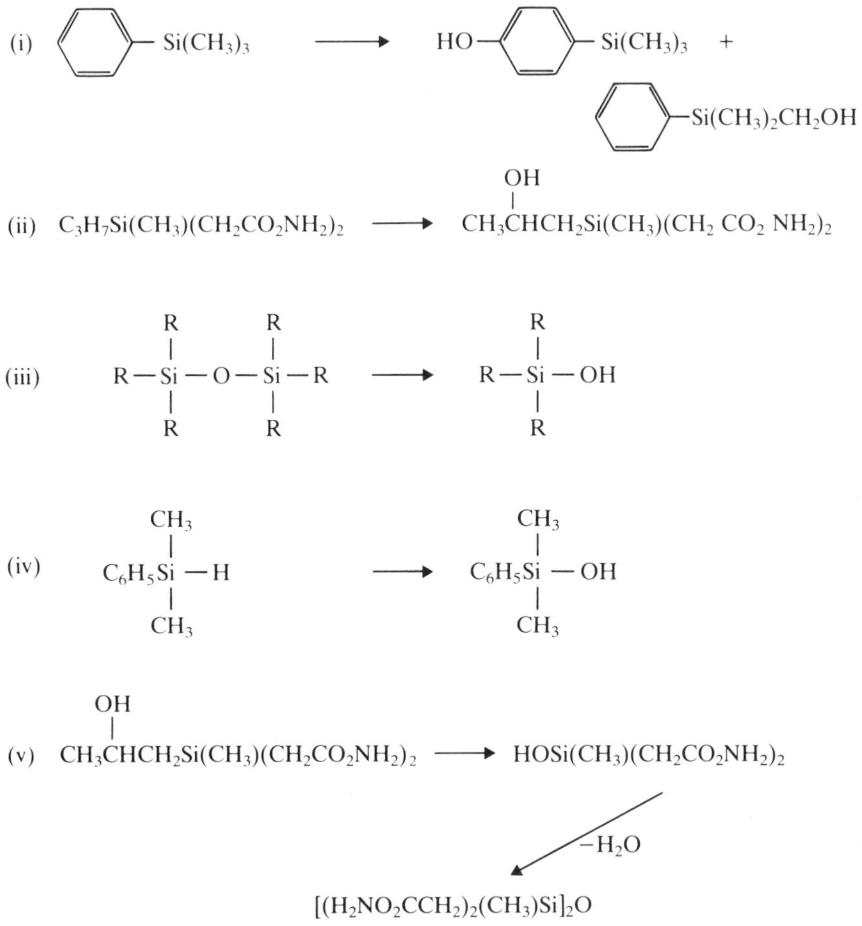

(i) ⟨benzene⟩—Si(CH₃)₃ ⟶ HO—⟨benzene⟩—Si(CH₃)₃ +

⟨benzene⟩—Si(CH₃)₂CH₂OH

(ii) C₃H₇Si(CH₃)(CH₂CO₂NH₂)₂ ⟶ CH₃ĊHCH₂Si(CH₃)(CH₂ CO₂ NH₂)₂

$$\text{(iii)} \quad R-\underset{\underset{R}{|}}{\overset{\overset{R}{|}}{Si}}-O-\underset{\underset{R}{|}}{\overset{\overset{R}{|}}{Si}}-R \quad \longrightarrow \quad R-\underset{\underset{R}{|}}{\overset{\overset{R}{|}}{Si}}-OH$$

$$\text{(iv)} \quad C_6H_5\underset{\underset{CH_3}{|}}{\overset{\overset{CH_3}{|}}{Si}}-H \quad \longrightarrow \quad C_6H_5\underset{\underset{CH_3}{|}}{\overset{\overset{CH_3}{|}}{Si}}-OH$$

(v) CH₃ĊHCH₂Si(CH₃)(CH₂CO₂NH₂)₂ ⟶ HOSi(CH₃)(CH₂CO₂NH₂)₂

⟍ −H₂O

[(H₂NO₂CCH₂)₂(CH₃)Si]₂O

Fig. 6.5 Hydroxylation reactions

pounds, named by Voronkow as silatranes, can best be described as caged structures. In these compounds, shown in Fig. 6.4, the silicon has some penta-coordinate and the nitrogen some tetracoordinate character.[40] The most prominent property of the silatranes is the exceptional activity exhibited by the 1-aryl derivatives. LD_{50} values for typical 1-substituted derivatives are shown in Table 6.1.

Some silatranes have greater toxicity to warm-blooded animals than cyanide or strychnine but are practically harmless to cold-blooded species. Commercial use has been made of this high and rather specific toxicity.[13,15,41] The 1-(p-chlorophenyl) silatrane has been marketed as a rodenticide. It has the advantage of high primary toxicity which declines rapidly as the material is hydrolysed within the body of the animal, so the bodies are then harmless to other animals such as domestic pets. The silatranes have the further advantage

Table 6.1   Toxicity* of 1-substituted silatranes $RSi(OCH_2CH_2)_3N$

R	$LD_{50}(mg\ kg^{-1})$	R	$LD_{50}(mg\ kg^{-1})$
$4\text{-}CH_3C_6H_4$	0.15	Cl_2CH	600
C_6H_5	0.33	$ClCH_2$	2800
$4\text{-}ClC_6H_4$	1.7	$C_6H_5CH_2CH_2$	3000
$BrCH_2CH_2CH_2$	5.0	$CH_2{=}CH$	3000
$4\text{-}CH_3OC_6H_4$	17	CH_3CH_2	3000
C_6H_{11}	150	CH_3	3000
$FCH_2CH_2CH_2$	223		

* Taken from Ref. 16 with permission

that they are not absorbed dermally by human beings. As shown in Table 6.1 the toxicity of the silatranes varies widely, being determined mainly by the nature of the 1-substituent. While 1-arylsilatranes are highly toxic the 1-alkyl, 1-vinyl, 1-ethynyl and 1-alkoxy compounds are practically non-toxic. The 1-chloromethyl and 1-ethoxy compounds, although non-toxic, are bioactive and are reported to have a remarkably large number of useful properties. They promote hair growth and wound healing, act as growth stimulants in young animals, and have antimicrobial, anticoagulant and antitumour activity.[13,15,39-43] The related silatrane 'mival' has been shown to be non-carcinogenic; it has an LD_{50} in the 2000 mg kg^{-1} range. Some silatranes have also been shown to affect plant growth and 'mival' has been marketed as a herbicide.

Only in the case of the siloxanes used in personal care products have detailed absorption and metabolic studies been carried out and very little is known of the metabolic fate of other types of organosilicon compounds, even the silatranes and those other organosilicon compounds which may be used as drugs. It is reported that the methylsilanes undergo hydroxylation *in vivo*. The phenylsilanes, specifically, do so in the *para* position (Fig. 6.5(i))[18,44] and the silaneprobamate in the propyl β-carbon position (Fig. 6.5(ii)).[45]

The hydroxylation reactions at carbon are common to non-silicon- and silicon-containing compounds (e.g. see Ch. 3). Metabolic breakdown involving attack at silicon also occurs.[18] Ingested siloxanes are broken down *in vivo* to silanol,[20] the silanes to silanols[44] (Fig. 6.5(iii) and-(iv)) and the hydroxylated silaneprobamate to a disiloxane (Fig. 6.5(v)).[45]

The *in vivo* reactions at silicon could be simple chemical reactions rather than enzyme-promoted processes. The formation of silanols from siloxanes occurs readily in vitro and is catalysed by both acids and bases (Fig. 6.6(i)). The *in vitro* hydrolysis of the silanes is usually conducted in basic solution but it may be acid catalysed (Fig. 6.6(ii)). The cleavage of the β-hydroxysilanes is an acid-catalysed reaction (Fig. 6.6(iii)). The *in vivo* versions of all these reactions could have taken place in the highly acidic environment of the stomach.

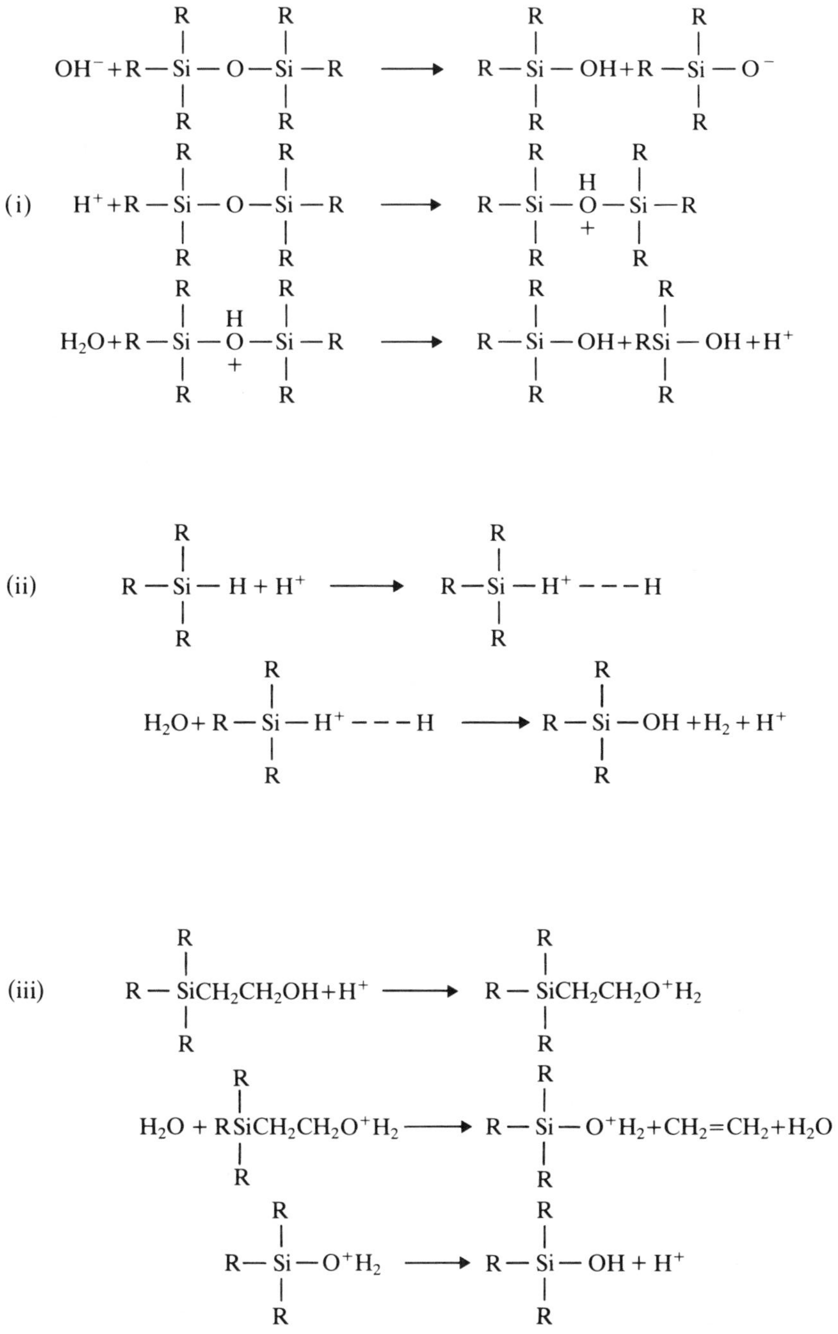

(i)

$$OH^- + R-\underset{\underset{R}{|}}{\overset{\overset{R}{|}}{Si}}-O-\underset{\underset{R}{|}}{\overset{\overset{R}{|}}{Si}}-R \longrightarrow R-\underset{\underset{R}{|}}{\overset{\overset{R}{|}}{Si}}-OH + R-\underset{\underset{R}{|}}{\overset{\overset{R}{|}}{Si}}-O^-$$

$$H^+ + R-\underset{\underset{R}{|}}{\overset{\overset{R}{|}}{Si}}-O-\underset{\underset{R}{|}}{\overset{\overset{R}{|}}{Si}}-R \longrightarrow R-\underset{\underset{R}{|}}{\overset{\overset{R}{|}}{Si}}-\underset{+}{\overset{H}{O}}-\underset{\underset{R}{|}}{\overset{\overset{R}{|}}{Si}}-R$$

$$H_2O + R-\underset{\underset{R}{|}}{\overset{\overset{R}{|}}{Si}}-\underset{+}{\overset{H}{O}}-\underset{\underset{R}{|}}{\overset{\overset{R}{|}}{Si}}-R \longrightarrow R-\underset{\underset{R}{|}}{\overset{\overset{R}{|}}{Si}}-OH + RSi-\underset{\underset{R}{|}}{}OH + H^+$$

(ii)

$$R-\underset{\underset{R}{|}}{\overset{\overset{R}{|}}{Si}}-H + H^+ \longrightarrow R-\underset{\underset{R}{|}}{\overset{\overset{R}{|}}{Si}}-H^+ ---H$$

$$H_2O + R-\underset{\underset{R}{|}}{\overset{\overset{R}{|}}{Si}}-H^+ ---H \longrightarrow R-\underset{\underset{R}{|}}{\overset{\overset{R}{|}}{Si}}-OH + H_2 + H^+$$

(iii)

$$R-\underset{\underset{R}{|}}{\overset{\overset{R}{|}}{Si}}CH_2CH_2OH + H^+ \longrightarrow R-\underset{\underset{R}{|}}{\overset{\overset{R}{|}}{Si}}CH_2CH_2O^+H_2$$

$$H_2O + R\underset{\underset{R}{|}}{\overset{\overset{R}{|}}{Si}}CH_2CH_2O^+H_2 \longrightarrow R-\underset{\underset{R}{|}}{\overset{\overset{R}{|}}{Si}}-O^+H_2 + CH_2=CH_2 + H_2O$$

$$R-\underset{\underset{R}{|}}{\overset{\overset{R}{|}}{Si}}-O^+H_2 \longrightarrow R-\underset{\underset{R}{|}}{\overset{\overset{R}{|}}{Si}}-OH + H^+$$

Fig. 6.6 Hydroxylation of siloxanes

The only *in vivo* reaction at silicon so far reported which may be enzyme catalysed is the direct transfer of methyl groups from silicon without prior oxidation.[46] Although analogous chemical processes exist, methyl is transferred from silicon to other metal and metalloidal species *in vitro* (equations 6.1 and 6.2).[47,48]

$$((CH_3)_2SiO)_n + GaCl_3 \rightarrow CH_3CaCl_2 + (CH_3SiClO)_n \qquad [6.1]$$

$$(C_2H_5)_4Si + HgCl_2 \rightarrow C_2H_5HgCl + (C_2H_5)_3SiCl \qquad [6.2]$$

Depending upon the nature of the acceptor there is the possibility of enzyme intervention.

Within the toxicity test results discussed above, it should be noted that some of the high polymerelastomeric organosilicon derivatives have recently been shown to give positive results in tests for mutagenicity and teratogenicity and have also been shown to have gonadotrophic and other properties towards some animals and in *in vitro* tests against some bacteria. Further details are given in Chapter 9, section 9.9.

6.3   Distribution of organosilicon compounds in natural water systems

Pellenbarg has measured the distribution and concentration of organosilicon compounds at a number of sites off the east coast of the USA.[49–51] It was assumed that silicones, being in many ways similar to hydrocarbons, would accumulate at phase boundaries and initial studies were concentrated in these areas. Samples[49] of solid, sludge and aqueous effluents, taken from a sewage works which processes much of the waste water from the Potomac River watershed, contained on average 36 μg g^{-1}, 96 μg g^{-1} and 5 μg dm^{-3} of organosilicon respectively. Surface sediment from the river itself contained on average 1.4 μg g^{-1}. The sediments with the highest silicon content were those close to the outlet from the waste treatment works and concentrations fell quite rapidly on moving downstream. The sediment of Delaware Bay, which is further from a source of anthropogenic input, displayed an average silicone content of about half that of the Potomac sediment. The bay sediment with the highest organosilicon levels was taken from an area of rapid sediment accumulation downstream from a tanker lightering depot. Core samples removed from Delaware Bay showed some interesting trends (Table 6.2).

Organosilicon was not detectable in the sandy layer from Core One (Table 6.2) and although the lower layers were apparently homogeneous, the silicone content decreased with depth. In contrast, in the second sample taken from a different area of the bay the deepest layer had the highest silicon content. Silicone levels in the surface waters of the bay were extremely low and very similar for both contaminated and non-contaminated areas. One sample, taken close to a marina which could serve as a source of silicone, contained

Table 6.2   Silicone in sediment cores from Delaware Bay*

	Depth (cm)	Organic Silicon (μg g^{-1})	Characteristics
Core One	3–8	0.39	distinct petroleum odour
	12–14	0.0	sandy layer
	30–33	0.45	grey-black gelatinous
	60–63	0.0	grey-black gelatinous
Core Two	4–6	0.12	tan brown, very fluid, small particle size
	15–17	0.0	gelatinous, no petroleum odour
	56–58	1.13	black, fibrous with plant fragments
	80–82	1.20	black, distinct petroleum odour

* Reprinted with permission from Pellenbarg R, 1979 *Environ Sci Technol* 13: 565–9. Copyright 1979 American Chemical Society

about 30 μg dm^{-3}; another, taken away from any source of contamination, contained about 34 μg dm^{-3}. In Chesapeake Bay,[50] a large open waterway with some anthropogenic input, organic silicon in the sediment ranged from 0.2 μg g^{-1} to 36 μg g^{-1}. In New York Bight,[51] a relatively small waterway which receives much of the municipal waste of New York City, the organosilicon in the sediment ranged from 50 μg g^{-1} to below detection limits. From the movement and distribution of silicones in these various waters and sediments it was concluded that silicone derivatives are affected by the same physico-chemical processes as the other components of the sediment system and are rapidly removed from the water column by flocculation–precipitation processes and tend to accumulate in areas of high sediment deposition. Similar observations are made in the case of mercury deposition in water–sediment systems (Ch. 2).

Since organic silicon is so rapidly transported from the water column to sediment, and since only minute amounts of silicone were detectable at the air/sea interface, even in the relatively highly contaminated Delaware Bay area,[49] it would not be expected that silicones would be detectable in areas remote from anthropogenic input. It is somewhat surprising, therefore, that PDMS has been cited as an important component of Arctic aerosols[52] and that the burning of organofluorine and organochlorine wastes at sea gives organosilicon fluoride products.[53] Considerable care must be exercised in the determination of organosilicon in environmental samples. There is no doubt

that PDMS concentrations will normally be higher in the laboratory environment than in the environment generally. Silicone materials are used in GC
column packings, heating baths, syringes, lubricants, hand lotions,
deodorants, etc. When measuring organic silicon at the $\mu g\ g^{-1}$ and ng g^{-1}
levels, particular attention must be paid to blank and control values, since the
observed silicone may be merely an artefact of the analytical procedure.
Hence all observations of low levels of organic silicon in unlikely places must
be viewed with caution.

6.4   Degradation of organosilicon compounds

The silicones are generally highly resistant to biodegradation. No biodegradation was observed[19] when a 15 per cent emulsion of a ^{14}C-labelled PDMS fluid
(300 cSt) was treated with activated sewage sludge over 70 days. In the same
time and under the same conditions ^{14}C[octadecane] was almost completely
converted into ^{14}CO$_2$ and polar products. In a general study of the activated
sludge degradability of organic substances, in which the biodegradability of
organic materials in water by activated sludge was evaluated from total oxygen demand (TOD) and chemical oxygen demand (COD) measurements, the
silicone surfactants proved to be among the most resistant materials to
degradation.[54] A comparison[55] of the biodegradability of detergents revealed
the following reactivity order: alkylbenzene sulphonates > ethoxyfatty acids
> silicone oils.

   Although not biodegradable, PDMS fluids can undergo chemical degradation under environmental conditions. Bush and Ingebriston[56] found that
PDMS derivatives in intimate contact with clays and clay-containing soils
underwent rearrangement and hydrolysis to give lower molecular weight
linear and cyclic products. The catalytic activity was dependent upon the type
of soil, the moisture level and, as would be expected for a heterogeneous reaction, surface area. Using as model compounds octamethylcyclotetrasiloxane
$(((CH_3)_2SiO)_4)$ and dodecamethylpentasiloxane $((CH_3)_{12}Si_5O_4)$ it was revealed
that soils high in humus or sand showed negligible rearrangement activity
whereas soils with a reasonable clay content were active towards both the
cyclic and linear materials but generally more active towards the linear than
the cyclic compounds. It was found that this redistribution reaction is affected
not only by the amount of clay present but also by the type. Kaolinite attacks
preferentially at dimethylsiloxy chain sites giving rise to volatile products
mainly hexamethylcyclotrisiloxane $(((CH_3)_2SiO)_3)$. In contrast, montmorillonite attacks at the trimethylsiloxy end sites giving oligomeric siloxanols
as the major products, together with some volatile material, mainly hexamethyldisiloxane as a minor product. Thus a kaolinite-based soil will promote
the formation of predominantly volatile rearrangement products whereas a
montmorillonite-based soil will give rise in the main to water-soluble
hydrolysis products. A moisture content as low as 1 per cent substantially

altered the clay's activity; montmorillonite was more affected by moisture than kaolinite. The effect of moisture is reversible, the catalytic activity of washed clays is restored on drying. Kinetic studies indicated that the half-life of PDMS fluids on soil can range from several minutes to weeks or more depending upon the soil type, hydration level, area of contact, etc. The dramatic effect of moisture on activity is obviously a negative factor from an environmental point of view. However, using labelled derivatives, it was found that PDMS compounds could move up and down in soils and that PDMS located in a wet area could migrate to a dry area and hence be degraded more rapidly.

The volatile and water-soluble products obtained from the soil degradation can be broken down further under the type of conditions which would prevail in normal atmospheric and aquatic systems.[23] When exposed to simulated atmospheric conditions volatile methylsilicon compounds undergo oxidative degradation at a rate comparable with that of ordinary hydrocarbons. Dimethylsilanediol ($(CH_3)_2Si(OH)_2$), which would formally be expected to be the ultimate hydrolysis product from PDMS derivatives, can be demethylated on irradiation with UV light in aqueous solution. Demethylation can also be achieved using natural ground level sunlight, and is enhanced by the addition of as little as 10 ng g^{-1} of sodium nitrate and nitrite. The nitrogen oxides undergo photochemical degradation to yield atomic oxygen and, in the presence of water, hydroxyl radicals. It is this latter highly reactive species which is believed to be responsible for the cleavage of the Si—C bond.[57] Further proof that the bond can be broken photolytically by natural levels of irradiation was obtained from a study of diatoms which require silicic acid for their skeletons. It was found that diatoms could be cultured in a medium in which the only source of silicic acid was that produced by sun lamp irradiation of dimethylsilanediol. Neither dimethylsilanediol itself nor methylsilanetriol could serve as a source of silicon for the diatoms affirming previous observations that the Si—C bond is not biodegradable.[19]

In the environment the silicone fluids are distributed over vast surfaces and are exposed to the action of catalytic clays and irradiation by sunlight. So these model studies suggest that the silicones are not going to survive in perpetuity. Time and exposure to soils and sunlight will convert them finally to carbon dioxide, water and silicic acid.

6.5 Silicon as a trace element

Although silicon is obviously an essential element for the primitive species which include it in their structure, there has always been some doubt whether higher organisms have a silicon requirement. More than a century ago Sachs,[58] in one of the earliest articles on plant nutrition, posed the questions 'whether silicic acid is an indispensable substance for those plants that contain silica, whether it takes part in nutritional processes, and what is the relationship that exists between the uptake of silicic acid and the life of the plant?' and then concluded in the light of evidence available to him at that time (from studies on *Zea* maize) that 'silicic acid is insignificant for the nutritional process of the maize plant'.

The ubiquity of silicon, its presence in atmospheric dust, water supplies, glass apparatus, etc. makes it difficult to set up silicon-free culture conditions to determine the effect of silicon deficiency on living material.[59] Early workers used paraffin coated glass apparatus[60] and asphalt painted iron containers.[61] The development of plastic apparatus simplified silicon deficiency studies. Several groups working independently[62-65] have found that species of rice, grain and equisetum grown in the absence of silicon develop necrosis of their leaves and show reduced growth, fertility and grain yield. In contrast another group found that the general appearance of rice plants grown with and without silicon was essentially the same.[66] Neither did the addition of silica affect the yields of various fruit and vegetable species including tomato, radish, onion and cabbage cultured in plastic containers.[67] On balance the evidence so far suggests that silicon is not required by most higher plant species but it is probably an essential element for plants such as rice and other grains which usually contain high concentrations of silicon when grown under normal conditions.

Although there has always been some lingering doubt whether higher plants have a silicon requirement, it has been assumed that for higher animals silicon is not an essential element. In an extensive review[12] of the biological properties of silicon published in 1967 it was stated that 'that although traces of silica are found in all animal tissues, there is no evidence that there is any biological need for silicon in higher animals'. Subsequently it was reported by Schwarz that rats kept in all-plastic isolators and fed on a silicon-deficient amino acid diet displayed reduced growth and developed bone deformities. Controls given dietary supplement of silicates showed none of these symptoms.[68-70] Similar effects were noted in chicks fed on silicon-deficient diet.[71,72] Studies using electron microprobe analysis[73] have revealed the localization of silicon in active calcification sites in young bone. In the primary stages of calcification, when the calcium content of the preosseus tissue is very low, there is a direct relationship between silicon and calcium content. In consequence of these findings it has been suggested that the silicon and calcium are associated in the early stages of bone formation.[74] *In vivo* experiments[75] with rats have also shown this relationship between calcium and silicon in bone formation. It was demonstrated that dietary silicon increases the rate of mineralization; this was particularly apparent on a low calcium diet. There also appears to be some relationship between silicon, magnesium and fluorine in bone formation in the chick.[76] Dietary deficiency studies have demonstrated that silicon also plays some role in connective tissue metabolism and structure. It has been suggested that silicon serves as a cross-linking agent by forming links between individual polysaccharide chains and joining polysaccharide chains to proteins.

In this way silicon helps in the development of the architecture of the fibrous elements of connective tissue and provides strength and elasticity to its structure.[77] This relationship of silicon with the elasticity of biological structures is a theme which runs through a number of studies, for example the decreased elasticity of the arterial walls in atherosclerosis has been related to the decreased silicon content of the arterial walls.[78]

6.6   Transport, bioconcentration and transalkylation of silicon in the environment

Inorganic silicon is ubiquitous in the environment. It is transferred from non-living to living systems by both passive and active processes. In plants with a low silicon content it is distributed rather uniformly between roots and shoots whereas in plants in which the overall silicon content is high the silicon is concentrated mostly in the aerial parts.[59,79] Within the plant tissues the silicon is concentrated in the epidermal cells and the cell walls. In higher animals the element is concentrated in the ground matrix and structural elements of connective tissue, cartilage skin and bone and appears to be associated specifically with glycosoaminoglycans and also collagens.[3,77]

As far as organosilicon compounds are concerned the PDMS fluids are the only compounds which could be of environmental importance since they are the only organosilicon compounds produced and released into the environment in any quantity. Compounds such as the arylsilatranes, which could be a potential hazard because of their excessively high toxicity, are normally only laboratory curiosities and their main commercial use, so far, as rodenticides was based on their emphemeracy.[13,15] All other organosilicon compounds known to have biological activity are, like the silatranes, hydrolytically unstable and/or produced in such low quantities that they could not possibly have any significant adverse ecological effects.

The environmental fate of the PDMS fluid will depend upon its chemical and physical properties. The more volatile silicones will be dispersed in the atmosphere where they can undergo photolytic oxidation[23] to give water-soluble products which can then be precipitated into soil and water systems with rainfall. At this point they then undergo further decomposition and the final products will be carbon dioxide, water and silicic acid, all of which are harmless.

The water-soluble siloxanols of the types used in personal care products, or produced from high molecular weight silicones, on clay surfaces can (like the materials dispersed in the atmosphere) undergo photolytic decomposition. The products can then partition into air and water depending upon their volatility until oxidation is completed.

The high molecular weight silicones are much less readily photolysed than their low molecular weight counterparts. The high molecular weight materials, when they enter waste water systems, will be absorbed on to other suspended material and end up in the sludge fraction at the sewage works. Their final destination will depend upon the fate of the sludge. Some will be consigned to landfills, others used as fertilizers, and others dumped at sea. The materials assigned to the land could in time undergo clay-catalysed decomposition[56] to give lower molecular weight products which might then break down further. According to Pellenbarg[51] the material received by the sea will be rapidly removed from the water column and precipitated on sediment. Thus silicones which escape degradation tend to end up imprisoned on sediments. Because of the levels at which they occur and their very low toxicity, they themselves are unlikely to produce any harmful effects and in these circumstances the only processes and reactions which might give rise to adverse ecological effects are bioconcentration and transalkylation.

Bioconcentration is a natural phenomenon by which materials present in low concentrations can build up to levels of significant toxicity within an organism. The ability of organic substances to bioconcentrate in aquatic media has been found to be dependent upon the partitioning of molecules between lipid and aqueous phases. It has been found for some types of organic compounds, in particular low molecular weight chlorinated hydrocarbons,[80] that bioconcentration factors can be estimated by measuring the related octanol/water partition coefficients (log P). This correlation does not hold for organosilicon compounds in general, mainly because the chemical and physical properties of this broad class of compounds are too disparate for such generalizations to be applicable. Although PDMS derivatives show very large log P values they do not exhibit significant bioconcentration.[81] The correlation does not hold in this case because, in order to reach the lipid phase, molecules must be able to pass through biological membranes. The few commercial silicones which are sufficiently small to pass through the cell membranes are generally so volatile that they will end up in the atmosphere rather than in the aquatic environment. The role of partitioning is further discussed in Chapter 1 (sect. 1.6.5)

Since both siloxanes and heavy metals are adsorbed into sediment it is possible that organic groups could be transferred to elements such as lead, mercury and arsenic. The organic derivatives are generally much more toxic than their inorganic counterparts and have in fact been the agents responsible for a number of major environmental catastrophes (Chs 2, 4, 5).[82,83]

It has been known for a very long time that organosilicon compounds can transfer alkyl and aryl groups to a variety of metal derivatives. Typical examples are shown in equations 6.3−6.5.[47,48,84−86]

$$CH_3C_6H_4Si(CH_3)_3 + HgCl_2 \rightarrow CH_3C_6H_4HgCl + (CH_3)_3SiCl \quad [6.3]$$

$$3(NH_4)_2(CH_2{=}CHSiF_5) + SbF_3 \rightarrow (CH_2{=}CH)_3Sb + 3(NH_4)_2SiF_6 \quad [6.4]$$

$$(NH_4)_2(CH_3SiF_5) + Pb(OAc)_4 \rightarrow CH_3PbF_3 + (NH_4)_2Si(OAc)_4F_2{}^* \quad [6.5]$$

* initial product

Recently it has been demonstrated that the species $(CH_3)_3Si(CH_2)_3SO_3^-$ and $(CH_3)_3Si(CH_2)_2CO_2^-$, frequently used as NMR standards in aqueous solution, react with mercuric salts to give methylmercury derivatives (equation 6.6).[87]

$$(CH_3)_3Si(CH_2)_3SO_3NA + Hg(NO_3)_2 \rightarrow ((CH_3)_2Si(CH_2)_3SO_3Na)NO_3 +$$

$$CH_3HgNO_3 \qquad\qquad [6.6]$$

Similar reactions have been observed with thallium(III) acetate, lead(IV) acetate and perhaps potassium hexachloroplatinate.[88] Mercuric nitrate, the most reactive mercurial electrophile, reacts with hexamethyldisiloxane to give products arising from methyl transfer,[89,90] but does not react with cyclodimethylsiloxanes – even after several months contact.

Only occasionally are high levels of organometallic compounds found in sediments. This is true even for organoleads which might be expected to be present in relatively large concentrations since they are continuously emitted from motor vehicle exhausts.[91] The organometal derivatives of heavy metals are photolysed or otherwise transformed in the environment[92,93] and their lifetime in natural systems is generally very short. One would predict that the yields of heavy metal derivatives formed by alkyl transfer from silicon would be very low and their rate of destruction relatively rapid. It is, therefore, unlikely that the silicon transalkylation reactions would be environmentally significant in normal circumstances. The only situation where there might be cause for concern would be in a spill where large concentrations of siloxane came into contact with large concentrations of heavy metal derivatives.

In conclusion, there is no doubt that PDMS derivatives are being released into the environment in relatively large quantities overall. They are non-biodegradable and only broken down chemically in rather special circumstances. In the light of their persistency it would be anticipated that they would accumulate in the environment and could become a cause for concern. This is probably not the case, for although lighter than water, they are not accumulating on surface waters or in the water column but are rapidly adsorbed on to sediments. They are detectable in river and coastal sediments but are generally not transported any distance from areas of anthropogenic input. The occurrence of siloxanes in sediments is unlikely to present any threat to benthic species. The PDMS derivatives have been shown to be virtually non-toxic and not to undergo bioconcentration and it is improbable that they will produce significant concentrations of more toxic species by reactions such as transalkylation.

6.7   Analysis of environmentally important silicon species

6.7.1   Analysis of silicon in organosilicon compounds

This subject has been reviewed by Belcher,[94] Sykes[95] and Crompton.[96]

Silicon in organosilicon compounds is determined quantitatively, usually by decomposing the organosilicon compound then measuring the silica or silicate produced. Fusion with sodium peroxide in a Parr-type bomb has generally been found to be the most rapid and cleanest method of decomposition for a wide variety of organosilicon compounds,[97-99] and is the technique most widely used. The silicate produced by fusion can be converted to molybdosilicic acid by treatment with ammonium molybdate, and determined colourimetrically[100] by reduction to molybdenum blue or volumetrically[97] or gravimetrically[98] as the oxine or quinoline salt of molybdosilicic acid. Fusion with alkali has also been used for the decomposition of organsilicon materials.[101] This method gave good results with high molecular weight compounds but rather less than 50 per cent recovery of silicon was achieved with volatile low molecular weight silicon derivatives unless an alcoholic alkali pretreatment was used. This pretreatment converted volatile siloxanes to involatile silanolates.

Various methods using mixtures of strong acids for the decomposition have been reported.[102-104] Commonly the organosilicon compound is digested in mixtures of concentrated nitric and sulphuric acids in a platinum crucible. The particular strength and blend of acids used depends upon the nature of the organosilicon compound. With low molecular weight alkylsilicon compounds certain precautions such as cooling the oxidizing acids in solid carbon dioxide or insertion of the material under the acid surface have to be observed to minimize volatilization. Following the digestion stage the crucible contents are heated slowly and then ignited and the silica produced determined volumetrically or gravimetrically. Kreshkow *et al.*[104] used ethanolic hydrofluoric acid to convert the silicon to hydrofluorosilicic acid. The hydrofluorosilicic acid can be estimated by a variety of methods.

The determination of silicon in organosilicon compounds using emission spectroscopy[105] and X-ray fluorescence[106] has been reported but such physical methods have not been widely used.

6.7.2   Analysis of organosilicon compounds

Organosilicon compounds may be determined qualitatively and quantitatively using a variety of chemical and physical procedures.[10,96,107,108] The particular method used is dependent upon the nature of the organosilicon compound. The silicone content of a range of natural sediment and water systems has recently been measured by Pellenbarg.[49] The silicones were extracted with organic solvents and the extracts analysed by nitrous oxide − acetylene flame AA. Very recently a highly sensitive method for the estimation of waterborne organosilicon compounds has been developed by the Dow Corning group.[109] Water-soluble organosilicon compounds are extracted and derivatized with hexamethyldisiloxane. The hexamethyldisiloxane-soluble derivatives are analysed by GC. Water-insoluble materials are digested with acid prior to extraction and derivatization. It is reported that the method is accurate in the mg dm^{-3} range and has the potential to be extended to measuring silicone levels in the low μg dm^{-3} range.

6.8 Commercial biologically active organosilicon compounds

$RSi(OCH_2CH_2)_3N$ Silatrane system (see Fig. 6.4)
$R = CH_2Cl$ ('Mival', herbicide); $R = $ p-ClC_6H_4 (rodenticide)

$ClCH_2CH_2Si(OCH_2CH_2OCH_3)_3$ 'Alsol' growth regulator

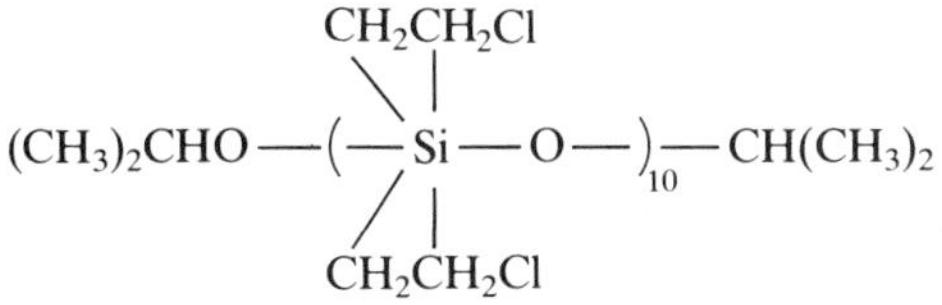

$$(CH_3)_2CHO-\left(-\overset{\overset{\displaystyle CH_2CH_2Cl}{|}}{\underset{\underset{\displaystyle CH_2CH_2Cl}{|}}{Si}}-O-\right)_{10}-CH(CH_3)_2$$

$ClCH_2CH_2Si(CH_3)(OR)_2$

Fig. 6.7 Biologically active organosilicon compounds

So far only the biologically active organosilicon compounds of the type shown in Fig. 6.7 have had sufficient potential to be considered for commercial use. They are shown here to exemplify structural features relevant to biological activity in this field. It is possible that as these organosilicon compounds have biological activity, other organosilicons having a greater environmental significance will be developed in the future.

References

1. Vernadskii V 1934 *Outline of Geochemistry*. ONT, Moscow; Leningrad, pp 89, 90, 128–31
2. Engel W 1958 *Planta* **41**: 358–90
3. Carlisle E M 1974 *Fed Proc* **33**: 1958–66
4. Schwarz K 1973 *Proc Natl Acad Sci US* **70**: 1608–12

5. Garson L, Kirchner L K 1971 *J Pharm Sci* **60**: 1113–27
6. Holzapul L 1942 *Naturwiss* **30**: 185–6; ibid **31**: 386; ibid **34**: 189
7. Holt P, Yates D 1953 *Biochem J* **54**: 300–5
8. Weiss A, Herzog A 1977 In *Biochemistry of Silicon and Related Problems* (eds Bendz G, Lindquist J). Plenum Press, New York; London, pp 109–25
9. Heinen W 1965 *Arch Biochem and Biophys* **110**: 137–49
10. Eaborn C 1960 *Organosilicon Compounds*. Butterworth, London
11. Voronkov M G, Lukevics E 1969 *Russ Chem Rev* **38**: 975–86
12. Fessenden R J, Fessenden J S 1967 *Adv Drug Res* **4**: 95–131
13. Voronkov M G, Zelchan G I, Lukevics E 1971 *Silicon and Life*. Zinatne, Riga
14. Voronkov M G 1973 *Chem Brit* **9**: 411–5
15. Voronkov M G, Lukevics E, Wannagat U, Levier R R, Chandler M L, Wender S R, Strinberg B 1977 In *Biochemistry of Silicon and Related Problems* (eds Bendz G, Lindquist I). Plenum Press, New York; London
16. Voronkov M G 1979 *Top Curr Chem* **84**: 77–135
17. Tacke R, Wannagat U 1979 *Top Curr Chem* **84**: 1–75
18. Fessenden R J, Fessenden J S 1981 *Adv Organometal Chem* **18**: 275–99
19. Hobbs E J, Keplinger M L, Calandra J C 1975 *Environ Res* **10**: 1397–406
20. Frye C L 1983 *Soap Cosmetic Chem Specialities*, August issue
21. Calandra J C, Keplinger M L, Hobbs E J, Tyler L J 1976 *Division of Polymer Chem Amer Chem Soc Polymer Reprints* **17**: 12–15
22. Kennedy G L, Keplinger M L, Calandra J C, Hobbs E J 1976 *J Toxicol and Environ Health* **1**: 909–20
23. Frye C L 1980 *J Organometal Chem Library* **9**: 253–60
24. Sipar – Pharm 1981 *Belg B E 889*, **453**: 1982 (*Chem Abstr* **96**: 168–759)
25. Magg P, Alzieu C 1977 *Sci Perche* **269**: 1–3
26. Mann H, Ollenschlaeger B 1977 *Fisch Uniwelt* **3**: 19–22
27. Wang S Y, Smith D M 1980 *Energy Res Abstr* **5**: 14377
28. Hilado H J, Casey C J, Chastensen D F, Lipowitz J 1978 *J Combust Toxicol* **5**: 130–40
29. Debata A, Murty K S 1981 *Oryza* **18**: 177–9
30. Anderson J E, Kreith F 1978 *Plant Soil* **49**: 161–73
31. Van der Post D C 1978 *Water Pollut Control* **77**: 520–4
32. Palazzolo R J, McHard J A, Hobbs E J, Fancher O E, Calandra J C 1972 *Toxicol Appl Pharmacol* **21**: 15–28
33. Hobbs E J, Fancher O E, Calandra J C 1972 *Toxicol Appl Pharmacol* **21**: 45–54
34. Bennett D R, Garzinski S J, Le Beau J E 1972 *Toxicol Appl Pharmacol* **21**: 55–67
35. Le Vier R R, Jankowiak M E 1972 *Toxicol Appl Pharmacol* **21**: 80–8

36. Thayer J S 1984 *Organometallic Compounds and Living Organisms.* Academic Press, New York
37. Balthais J J, Voronkov M G, Zelchan G I 1964 *Izv Akad Neut Latv SSR*: 102–6; 1964 (*Chem Abstr* **61**: 9932)
38. Schwarz M, Beiter S, Kaplin S, Loeffler O, Martino M, Damle S 1970 *Ger Offen 2, 009, 864*; 1970 (*Chem Abstr* **73**: 120, 748)
39. Irkutsk Inst of Org Chem Siberian Dept Acad Sci USSR 1977 *British Patent* 1465, 455; 1977 (*Chem Abstr* **87**: 73, 369)
40. Voronkov M G, Platonova A T, D'yakov V M, Katrush K M, Kazakul A J, Kuznetsov I G, Mansurova L A 1977 *USSR Patent* 541, 473; 1977 (*Chem Abstr* **87**: 83, 547)
41. Katrush K M, Voronkov M G, D'yakov V M 1977 *Biol Atk Soedin Elem IV B Gruppy*: 162–9; 1978 (*Chem Abstr* **89**: 58, 833)
42. Simkovich B Z, Lukevics E, Zelchan G I, Zamaraeva T V, Mazurov V D 1977 Biokhimya **42**: 1128–9; 1977 (*Chem Abstr* **87**: 79, 039)
43. Shevchenko S G, Platonova A T, Sadakh V V, Voronkov M G 1977 *Biol Akt Soedin Elem IV B Gruppy*: 95–8; 1978 (*Chem Abstr* **89**: 70, 791)
44. Fessenden R J, Hartman R A 1970 *J Med Chem* **13**: 52–4
45. Fessenden R J, Ahlfors C 1967 *J Med Chem* **10**: 810–2
46. Hughes G S 1972 MS Thesis, University of Montana, Missoula Montana, USA
47. Schmidbaur H, Findeiss W 1964 *Angew Chem Intern Edn* **3**: 696
48. Manulkin Z M 1949 *J Gen Chem USSR* **18**: 299–305; 1947 (*Chem Abstr* **41**: 90)
49. Pellenbarg R 1979 *Environ Sci Technol* **13**: 565–9
50. Pellenbarg R 1979 *Marine Poll Bull* **10**: 267–9
51. Pellenbarg R 1982 *Marine Poll Bull* **13**: 427–9
52. Weschler C J 1981 *Atmos Environ* **15**: 1365–9
53. Compaan H, Van der Berg A A, Timner J M, Van Leenwen P 1982 Emission Measurements on Board the Incineration Ship *Vulcanus* Report No CL 81/108 Order No 30057
54. Matsui S, Murakami T, Sasaki T, Hirose Y, Iguma Y 1975 *Prog Water Technol* **7**: 645–59
55. Hellman H 1982 *Z Wasser Abwasser Forsch* **15**: 229–32; 1983 (Chem Abstr **98**: 21968)
56. Bush R R, Ingebriston D N 1979 *Environ Sci Technol* **13**: 676–9
57. Heicklen J 1976 *Atmospheric Chemistry.* Academic Press, London, p 282
58. Sachs J 1862 *Flora* **20**: 31–8, 49–55, 65–71
59. Lewin J C, Reimann B E F 1969 *Ann Rev Plant Physiol* **20**: 289–304
60. Wagner F 1940 *Phytopathol Z* **12**: 427–79
61. Raleigh G J 1945 *Soil Sci* **60**: 133–5
62. Yoshida S, Ohnishi Y, Kitagishi K 1962 *Soil Sci Plant Nutr (Tokyo)* **8**: No 3 15–21

63. Mitsui S, Takatoh H L1963 *Soil Sci Plant Nutr (Tokyo)* **9**: 7–11
64. Chen C, Lewin J 1969 *Can J Botany* **47**: 125–31
65. Vlamis J, Williams D E 1967 *Plant Soil* **27**: 131–40
66. Tanaka A, Park V D 1966 *Soil Sci Plant Nutr (Tokyo)* **12**: 23–8
67. Wooley J T 1957 *Plant Physiol* **32**: 317–21
68. Schwarz K, Milne D B 1972 *Nature* **239**: 333
69. Schwarz K 1974 *Fed Proc* **33**: 1748–57
70. Milne D B, Schwarz K, Sognnaes R 1972 *Fed Proc* **31**: 700
71. Carlisle E M 1972 *Science* **178**: 619–21
72. Carlisle E M 1972 *Fed Proc* **31**: 700
73. Carlisle E M 1969 *Fed Proc* **28**: 374
74. Carlisle E M 1970 *Science* **167**: 279–80
75. Carlisle E M 1970 *Fed Proc* **29**: 565
76. Carlisle E M 1971 *Fed Proc* **30**: 462
77. Schwarz K 1977. In *Biochemistry of Silicon and Related Problems* (eds Bendz G, Lindquist I). Plenum Press, New York; London, p 207
78. Loeper J, Fragry M 1977. In *Biochemistry of Silicon and Related Problems* (eds Bendz G and Lindquist I). Plenum Press, New York; London, p 281
79. Yoshida S, Ohnishi Y, Kitaqishi K 1962 *Soil Sci Plant Nutr (Tokyo)* **8**: No 1 30–5; ibid 36–41; ibid No 2 1–5
80. Neeley W B, Branson D R, Blan G E 1974 *Environ Sci Technol* **8**: 1113–15
81. Frye C L 1984 Private Communication
82. Challenger F 1945 *Chem Rev* **36**: 315–61
83. Chau Y K, Wong P T S 1978. In *Organometals and Organometalloids: Occurrence and Fate in the Environment* (eds Brinckmann F E, Bellama J M). ACS Symp Ser No 82, ACS Washington DC p 39
84. Manulkin Z 1946 *J Gen Chem USSR* **16**: 235–42
85. Muller R, Dathe C 1966 *Chem Ber* **99**: 1609–15
86. Eaborn C, Bott R 1968. In *Organometallic Compounds of the Group IV Elements* Vol 1, Pt 1 (ed McDiarmid A G). M Dekker, New York, p 436
87. De Simone R E 1972 *J Chem Soc Chem Commun* 780–1
88. Thayer J S 1978 In *Organometals and Organometalloids Occurrence and Fate in the Environment* (eds Brinckmann F E, Bellama J M) ACS Symp Ser No 82, ACS Washington DC, p 157
89. Thayer J S 1978 *Synth React Inorg Met Org Chem* **8**: 371–9
90. Nies J D 1978 PhD Thesis, University of Maryland, USA
91. Potter H R, Jarvie A W P, Markall R N 1977 *Water Pollut Contr* **76**: 123–8
92. Jarvie A W P, Markall R N, Potter H R 1981 *Environ Res* **25**: 241–9
93. Jarvie A W P, Whitmore A P, Markall R N, Potter H R 1983 *Environ Pollut Ser B* **6**: 69–79
94. Belcher R, Gibbons D, Sykes A 1952 *Mikrochem*: 76–103

95. Sykes A 1956 *Mikrochim Acta*: 1155–68
96. Crompton T R 1974 *Chemical Analysis of Organometallic Compounds* Vol 2. Academic Press, London
97. McHard J A, Servais P C, Clark H A 1948 *Anal Chem* **20**: 325–8
98. Wilson H 1949 *Analyst* **74**: 243–8
99. Jean M 1955 *Chim Anal* **38**: 37–49
100. Ringbom A, Ahlers P E, Siitonen S 1959 *Anal Chim Acta* **20**: 78–83
101. Wetters J H, Smith R C 1969 *Anal Chem* **41**: 379–81
102. Sir Z, Komers R 1956 *Coll Czech Chem Commun* **21**: 873–9
103. Smith A L, Brown L H, Tyler L J, Hunter M J 1957 *Ind Eng Chem* **49**: 1903–6
104. Kreshkow A P, Myshlyaeva L V 1963 *Zavod Lab* **29**: 924–6; 1963 *Chem Abstr* **59**: 121896
105. Radell J, Hunt P D 1958 *Anal Chem* **30**: 1280–1
106. Chan F L 1965. In *Proceedings of the Society for Analytical Chemistry Conference*, Nottingham. W Heffer and Sons Ltd, Cambridge pp 89–101
107. Bellamy L J 1958 *The Infrared Spectra of Complex Molecules* 2nd edn. Methuen, London
108. Ebsworth E A V 1968. In *Organometallic Compounds of the Group IV Elements* Vol 1, Pt 1. Marcel Dekker Inc, New York
109. Mahone L G, Garner P J, Buch R R, Lane T H, Tatera J F, Smith R C, Frye C L 1983 *Environ Toxicol Chem* **2**: 307–13

Chapter 7

Organic Group VI elements in the environment

7.1 Introduction

Microbial transformation of Group VI elements, notably selenium and tellurium, has been known since 1902 when Rosenheim[1] observed the generation of unpleasant odours upon growth of *Scopulariopsis brevicaulis* on sterilized breadcrumbs in the presence of inorganic compounds of selenium and tellurium. The gas evolved from the cultures with selenium compounds was later identified by Challenger and North[2] as dimethyl selenide. A few years later, the gas liberated by the mould in the tellurium experiments was identified as dimethyl telluride by Challenger.[3]

Organic forms of sulphur compounds constitute a large and diverse group of compounds which are closely associated with life. Organosulphur compounds are ubiquitous in cell compartments of life and the globe. Many of the organosulphur compounds are biogenic and their formation often involves biochemistry of living matter. The chemistries of sulphur and selenium, and their interactions with all forms of life are extremely complex. It is intended to limit the review to organo Group VI compounds of environmental interest and significance, and their occurrence and formation including toxicity of these compounds only to aquatic biota. Without these limitations the overwhelming amount of literature available, even on one element, makes it an impossible task to complete in a single chapter. Industrial uses of inorganic forms are noted in order to give an indication of release to the environment prior to possible formation of organic derivatives therein.

7.2   Industrial use of Group VI elements and environmental effects

7.2.1   Sulphur

Sulphur is an extremely versatile element in its usage. Its major product, sulphuric acid, is involved extensively in many chemical industries. In recent years, the largest demand for sulphuric acid has been for agricultural purposes in the processing of phosphate fertilizers and ammonium sulphate for fertilizers. Other uses of sulphur are in petroleum refining, metal mining and processing, production of organic and inorganic chemicals, paints and pigments, paper and pulp, plastic and synthetic materials, and in many other industrial uses. The use of sulphur is rather unusual as compared with other elements in that the largest portion of it is used as chemical reagents rather than as a component of a finished product. Because of this the sulphur is not retained in the final product, but is more often discarded as a waste product. The use of sulphuric acid in the manufacture of phosphate fertilizers is a typical example.

Table 7.1 Agricultural and industrial use of some organic sulphur and organic phosphorus−sulphur compounds

Compound	Formula	Uses
Sodium lauroyl isethionate	$CH_3(CH_2)_{10}COOCH_2CH_2SO_3Na$	Mild cosmetic detergent
Octyl mercaptan	$CH_3(CH_2)_6CH_2SH$	Silver polish
Parathion (methyl and ethyl)	$(CH_3O)_2PSO\text{-}C_6H_4\text{-}NO_2$	Cotton pest control
Zinc dialkyl phosphorodithioate	$Zn\,[SPS(OR)_2]_2$	Anti-wear in lubricating oil
Malathion	$(CH_3O)_2PS_2\text{-}CH(CH_2COOEt)\text{-}COOEt$	Insecticide
Thimet	$(C_2H_5O)_2PS_2CH_2SC_2H_5$	Insecticide
Nabam	$(NaSCNHCH_2\text{-})_2$ with $\overset{\|}{S}$	Pesticide
Captan		Pesticide

Some synthetic organic phosphorus—sulphur compounds have agricultural and industrial applications. For example, methyl- and ethylparathions are used extensively to control cotton pests. Zinc dialkylphosphorodithioate is used as an anti-wear additive in lubricating oil; malathion and thimet are used as insecticides. These compounds, although classified as organophosphorus compounds, do contain sulphur in their formulation. Some examples are given in Table 7.1.

7.2.2   Selenium and tellurium

The unique photoelectrical and semiconducting properties of selenium have resulted in extensive use of this element in photocell devices and in xerography, solar batteries, specialty transformers and rectifiers. Its major use is in coloured glass manufacturing both for decorative purposes and for cutting down glare and heat transfer. A summary of its use pattern and relative consumption is summarized in Table 7.2.[4] The principal commercial selenium compounds are mostly inorganic salts. Organoselenium compounds are mostly formed in living systems such as the protein derivatives, selenocystine $((-SeCH_2CH(NH_2)COOH)_2)$ and   selenomethionine $(CH_3SeCH_2CH_2CH(NH_2)COOH)$ which have no significant commercial applications. There is much interest in synthetic organoselenium compounds for potential use as chemotherapeutic agents,[5] such as antibacterial agents,

$$\left(H_2N-\!\!\left\langle\bigcirc\right\rangle\!\!-Se- \right)_2$$

hypnotics, 2-selenobarbituric acid,

and other compounds that affect the autonomic nervous system, including *o*-carboxybenzeneseleninic acid,

and anti-inflammatory compounds such as 5,5'-selenobiasalicylic acid,

Table 7.2 Use pattern of selenium

Application	Per cent of total	Purpose
Glass manufacture	35	– Small amounts of selenium added to glass melts to decolourize the green tint caused by iron impurities
		– Large amounts used to produce a ruby red glass in tableware, signal, and decorative uses
		– Used in the manufacture of dark coloured glass in building and vehicles to reduce glare and rate of heat transfer
Electronic application	20	– Selenium rectifier, which changes alternating current to direct current, photographic exposure meter, solar batteries
Photocopying	25	– The amorphous selenium is used to transfer photographic image by static electricity
Pigments	15	– Cadmium sulphoselenide pigments (orange–red–maroon) for use in plastics, paints, enamels, inks and rubber
Other	5	– Additives to poultry and swine feed. Supplementation for selenium-deficient areas
		– Selenium sulphide as fungicide for the control of dandruff and dermatitis

(Per cent breakdown use pattern in 1972 taken from US Environmental Protection Agency Publication No. EPA-560/2-75-005D (1975) – see Ref. 4.)

Organoselenium compounds are also used in polymers as semiconducting materials.[6] Radioactive ^{75}Se[selenomethionine] is used in human pancreatic scanning.[7] Use patterns for selenium are given in Table 7.2.[4]

Tellurium has similar applications in the semiconductor industry and electronics, in the production of thermoelements, photoelements and other devices in automation equipment. The increasing demand for semiconductor use necessitates research work on the application of various tellurium compounds as semiconductors. Tellurium is also used for the production of detectors and thermocouples.

In metallurgy, selenium and tellurium are used as additives to improve the machinability and stability of stainless steel and copper alloys. Additions of tellurium increase the anticorrosive properties of copper and lead. Lead telluride is added to acid-resistant materials used in sulphuric acid manufacturing for anticorrosive purposes.

In the chemical industry, selenium and tellurium are used as catalysts in petroleum processing and in organic syntheses. They are also used in the rubber industry to increase the wear resistance and elasticity of rubber. The commonly used selenium compounds and their applications are listed in Table 7.3.[8]

7.2.3   Environmental contamination

Anthropogenic inputs of Group VI elements into the environment through their uses, production and other sources, contribute to contamination of the environment. The Group VI elements can be transformed from inorganic compounds to organometalloids through microbial activities. The resultant methylated compounds are more lipophilic and may be more volatile, thus changing the pattern of transport and possibly the toxicological behaviour. The ecological consequences of such processes are not known. It should be pointed out, however, that the methylation of inorganic selenium oxyanion salts does not pose the same kind of environmental impact as the methylation of inorganic mercury because dimethyl selenide is much less toxic than the inorganic selemium oxyanion salts.[9]

7.2.3.1   Contamination from use

Selenium and its salts used in glass may be lost to the environment on handling and batch processing. A large amount of elemental selenium used is volatilized during melting. The glass industry in the USA alone estimated that about 184 tonnes of selenium were emitted to the atmosphere in 1969.[10]

In the photocopying process, selenium is used in an enclosed environment except in the manufacture of copying equipment. The industry estimates that about 1 kg per tonne of selenium processed, or about 135 kg in 1972, escaped into the atmosphere.[10]

The processes involved in the use of selenium in pigments in paints results in minimal losses to the atmosphere. Use pattern of tellurium is similar to selenium in many ways, but probably in smaller quantities.

Sulphur is a more widespread element through its many uses in chemical industries and agriculture. Loss of man-made sulphur dioxide through emissions alone is the single largest input into the atmosphere (see sect. 7.2.3.3).

7.2.3.2   Contamination from production

The affinity of selenium and tellurium for copper and sulphur is very marked. The accumulation of copper in ores is accompanied by high concentrations of

Table 7.3 Some commonly used selenium compounds and their applications

Aluminium selenide, Al_2Se_3	In preparation of hydrogen selenide; in semiconductor research
Ammonium selenite, $(NH_4)_2SeO_3$	In manufacture of red glass; as reagent for alkaloids
Arsenic hemiselenide, As_2Se	In manufacture of glass
Bismuth selenide, Bi_2Se_3	In semiconductor research
Cadmium selenide, CdSe	In photoconductor, semiconductors, photo-electric cells, and rectifiers; in phosphors
Calcium selenide, CaSe	In electron emitters
Cupric selenate, $CuSeO_4$	In colouring Cu or Cu alloys black
Cupric selenide, CuSe	As catalyst in Kjeldahl digestions; in semiconductors
Indium selenide, InSe	In semiconductor research
Potassium selenate, K_2SeO_4	As reagent
Selenium disulphide, SeS_2	In remedies for eczemas and fungus infections in dogs and cats; as antidandruff agent in shampoos for human use; usually employed as a mixture with the monosulphide
Selenium hexafluoride, SeF_6	As gaseous electric insulator
Selenium monosulphide, SeS	Topically against eczemas, fungus infections, demodectic mange, flea bites in small animals; usually employed as a mixture with the disulphide
Selenium dioxide, SeO_2	In the manufacture of other selenium compounds; as a reagent for alkaloids
Sodium selenate, Na_2SeO_4	As veterinary therapeutic agent
Sodium selenite, Na_2SeO_3	In removing green colour from glass during its manufacture; as veterinary therapeutic agent.

(From: *Selenium*, National Academy of Sciences, Washington DC (1976) – see Ref. 8.)

selenium and tellurium. In the USA recovery of selenium and tellurium is principally a by-product of copper refining. The main sources of selenium and tellurium in industry are the sludges of copper electrolytic plants. In the course of recovery of selenium and precious metals, selenium is lost in three processes; melting operation, drying oven and the bisulphate fusion. Selenium is present in the hot exhaust gases from the reverberatory furnaces and convectors in the copper smelter.

Sulphur dioxide emission, as a result of sulphur production, is more serious than selenium and tellurium together. However, all these sources when compared to the sulphur dioxide emission in burning fossil fuels, are utterly insignificant.

7.2.3.3   Contamination from fossil fuels

Much of the information about sulphur pollution is from its association with acid rain. It is estimated that some 60 million tonnes of man-made sulphur dioxide emissions are given off each year. This acid gas will be transported by the prevailing winds to remote areas and will fall back on earth as acid rain. The acid rain problems are beyond the scope of this chapter, but its magnitude does help an appreciation of the ecological impacts of Group VI elements, in particular, of sulphur.

Contributions of selenium to the environment through the burning of fossil fuels are also serious. A study indicated that zooplankton in Lake Michigan contained 7 μg g^{-1} of selenium, whereas selenium in the sediment of this part of the lake was less than 0.5 μg g^{-1}.[11] Such an observation indicates that selenium is quickly taken up by organisms and does not accumulate at the bottom. Selenium concentration in zooplankton increases as one approaches the city of Chicago. This is explained by the prevailing winds over Chicago from the south-west so that most of the smoke from fossil fuel combustion is blown out over the southern end of the lake. These environmental aspects of selenium require further study. Table 7.4 indicates the amount of selenium entering the atmosphere through various routes due to anthropogenic activities. The major source of selenium in the environment is still the weathering of rocks and minerals.

Table 7.4 Estimation of anthropogenic inputs of selenium to the environment

Source	Amount entering environment (tonnes yr^{-1})	Route
Manufacture of glass	200	to atmosphere
Fuel oil combustion	30	to atmosphere
Coal combustion	573	to atmosphere
	2500	to land
Other	200	
Total	3503 tonnes yr^{-1}	

(From US Environmental Protection Agency Publication no. EPA-56012-75-005D (1975) − see Ref. 4.)

Little information is available on environmental aspects of tellurium. On account of its similar properties and occurrence to selenium, it may be assumed that the above estimates may be valid for tellurium.

7.3 Natural formation of organometalloidal compounds – biomethylation of Group VI elements

7.3.1 Sulphur

The first observation related to sulphur methylation (by Challenger and Simpson[12]) was that dimethyl sulphide ($(CH_3)_2S$) was evolved by the red alga *Polysiphonia fastigiata* on exposure to air and especially on treatment with alkali. The evolution of dimethyl sulphide was also observed in soils, seawater, and in a few samples of marine algae.[13] Its concentration in pond water has been found to exceed that of hydrogen sulphide.[14] So far, however, there has been no affirmative proof to support microbial methylation of sulphur.

Challenger[15] concluded that a normal metabolism of the algae is most likely responsible for the emission of dimethyl sulphide, since the test alga exerts no methylating action on sodium arsenite and arsenate or potassium tellurite when suspended in artificial seawater containing these compounds. Production of dimethyl sulphide still occurred when the weed was placed in water saturated with toluene; therefore, it is unlikely that bacterial action is involved. It seems likely that the dimethyl sulphide produced in aquatic systems is due to the decomposition of dimethyl-β-propiothetin ($(CH_3)_2S^+CH_2CH_2COO^-$), which has been widely found in marine and freshwater algae.[15,16]

Laboratory studies with pure cultures have demonstrated that many species of bacteria such as *Aerobacter aerogenes* and *Pseudomonas putrefaciens*, and fungi such as *Schizophyllum commune*, *Penicillium notatum* and *Aspergillus niger* can produce volatile organic sulphur compounds such as methyl mercaptan, dimethyl sulphide and dimethyl disulphide from different sulphur-containing substrates.[17,18]

Moje *et al.*[19] detected emissions of carbonyl sulphide (COS) through microbial decomposition of a pesticide called Nabam (disodium ethylene-bis-dithiocarbamate). Similar production was observed[20] from Captan (pesticide) by axenic cultures of *Neurospora Crassa*, and in the gases evolved from amended and unamended soils under aerobic or waterlogged conditions.[21-23]

Carbon disulphide has also been observed in the evolved gases from animal manures under anaerobic conditions,[22] and also in amended and unamended soils under aerobic or waterlogged conditions[22-24] as a result of microbial decomposition of cysteine ($HSCH_2CH(NH_2)COOH$) and cystine ($(-SCH_2CH(NH_2)COOH)_2$). Such processes may also account for the wide distribution of carbon disulphide in coastal and ocean waters.[13]

Table 7.5 summarizes some volatile organic sulphur compounds produced by micro-organisms.[25-37] It should be noted that very little is known about the mechanisms and biochemistry of the reactions through which micro-organisms produce volatile organosulphur compounds. Recently the effects

Table 7.5 Volatile organic sulphur compounds produced by micro-organisms

Compound	Formula	Reference
Carbon disulphide	CS_2	21–25
Carbonyl sulphide (carbon oxysulphide)	COS	19, 20, 26
Methyl mercaptan (methanethiol)	CH_3SH	27–30
Ethyl mercaptan (ethanethiol)	CH_3CH_2SH	24
n-Propyl mercaptan (1-propanethiol)	$CH_3CH_2CH_2SH$	31
iso-Propyl mercaptan (1-methyl-1-ethanethiol)	$(CH_3)_2CHSH$	32
Allyl mercaptan (2-propene-1-thiol)	$CH_2=CHCH_2SH$	31, 33
n-Butyl mercaptan (1-butanethiol)	$CH_3CH_2CH_2CH_2SH$	32, 34
iso-Butyl mercaptan (2-methyl-1-propanethiol)	$(CH_3)_2CHCH_2SH$	32
Dimethyl sulphide (methylthiomethane)	CH_3SCH_3	13, 17, 32, 35
Dimethyl disulphide (methyldithiomethane)	CH_3SSCH_3	21, 24, 32, 36
Dimethyl trisulphide (methyltrithiomethane)	CH_3SSSCH_3	37
Ethyl methyl sulphide (methylthioethane)	$CH_3CH_2SCH_3$	24
Diethyl sulphide (ethylthioethane)	$CH_3CH_2SCH_2CH_3$	2
Diethyl disulphide (ethyldithioethane)	$CH_3CH_2SSCH_2CH_3$	24
Dipropyl disulphide (propyldithiopropane)	$CH_3CH_2CH_2SSCH_2CH_2CH_3$	31
Diallyl sulphide (allyl sulphide)	$CH_2=CHCH_2SCH_2CH=CH_2$	31
Diallyl disulphide (allyl disulphide)	$CH_2=CHCH_2SSCH_2CH=CH_2$	31
Methyl allyl sulphide	$CH_3SCH_2CH=CH_2$	33

(From: Bremner J M, Steele C G (1978) – see Ref. 80.)

of sulphur source, sulphate concentration and salinity on the biosynthesis of dimethyl sulphide and dimethyl propiothetin by marine phytoplankton have been investigated in detail.[38] Experimental evidence has been obtained to support the postulate that the production of dimethyl propiothetin is a biological function in osmotic regulation in the cell.

7.3.2   Selenium

The methylation of selenium as well as tellurium by organisms has been stated repeatedly in the literature since the early part of this century. It is known that several genera of fungi produce dimethyl selenide from inorganic selenium compounds.[39,40] Rats fed with selenate or selenite salts exhale dimethyl selenide[41] and dimethyl diselenide.[42] In addition to these volatile selenium compounds, trimethylselenonium ion is also excreted in the urine.[43] Other studies conducted with higher plants grown in selenite-enriched soils, identify dimethyl selenide as the single product of a non-accumulator species – cabbage,[44] and dimethyl diselenide as the product of an accumulator species – Astragalus.[45] Incubation of soils amended with glucose and selenite produces dimethyl selenide.[34,46] A strain of *Penicillium* which produced dimethyl selenide has been isolated from raw sewage.[39]

Apart from the dimethyl selenide and dimethyl diselenide, other volatile selenium compounds have been detected and identified. In a study with lake sediment enriched with inorganic selenium compounds, an unknown volatile selenium was detected in addition to dimethyl selenide and dimethyl diselenide.[47] This unknown species was also produced in soils and sewage sludge, and was identified as dimethyl selenone ($(CH_3)_2SeO_2$).[48] It is an intermediate product in the methylation sequence in the formation of dimethyl selenide originally proposed by Challenger.[15] The erratic appearance of dimethyl selenone in the lake sediment methylation could be due to kinetic differences of the intermediate and the final products, or to the differences in micro-organisms which are responsible for the formation of each selenium species.[48]

Not much is known about the methylation of selenium *in situ* in the environment. In experiments to simulate *in situ* methylation of selenium in the aquatic environment, Chau *et al.*[47] incubated various lake sediments and lake water enriched with inorganic and organic selenium compounds, and analysed the head space gases for volatile selenium species. Production of dimethyl selenide and dimethyl diselenide was observed from soil and sediment enriched with the following selenium compounds at 5 mg dm^{-3} in the culture substrate: sodium selenite, sodium selenate, selenocystine, selenourea and seleno-DL-methionine. In many instances, an unknown volatile selenium compound, now identified as dimethyl selenone, was also produced. Certain lake sediments produced the volatile species upon incubation without the addition of selenium compounds. This is the best laboratory simulation to date for the real environment. All the volatile selenium production so far reported is directly related to microbial growth, and hence attributed to biomethylation processes.

The production of volatile selenium in sediment is temperature dependent. The production of dimethyl selenide from selenite is reduced to 75 per cent when the temperature is lowered from 20 to 10 °C. At 4 °C the production was lowered to about 10 per cent.[47] Conversion of selenate to volatile selenium is even more influenced by temperature. The production at 10 °C is about 15 per cent of that at 20 °C. At 4 °C virtually no volatile selenium is produced.

The conversion of both selenite and selenate in sediment was observed under aerobic and anaerobic conditions.[47] Under environmental conditions, where the sediment temperature is low, biomethylation of selenium may be too slow to produce detectable amounts of selenium metabolites. Since these metabolites are volatile, their detection in the atmosphere may present further difficulties. Indeed, there is a scarcity of data available on environmental occurrence of alkyl selenides.

Certain fungi, among them *Scopulariopsis brevicaulis, Aspergillus niger* and *Pencillium notatum* [15] and bacteria, which were not considered to be able to methylate selenium and tellurium,[3] are now known to methylate both selenite and tellurite. A strain of *Corynebacterium*,[49] *Aeromonas* sp., *Flavobacterium* sp. and a tentatively identified *Pseudomonas* sp. have been demonstrated to methylate selenium in sediment.[47]

7.3.3   Tellurium

Microbial transformations for tellurium follow similar pathways as for selenium, which include methylation and reduction. Fungi can also produce volatile dimethyl telluride from tellurium salts.[50] A strain of *Penicillium*, which produced dimethyl selenide from inorganic selenium, was isolated from raw sewage. Dimethyl tellurite was also produced by this organism from several tellurium salts ($TeCl_4$; H_2TeO_3; H_6TeO_6), but this product was synthesized only in the presence of both tellurium and selenium.[39] Thus the methylation of tellurium may not be a direct biological methylation, but rather a transmethylation from biologically formed methyl selenide compounds.[51] Since the occurrence of tellurium in nature is rare, its environmental impact due to methylation is not expected to have serious effects.

7.3.4   Biomethylation mechanisms for selenium

7.3.4.1   Mechanism proposed by Challenger[15]

The mechanism suggested by Challenger (1951) for the methylation of selenium by fungi is as shown in equation 7.1:

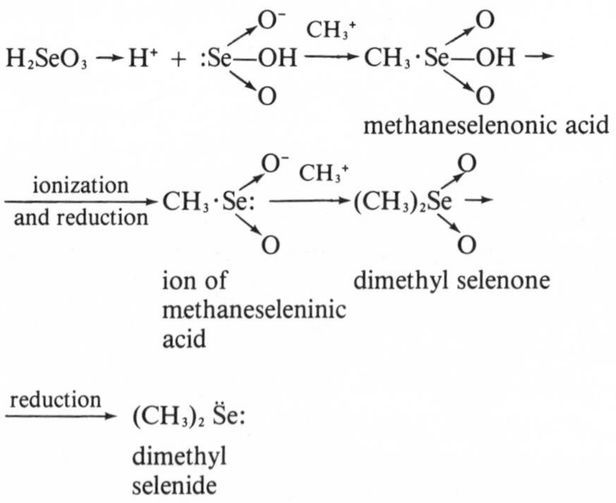

$$H_2SeO_3 \rightarrow H^+ + \;:\!\underset{\diagdown O}{\overset{\diagup O^-}{Se}}\!-OH \xrightarrow{\;CH_3^+\;} CH_3\cdot\underset{\diagdown O}{\overset{\diagup O}{Se}}\!-OH \rightarrow$$

methaneselenonic acid

$$\xrightarrow[\text{and reduction}]{\text{ionization}} CH_3\cdot\underset{\diagdown O}{\overset{\diagup O^-}{Se}}\!: \xrightarrow{\;CH_3^+\;} (CH_3)_2\underset{\diagdown O}{\overset{\diagup O}{Se}} \rightarrow \qquad\qquad [7.1]$$

ion of   dimethyl selenone
methaneseleninic
acid

$$\xrightarrow{\text{reduction}} (CH_3)_2\,\ddot{S}e\!:$$

dimethyl
selenide

The postulated intermediate selenium compounds were not detected in culture media, and it was concluded that the biological methylation process was confined to the mould cells and did not take place in the medium. The mechanism is basically a transfer of a methyl group as a cation. The mechanism also requires the methyl acceptor to be nucleophilic, i.e. in the form of negative ion which contains unshared electrons for coordination with the positive methyl group. This could then undergo reduction and ionization, followed by further coordination of a CH_3^+ group.

7.3.4.2 Mechanisms proposed by Lewis *et al.*[52]

Based on a review of the literature, and on the results of their experiments with cabbage, pathways for the production of volatile selenium by non-accumulator and accumulator species of plants, under normal physiological conditions and non-toxic levels of selenium in the growth substrates have been proposed by Lewis *et al.*[52]

For non-accumulator species (cabbage) the pathway in equation 7.2 was proposed:

$$SeO_3^{2-}$$
$$SeO_4^{2-} \xrightarrow[\text{reduction to Se(II)}]{\text{assimilation and}} CH_3SeCH_2CH_2CHCOOH \rightarrow$$
$$\underset{NH_2}{|}$$

selenomethionine

$$\xrightarrow[CH_3^+]{\text{methylation}} \overset{CH_3}{\underset{CH_3}{\diagdown}} Se^+ - CH_2CH_2CHCOOH \xrightarrow[\text{enzymatic cleavage}]{H_2O}$$
$$\underset{NH_2}{|}$$

Se-methylselenomethionine ion

$$CH_3SeCH_3 + CH_2CH_2CHCOOH + H^+ \qquad\qquad [7.2]$$
$$\underset{OH}{|} \quad \underset{NH_2}{|}$$

dimethyl
selenide   homoserine

The intermediate product, selenomethionine, has been found in rye grass, wheat, and clover, all non-accumulators. Se-methylselenomethionine has been identified as the predominant soluble organic selenium compound in four non-accumulator species of *Astragulus*. This compound has also been found in cabbage, a non-accumulator grown on selenite- or selenate-enriched media.[44,53]

Evidence that the enzymatic cleavage reaction takes place in intact plants has been inferred from the studies discussed previously.[44,53]

For accumulator species (*Astragulus*) an alternative pathway has been proposed (equation 7.3):

$$\text{SeO}_3^{2-} \atop \text{SeO}_4^{2-} \quad \xrightarrow[\text{reduction to Se(II)}]{\text{assimilation and}} \quad \begin{array}{cc} \text{CH}_2 - \text{Se} - \text{CH}_2 \\ | \qquad\qquad | \\ \text{CH}_2 \qquad \text{CHNH}_2 \\ | \qquad\qquad | \\ \text{CHNH}_2 \qquad \text{COOH} \\ | \\ \text{COOH} \end{array} \longrightarrow \qquad [7.3]$$

(a) selenocystathionine

$$\begin{array}{ccc} \text{CH}_3\text{SeCH}_2\text{CHCOOH} & \longrightarrow & \text{CH}_3\text{SeCH}_2\text{CHCOOH} \\ | & & \downarrow \quad | \\ \text{NH}_2 & & \text{O} \quad \text{NH}_2 \end{array}$$

(b) Se-methylselenocysteine   (c) Se-methylselenocysteine selenoxide

$$\longrightarrow \text{CH}_3\text{SeSeCH}_3$$

(d) dimethyl diselenide

The intermediate, selenocystathionine (a) has been formed from selenate and selenite in accumulator species.[54] The formation of S-methylcysteine in plants results from the methylation of cysteine.[55] Analogous pathways for the formation of Se-methylselenocysteine (b) in *A. bissulcatus* have been suggested based on ^{75}Se-selenomethionine experiments.[56] The formation of Se-methylselenocysteine from selenocystathionine in accumulator species fed with selenite or selenate required further elucidation. However, Se-methylselenocysteine has been found in accumulator plant species,[57] and has been suggested as a biochemical means for distinguishing accumulators from non-accumulators. No direct evidence is yet available for the formation of Se-methylselenocysteine selenoxide (c) other than for the sulphur analogue. The natural ($+$) isomer of S-methyl-L-cysteine sulphoxide was formed from S-methyl-L-cysteine in broccoli leaves. It has been known that acid hydrolysis of S-methyl-L-cysteine sulphoxide yields dimethyl disulphide.[58] Analogous reactions may be expected for selenoxide. Dimethyl diselenide (d) has indeed been identified in volatile materials from the accumulator plant *A. racemosus*.[45]

The two pathways proposed by Lewis *et al.* are also pathways involving transfer of methyl carbonium groups.

7.3.4.3   Mechanism proposed by Reamer and Zoller[48]

The mechanism proposed by Reamer and Zoller[48] is basically a methyl carbonium pathway similar to that proposed by Challenger.[15] They proposed a reduction pathway for the intermediate $\text{CH}_3\text{SeO}_2^-$ to become dimethyl diselenide by reduction. The identification of dimethyl selenone as a volatile metabolite supports Challenger's mechanism (see sect. 7.3.4.1). The present

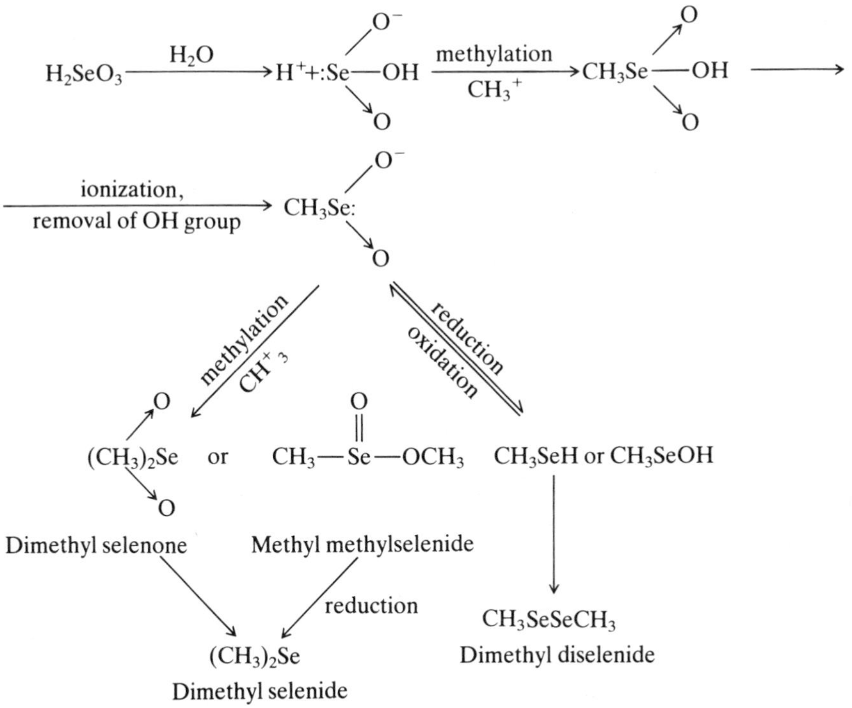

Fig. 7.1 Biomethylation mechanism of Reamer and Zoller for selenium

proposed mechanism does not involve the pathways of selenomethionine and Se-methylselenomethionine mentioned in Lewis's suggestions.[52] Details of the mechanism are given in Fig. 7.1.

7.3.5 The role of sulphur in transmethylation processes

Sulphide ion has been known to catalyse further methylations of trimethyllead salts[59] by forming a sulphide complex $((CH_3)_3Pb)_2S$ which subsequently disproportionates to tetramethyllead. Its involvement in the methylation of methylmercury to dimethylmercury has also been elaborated.[60] Thus a potentially important cycling process for organomercury compounds can be accomplished by the role of hydrogen sulphide in the mobilization of methylmercury as volatile dimethylmercury in the environment. Similar processes also occur for tin and they are discussed in the relevant chapters (Chs 2−4).

Other biogenic organosulphur compounds such as methionine, *S*-methylmethionine, *S*-adenosylmethionine, and coenzyme M cofactor are known biological methyl transfer agents acting as sources of methyl carbonium ion (CH_3^+).[61] This type of methyl donor favours nucleophilic transmethylation reactions in which the methyl acceptor must be nucleophilic. Methylation of arsenic[3,62] and selenium[15,48,52] follows this pathway. The structures of some biogenic sulphur compounds, which are useful as methyl donors, are as follows:

$$CH_3SCH_2CH_2CHNH_3^+COO^-$$

methionine

$$(CH_3)_2\overset{+}{S}CH_2CH_2\overset{|}{\underset{NH_2}{C}HCOOH}$$

S-methylmethionine ion

$$\overset{CH_3}{\underset{Aden}{|}}\,CH_2SCH_2CH_2\overset{+}{\underset{NH_2}{C}HCOOH}$$

$$HOSO_2CH_2CH_2SH$$
coenzyme M

S-adenosylmethionine ion
(Aden = adenosyl group)

Other sulphur derivatives that are known to be methyl donors are trimethylsulphonium salts, such as: trimethylsulphonium iodide $((CH_3)_3S^+I^-)$, trimethylsulphoxonium iodide $((CH_3)_3SO^+I^-)$ and trimethylsulphoxonium hexachloroantimonate $((CH_3)_3SO^+SbCl_6^-)$. These compounds, to a various extent, have been demonstrated to methylate Pb(II) in aqueous solutions as carbonium donors.[63] Further details are given in Chapter 4.

Sulphydryl-containing amino acids, such as homocysteine $(HSCH_2CH_2CH(\overset{+}{N}H_3)COO^-)$ and cysteine $(HSCH_2CH(\overset{+}{N}H_3)COO^-)$ complex with metals such as Hg(II), Cd(II) and Pb(II). Their mercury complexes and implications in the methylation of mercury and other metals have been investigated.[64]

7.4   Natural occurrence of organic Group VI elements in the environment

7.4.1   Organosulphur compounds

There is a diverse group of organosulphur compounds occurring naturally in soil, most of which are related to microbial activities. Dimethyl sulphide is one of the most significant volatile sulphur compounds in the environment and in the sulphur cycle. It is widely distributed both in the atmosphere and in marine and freshwater systems. The highest concentration of dimethyl sulphide is found in surface water, particularly on the continental shelves, in estuaries and in upwelling regions, where its concentration correlates significantly with *chlorophyll a*.[65] The vertical distribution of dimethyl sulphide is closely related to the distribution of primary productivity as indicated by chlorophyll and nutrient concentrations. Therefore, marine primary producers are the dominant source of dimethyl sulphide to seawater and consequently to the atmosphere.

Other biogenic sulphur compounds, such as dimethyl sulphoxide ($(CH_3)_2SO$), the oxidation product of dimethyl sulphide, have also been found ($19-109$ nmol dm^{-3}) in surface ocean water and in rivers and lakes but are not detected below the euphotic zone.[66] The occurrence of dimethyl sulphoxide is associated with phytoplankton activity in the surface waters. Another organosulphur compound, carbonyl sulphide (COS), has been detected in surface seawater near Delaware Bay, ranging from 6 to 22 ng dm^{-3}. The production of carbonyl sulphide appears to be independent of salinity, photosynthetic activity and microbial activity.[67] Table 7.6 shows the concentrations of dimethyl sulphide in seawater and some ecological zones.

Other organosulphur compounds are formed as a result of biochemical reactions in living systems (see sect. 7.3.1). These compounds have not been detected in environmental samples, except in confined experimental or localized biological systems. It is beyond the scope of this environmental work to include them in the discussion.

7.4.2 The sulphur cycle

Sulphur exists in several oxidation states and also forms numerous compounds of organic and inorganic nature. Micro-organisms play a distinctive role in the conversions between them. The general cycle of sulphur in the environment is illustrated in Fig. 7.2.

The largest quantity of element flow is in the burning of fossil fuels, which releases sulphur dioxide gas into the atmosphere, which eventually returns to land and water. The soluble form of sulphur in soil is the sulphate which can be readily leached out. The organic bound sulphur represents about 75 per cent of the total, from which sulphur can be slowly released by microbial decomposition. It should be noted that wide distribution of dimethyl sulphide has been observed in oceans and lakes.[68] Soil treated with plant materials,

Table 7.6 Occurrence of dimethyl sulphide in the environment

Samples	DMS (ng dm^{-3} as S)	Reference
North Sea	58.0	68
Dover Strait	76.6	68
English Channel	116.4	68
North Atlantic (open ocean surface water)	62.5	68
South Atlantic (open ocean surface water)	37.1	68
Open waters Straits of Florida	$25-60$	65
Seagrass bed, south of Long Key	390	65
Reef break, south of Long Key	88	65

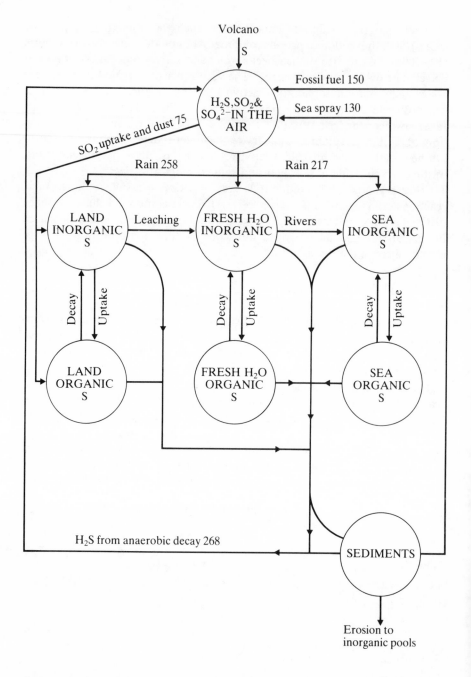

Fig. 7.2 The sulphur cycle in the biosphere. The flow rates between pools are 10^6 tonnes (as S) year^{-1}. From Campbell, R 1977 – see Ref. 69 and Kellogg, W W *et al.*, 1972 – see Ref. 70

sulphur-containing amino acids, proteins, manures and sludges, evolves dimethylsulphide as a major volatile sulphur compound. It is, therefore, considered as an important transport species for the transfer of sulphur from the sea through the air to the land.[13] Sulphur in the air is mostly sulphate, sulphur dioxide, dimethyl sulphide and hydrogen sulphide. The concentrations of these components are variable from place to place. Unpolluted air contains about 1–5 ng dm^{-3} of each of these components, although dimethyl sulphide could be more abundant in areas near the sea.

Seawater contains more sulphur than freshwater. Concentrations of sulphur in seawater average 2700 mg dm^{-3}, whereas concentrations in freshwater are 3 to 30 mg dm^{-3}. Major inputs to the aquatic system are atmospheric through rain and precipitation and land leaching.[69] Within each compartment, the conversions between organic and inorganic forms are carried out mainly by micro-organisms.

7.4.3 Organoselenium compounds in organisms

It has long been known that a high proportion of selenium in organisms is closely associated with protein and other organic fractions in the cells. A number of organoselenium compounds reported in organisms is listed in Table 7.7. The existence of these compounds in animals, plants and micro-organisms is quite well documented.[70–79] Some of these compounds have been isolated from the organisms in crystalline forms and authenticated.[73,74] The production of dimethyl diselenide and dimethyl selenide by plants has been suggested as by-products of normal, biochemical reactions within the plants, where selenium is metabolized along the pathways of sulphur.[52] However, the exhalation of dimethyl selenide by animals is apparently a detoxification mechanism.[75]

7.4.4 Organoselenium compounds in the environment

Selenium is a dispersed element and occurs in minute amounts in almost all substances. It is present in all living matter. In nature, it rarely occurs alone, but mainly associates with sulphur.[80] Organic selenium compounds have been synthesized specifically as analogues of known sulphur compounds. So far there is only one report which identifies the following compounds: dimethyl selenide, dimethyl diselenide and dimethyl selenone in the atmosphere over lake and ocean near shore[81] (Table 7.8). Production of volatile dimethyl selenide, dimethyl diselenide and dimethyl selenone has been previously observed in lake sediment[47] and in soils[40] after incubation in the laboratory.

The occurrence of organoselenium fractions in natural water has been a subject of debate.[82–84] In an anoxic basin, Cutter[85] found significant concen-

Table 7.7 Organoselenium compounds reported in organisms

Compound*	Organism	Reference
Adenosine phosphoselenate	*Saccharomyces cerevisiae* (yeast)	71
	Desulfovibrio desulfuricans (bacterium)	72
Selenomethionine	*Saccharomyces cerevisiae* (yeast)	73
	Escherichia coli (bacterium)	74
	Allium cepa (onions)	75
Se-adenosylselenomethionine	*Saccharomyces cerevisiae* (yeast)	76
Selenocystine	*Saccharomyces cerevisiae* (yeast)	73
	Escherichia coli (bacterium)	77
	Allium cepa (onions)	75
Dimethyl selenide	*Aspergillus niger* (fungus)	15
	Penicillium sp. (fungus)	39
	rats	41
	Brassica oleracea (cabbage)	53
Dimethyl diselenide	*Astragalus racemosus* (plant)	45
Trimethyl selenonium salts	rats	78
Se-methylselenomethionine	*Astragalus* sp. (plant)	57
Se-methylselenocysteine	*Astragalus bisulcatus* (plant)	79
Selenocystathionine	*Stanleya pinnata* (plant)	57

*Formulae in text.

trations of organoselenium compounds ($\sim$ 1 nmol dm^{-3}) which correlates to the concentrations of total amino acids. Similar distributions have been observed in near surface waters of the Pacific.[86,87] Using different methodology, Takayanagi and Wong[88] differentiated organic, colloidal and inorganic selenium compounds in river, estuarine and coastal seawater at Chesapeake Bay by ultrafiltration followed by ultraviolet irradiation of the sample. Organic selenium was found in all three types of water with values of 0.85 nmol dm^{-3} and 0.30 nmol dm^{-3} for dissolved and colloidal organic selenium respectively in seawater. Unfortunately, the forms of organic selenium were not characterized or identified in these studies. The term 'organic selenium' is simply operationally defined because this fraction has a behaviour pattern similar to that expected for organic materials during the analysis. Measures *et al.,*[89] however, did not detect these compounds and

Table 7.8 Occurrence of organic selenium compounds in the environment

Sample	$(CH_3)_2Se$	$(CH_3)_2Se_2$	$(CH_3)_2SeO_2$	Reference
(a) Atmosphere (ng m^{-3})				
Lake Wilrijk (Belgium)	0.47–0.84	0.35–0.62	<0.10–0.43	81
River Scheldt (Belgium)	<0.15–0.85	<0.30	<0.10	81
North Sea shore (Ostend)	0.15	<0.30	<0.10	81
(b) Sediment*				
Elbow Lake (Canada)	34	ND	3.3	47
Kelley Lake (Canada)	2.7	3.3	2.3	47
Johnnie (Canada)	6.3	ND	3.7	47

*Sediment 50 g incubated for 7 days, volatile alkyl selenides produced in ng Se.
ND = not detectable.

disputed the findings of other workers because of the lack of positive identification of organoselenium and the uncertainty of the analytical methods used by other workers.

No information is available regarding the forms of organoselenium compounds in sediment and soils, although much has been known about the release of dimethyl selenide and dimethyl diselenide from these materials.[47,48]

7.5 Toxicity of organic Group VI compounds to aquatic species

While a number of reviews[4,90,91] have been published on effects of organo Group VI compounds on mammalian systems, information on toxicity of these compounds on aquatic biota is rare. There has not been any systematic toxicological study of these compounds. Many of the volatile sulphur compounds evolved through microbial processes (Table 7.5) have been found to affect the growth of plants or micro-organisms. Information, however, is qualitative without any quantitative measurements. For example, carbonyl sulphide and carbon disulphide are toxic to fungal species and carbon disulphide inhibits oxidation of ammonium ion in the nitrification process in soils.[92] Methyl mercaptan, dimethyl sulphide, and dimethyl disulphide retard the growth and prevent germination of spores of *Aphanomyces euteiches*.[53] Nothing is known of their toxicity to aquatic biota.

Selenium toxicity is influenced by its chemical form. Few data are available on the toxicity of inorganic selenium compounds to aquatic organisms and even less on the toxicity of organic selenium compounds.

Inorganic selenium compounds are generally toxic to aquatic organisms at selenium concentrations greater than 50 μg dm^{-3}.[93] There is only one report that suggests organic selenium compounds are very much more toxic than inorganic forms. Niimi and LaHam (1976),[94] testing the toxicity of several different forms of selenium to zebra fish (*Brachydanio rerio*), showed that selenites were more toxic than selenates. There was a marked difference in toxicity between the selenides, with selenomethionine being the most toxic compound examined, and selenocystine being similar to the selenates. Reliable concentrations toxic to the fish could not be obtained for organic selenides due to the limited solubility of selenocystine and the significant losses of selenomethionine from the test solution attributable to the microbial methylation.

These differences in toxicity among the selenium compounds have also been demonstrated in phytoplankton.[95] The toxic effects of selenite, selenate and selenomethionine, on the different phases of growth of two species of blue-green algae (*Anacystis nidulans* and *Anabaena variabilis*) were compared. Selenomethionine and selenite were more toxic than selenate. These compounds were less toxic in culture medium containing sulphate, than in sulphur-free medium, thus suggesting a protective role of sulphur against selenium toxicity. Similar results were found in the study with a green alga, *Chlorella vulgaris*.

Both selenate and selenomethionine were observed to exert toxic effects by blocking the metabolic uptake of sulphate and the production of methionine respectively.[96,97] There appears to be a wide variation in toxicity among the organoselenium compounds in mammals, similar to that in the aquatic organisms.[98] In addition, organisms show great species and strain variation in their sensitivity to selenium compounds. For example, selenomethionine had no inhibitory effect on *Chlorella pyrenoidosa* at concentrations up to 1×10^{-3} mol dm^{-3},[99] but such concentrations inhibited the cell division of *Chlorella vulgaris*.[96]

No information is available on the aquatic toxicity of organic tellurium compounds.

Mammalian toxicities for organo Group VI compounds are discussed elsewhere.[4,90,91] The role of selenium in reducing methylmercury toxicity is discussed in Chapter 2 (sect. 2.7).

7.6   Concluding remarks

Organo Group VI compounds are not widely used industrially and commercially as compared with other organometals, such as tin and lead. Because of their limited usage, environmental contamination is not a serious problem. However, inorganic sulphur compounds are implicated in causing acid precipitation. Environmental occurrence of organic selenium and tellurium compounds is still not well established due to the lack of suitable analytical methodology. Biomethylation of selenium and tellurium has been well documented. However, there is still no mechanistic evidence to substantiate biomethylation of inorganic sulphur compounds, but of course methyl forms do occur. Environmental toxicological information of organo Group VI elements is extremely limited. No systematic or quantitative studies have been carried out, and further research should be addressed to the above areas.

References

1. Rosenheim O 1902 *Proc Chem Soc:* 138–9
2. Challenger F, North H E 1934 *J Chem Soc:* 68–71
3. Challenger F 1945 *Chem Rev* 36: 315–61
4. US Environmental Protection Agency Publication 1975 No. EPA-560/2-75-005 D Preliminary investigation of effects on the environment of boron, indium, nickel, selenium, tin, vanadium and their compounds.
5. Klayman D L 1973 Selenium compounds as potential chemotherapeutic agents. In *Organic Selenium Compounds: Their Chemistry and Biology* (eds Klayman D L, Günther W H H). Wiley Interscience, New York. pp 727–61
6. Mortillaro L, Russo M 1973 Selenium-containing polymers. In *Organic Selenium Compounds: Their Chemistry and Biology* (eds Klayman D L, Günther W H H). Wiley Interscience, New York, pp 815–33
7. Rosenfield I, Beath O A 1964 *Selenium*. Academic Press, New York, p 328
8. National Academy of Sciences 1976 *Selenium*. Washington, DC, p 34
9. McConnel K P, Portman O W 1952 *Proc Soc for Exp Biol Med* 79: 230–1
10. Davies W E and Assoc 1972 National inventory of sources and emissions Ba, B, Cu, Se and Zn 1969 Section IV US Environ Protection Agency Contract No. 68-02-0100 (NTIS-PB 210-679)
11. Copeland R 1970 *Limnos* 3: 7–9
12. Challenger F, Simpson M I 1948 *J Chem Soc:* 1591–97
13. Lovelock J E, Maggs R J, Rasmussen R A 1972 *Nature London* 237: 452–3
14. Bechard M J 1974 Emission of volatile organic sulphur by freshwater algae. M.S. Thesis, Washington State Univ, Pullman
15. Challenger F 1951 Biological methylation. In *Advances in Enzymology and Related Areas of Molecular Biology*, vol 12. Wiley Interscience, New York, pp 429–91
16. Richmond D V 1973 *Sulphur Compounds in Phytochemistry*, vol III. Van Nostrand Reinhold, New York, pp 41–73
17. Kadota H, Ishida Y 1972 *Ann Rev Microbiol* 26: 127–38
18. Alexander M 1974 *Adv Appl Microbiol* 18: 1–73

19. Moje W, Munnecke D E, Richardson L T 1964 *Nature (London)* **202**: 831–2

20. Somers E, Richmond D V, Pickard J A 1967 *Nature (London)* **215**: 214

21. Banwart W L, Bremner J M 1975 *J Environ Qual* **4**: 363–6

22. Banwart W L, Bremner J M 1976 *Soil Biol Biochem* **8**: 19–22

23. Banwart W L, Bremner J M 1976 *Soil Biol Biochem* **8**: 439–43

24. Banwart W L, Bremner J M 1975 *Soil Biol Biochem* **7**: 359–64

25. Banwart W L, Bremner J M 1974 *Soil Biol Biochem* **6**: 113–15

26. Elliott L E, Travis T A 1973 *Proc Soil Sci Soc Amer* **37**: 700–2

27. Challenger F, Charlton P T 1947 *J Chem Soc:* 424–9

28. Frederick L R, Starkey R L, Segal N 1957 *Proc Soil Sci Soc Amer* **21**: 287–92

29. Segal W, Starkey R L 1969 *J Bacteriol* **98**: 908–13

30. Sharpe M E, Law B A, Phillips B A 1976 *J Gen Microbiol* **94**: 430–5

31. Ling J E, Coley-Smith J R 1969 *Ann Appl Biol* **64**: 303–14

32. Jenkins D, Medsker L L, Thomas J F 1967 *Environ Sci Tech* **9**: 731–5

33. Challenger F, Greenwood D 1949 *Biochem J* **44**: 87–91

34. Francis A J, Duxbury J M, Alexander M 1974 *Appl Microbiol* **28**: 248–50

35. Herbert R A, Shewan J M 1976 *J Sci Food Agric* **27**: 89–94

36. Lewis J A, Papavizas G C 1970 *Soil Biol Biochem* **2**: 239–46

37. Miller A, Scanlan R, Lee J S, Libbey L M 1973 *Appl Microbiol* **26**: 18–21

38. Vairavamurthy A, Andreae M O, Iverson R L 1985 *Limn Oceanogr* **30**: 59–70

39. Fleming R W, Alexander M 1972 *Appl Microbiol* **24**: 424–9

40. Barkes L, Fleming R W 1974 *Bull Environ Contam Toxicol* **12**: 308–11

41. McConnell K P, Portman O W 1952 *J Biol Chem* **195**: 277–82

42. Vlasakova V, Benes J, Parizek J 1972 *Radiochem Radioanal Letts* **10**: 251–8

43. Palmer I S, Fischer D D, Halverson A W, Olson O E 1969 *Biochem Biophys Acta* **177**: 336–42

44. Lewis B G, Johnson C M, Broyes T C 1974 *Plant and Soil* **40**: 107–18

45. Evans C S, Asher C J, Johnson C M 1968 *Aust J Biol Sci* **21**: 13–20

46. Abu-Erreish G M, Whitehead E I, Olson O E 1968 *Soil Science* **106**: 415–20

47. Chau Y K, Wong P T S, Silverberg B A, Luxon P L, Bengert G A 1976 *Science* **192**: 1130–1

48. Reamer D C, Zoller W H 1980 *Science* **208**: 500–2

49. Doran J W, Alexander M 1977 *Appl Environ Microbiol* **33**: 31–7

50. Bird M L, Challenger F 1939 *J Chem Soc:* 163–8

51. Jernelöv A, Martin A-L 1975 *Ann Rev Microbiol* **29**: 61–77

52. Lewis B A G 1976 Selenium in biological systems and pathways for its volatilization in higher plants. In *Environmental Biochemistry*, vol. 1. *Carbon, Nitrogen, Phosphorus, Sulfur and Selenium Cycles* (ed Nriagu

J O). Ann Arbor Sci, Michigan, pp 389–409

53. Lewis J A, Papavizas G C 1971 *Phytopathology* **61**: 208–14
54. Peterson P, Butler G 1967 *Nature (London)* **213**: 599–600
55. Thompson J 1967 *Ann Rev Plant Physiol* **18**: 59–84
56. Chow C, Nigam S, McConnell W 1972 *Biochem Biophys Acta* **273**: 91–6
57. Virupaksha T K, Shaift A 1965 *Biochim Biophys Acta* **107**: 69–80
58. Ostermayer F, Tarbell D 1959 *J Amer Chem Soc* **82**: 3752–5
59. Jarvie A W P, Markall R N, Potter H R 1975 *Nature (London)* **255**: 217–18
60. Craig P J, Bartlett P D 1978 *Nature (London)* **275**: 635–7
61. Ridley W P, Dizikes L J, Wood J M 1977 *Science* **197**: 329–32
62. McBride B C, Merilees H, Cullen W R, Pickett W 1978 Anaerobic and aerobic alkylation of arsenic. In *Organometalloids: Occurrence and Fate in the Environment* (eds Brinckman F E, Bellama J M). ACS Symp Ser No 82, Washington DC, pp 94–115
63. Ahmad I, Chau Y K, Wong P T S, Carty A J, Taylor L 1980 *Nature (London)* **287**: 716–17
64. Carty A J 1978 Mercury, lead and cadmium complexation by sulfhydryl-containing amino acids. Implications for heavy-metal synthesis, transport and toxicology. In *Organometals and Organometalloids – Occurrence and Fate in the Environment* (eds Brinckman F E, Bellama J M). ACS Symp. Ser No 82, Washington DC, pp 339–58
65. Andreae M O, Barnard W R, Ammons J M 1983 The biological production of dimethyl sulfide in the ocean and its role in the global atmospheric sulfur budget. In *Environmental Biochemistry* (ed Hallberg R). *Ecol. Bull. Stockholm* **35**: 167–77
66. Andreae M O 1980 *Limnol Oceanogr* **25**: 1054–63
67. Ferek R J, Andreae M O 1984 *Nature (London)* **307**: 148–50
68. Andreae M O, Barnard W R 1983 *Anal Chem* **55**: 608–12
69. Campbell R 1977 Microbial ecology. In *Basic Microbiology* (ed Wilkinson J F). Blackwell Scientific Publications, London, p 52
70. Kellogg W W, Cadle R D, Allen E R, Lazarus A L, Martell E A 1972 *Science* **175**: 587–96
71. Wilson L G, Bandurski R S 1958 *J Biol Chem* **233**: 975–81
72. Akagi J M, Campbell L L 1962 *J Bacteriol* **84**: 1194–201
73. Blau M 1961 *Biochim Biophys Acta* **49**: 389–90
74. Tuve T, Williams H H 1961 *J Biol Chem* **236**: 597–601
75. Spare C G, Virtanen A I 1964 *Acta Chem Scand* **18**: 280–2
76. Mudd S H, Cantoni G L 1957 *Nature (London)* **180**: 1052
77. Weiss K-F, Ayres J C, Kraft A A 1975 *J Bacteriol* **90**: 857–62
78. Byard J L 1969 *Arch Biochem Biophys* **130**: 556–60
79. Trelease S F, Di Somma A A, Jacobs A L 1960 *Science* **132**: 618
80. Bremner J M, Steele C G 1978 Role of microorganisms in the atmos-

pheric sulfur cycle. In *Adv Microbiol Ecology* (ed Alexander M). Plenum Press, New York, pp 155–201

81. Jiang S 1983 *Atm Environ* **17**: 111–14
82. Ganther H E 1974 Biochemistry of selenium. In *Selenium* (eds Zingaro R A, Cooper W C). Van Nostrand Reinhold, New York, pp 546–614
83. Shrift A 1973 Selenium compounds in nature and medicine – metabolism of selenium by plants and microorganisms. In *Organic Selenium Compounds: Their Chemistry and Biology* (eds Klayman D L, Günther W H H). John Wiley, New York, pp 763–814
84. Ganther H, Levander O, Baumann C 1966 *J Nutr* **88**: 55–60
85. Cutter G A 1982 *Science* **217**: 829–31
86. Suzuki Y, Sugimura Y, Mijake Y 1980 *Proc 4th Symp Coop Study Kuroshio and Adj Regions*. Japan Acad Tokyo 1979 Saikon Publ Co, Tokyo, pp 369–414
87. Cutter G A, Bruland K W 1984 *Limn Oceanogr* **29**: 1179–92
88. Takayanagi K, Wong G T F 1984 *Marine Chem* **14**: 141–8
89. Measures C I, Grant G, Khadem M, Lee D S, Edmond J M 1984 *Earth and Planet Sci Lett* **71**: 1–12
90. Diplock A T 1976 *Rev Toxicol* **4**: 271–329
91. Johnson C M 1976 *Residue Rev* **62**: 102–30
92. Ashworth J, Briggs G G, Evans A A 1975 *Chem Ind* 1975: 749–50
93. Hart B T 1982 Australian water quality criteria for heavy metals. In *Aust Water Res Council Tech Paper No. 77,* 1–282
94. Niimi A J, LaHam Q N 1976 *Can J Zool* **54**: 501–9
95. Kumar H D, Prakash G 1971 *Ann Bot* **35**: 697–705
96. Shrift A 1954 *Ann J Bot* **4**: 223–30
97. Shrift A 1958 *Bot Rev* **24**: 550–83
98. Cooper W C 1967 Selenium Toxicity in man. In *Symposium: Selenium in Biomedicine* (eds Muth O H, Oldfield J E, Weswig P H). Avi Publish. Co. Westport, Connecticut, pp 185–99
99. Shrift A, Nevyas J, Turndorf S 1961 *Plant Physiol* **36**: 502–9

Chapter 8

Methyl transfer reactions of environmental significance involving naturally occurring and synthetic reagents

8.1 Introduction

8.1.1 Scope of review and importance of topic

This chapter summarizes methyl transfer reactions of environmental significance from organic reagents, methylcobalamin (CH_3CoB_{12}) and synthesized metal complexes to a variety of metals and metalloids. The term 'environmental significance' means that most discussion is of methyl transfer reactions in water, reflecting environmental observations. Some non-methyl alkyl transfer reactions in non-aqueous media are described when they are of environmental interest. The abbreviations and structures (Fig. 8.1) show the structure of the synthetic chelates, and structures of methylcobalamin and methylcobinamide are presented in Chapter 1. Methyl acceptors include compounds of mercury, tin, lead, and miscellaneous other metals and metalloids. Only laboratory experiments are described here and reactions of natural water and sediment samples are excluded. Such reactions are covered in appropriate chapters elsewhere in this work.

The reactions described in this chapter are important because organometallic compounds are much more toxic than their inorganic counterparts (arsenic being a notable exception). The enhanced toxicity arises from the ability of neutral organometallic compounds to permeate cell membranes and modify intracellular chemistry. Studies of organometallic compound formation by reactions of metals or metalloids with naturally occurring metabolites or known biochemical methylating agents in aqueous media under laboratory conditions are also important. These studies are steps toward the goal of understanding such processes in the environment and distinguishing between chemical and enzymatic methylation of metals and metalloids.

1a.(DH)$_2$ bis(dimethylglyoximato)
1b.(GH)$_2$ bis(glyoximato)
1c.(DpH)$_2$ bis(diphenylglyoximato)
1d.(DBF$_2$)$_2$ bis(boron difluoride)
                     derivative of (DH)$_2$

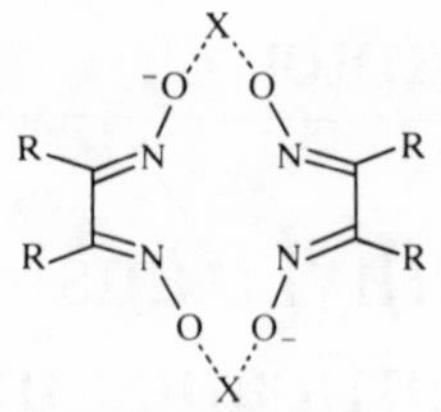

1a. R=CH$_3$, X=H
1b. R=H, X=H
1c. R=C$_6$H$_5$, X=H
1d. R=CH$_3$, X=BF$_2$

2. (DcH)$_2$)   bis(cyclohexyl glyoximato)

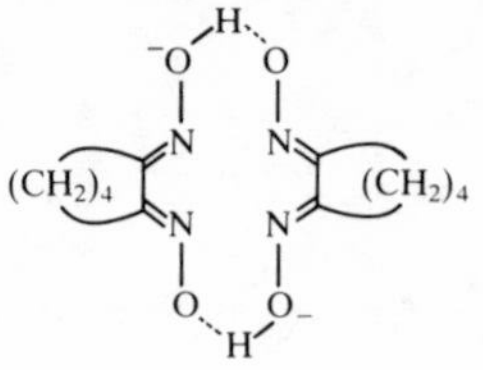

3a. salen   bis(salicylaldehyde)-
                     ethylenediamine

3b. 7,7^1-(CH$_3$)$_2$-salen   7,7^1-dimethylbis-
                                  (salicylaldehyde)-
                                  ethylenediamine

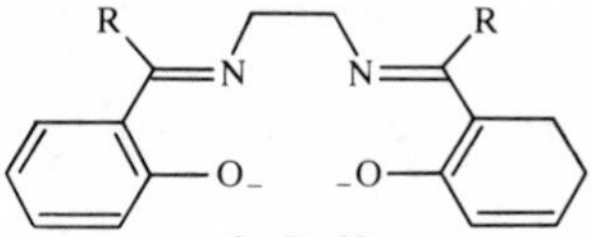

3a. R=H
3b. R=CH$_3$

4. bae   bis(acetylacetone)-
                     ethylenediamine

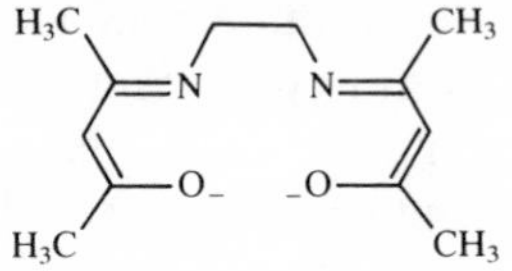

5a. (DO)(DOH)pn   2,3,9,10-tetramethyl-1,4,8,11,-
                     tetraazaunadeca-1,3,8,10-
                     tetraen-11-ol -1-olato

5b. (DO)(DOH)en   2,3,8,9-tetramethyl-1,4,7,10-
                     tetraazadeca-1,3,7,9-
                     tetraen-10-ol-1-olato

5c. (DO)(DOBF$_2$)pn   boron difluoride
                        derivative of (DO)(DOH)pn

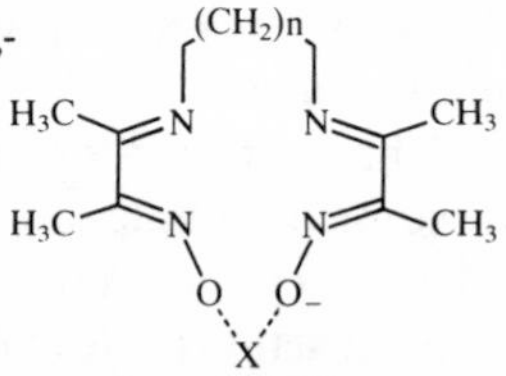

5a. n=3, X=H
5b. n=2, X=H
5c. n=3, X=BF$_2$

6. tim   2,3,9,10-tetramethyl-
                     1,4,8,11-tetraazacyclo-
                     tetradeca-1,3,8,10-tetraene

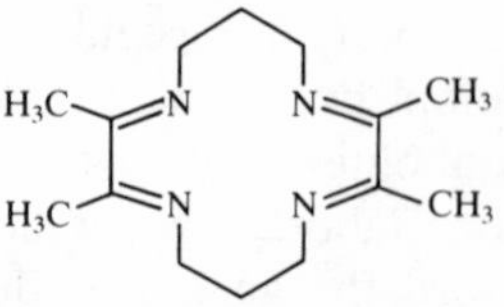

Fig. 8.1   Abbreviations and structures

8.1.2 Summary of recent reviews

Several recent reviews contain material relevant to this chapter. Toscano and Marzilli[1] emphasize synthesis, characterization and reactions of compounds with cobalt—carbon bonds. The cobalt—carbon bond cleavage reactions described include those of methylcobalamin (CH_3CoB_{12}). Craig,[2] who reviewed environmental organometallic chemistry, describes anthropogenic sources of organometallic compounds as well as their decomposition, formation and detection in the environment. Saxena and Howard[3] emphasize environmental transformations of alkylmercury, arsenic, and lead compounds. Davies and Smith,[4] in a general review of organotin chemistry, include a section on reactions of environmental significance. Two reports[5,6] describe methyl transfer reactions of CH_3CoB_{12}. This molecule has been identified as a methyl donor for mercury and hence routes to cobalt—carbon bond cleavage in this molecule are of environmental importance.

8.1.3 General aspects of cobalt—carbon bond cleavage

Cleavage of the cobalt—carbon bond occurs by a homolytic or heterolytic process.[7,8] Homolytic fission, which can occur by irradiation or reactions of metal ions such as chromium(II), results in a cobalt(II) product (equation 8.1).

The metal ion is oxidized by one electron, e.g. from chromium(II) to chromium(III):

$$CH_3Co(III)(chel) + Cr^{2+} \rightleftharpoons Co(II)(chel)^+ + CH_3Cr(III)^{2+} \qquad [8.1]$$

These formal oxidation state changes occur because of the organometallic chemists' method of counting electrons. Methyl bonded to metals is always considered to be a carbanion (CH_3^-). In homolytic bond cleavage and transfer of a methyl radical, the radical ($CH_3^\bullet$) leaves one electron behind on cobalt, resulting in its one-electron reduction, and it gains one from chromium(II) to form a $CH_3Cr(III)$ species (equation 8.1).

In contrast to homolytic bond breaking, heterolytic processes result in two-electron reduction and oxidation, or to no redox changes. Attack of $CH_3Co(III)$ (chel) by electrophiles such as mercury(II) exemplifies the latter case. Reaction products retain the trivalent oxidation state of cobalt and the divalent one of mercury(II) (equation 8.2).

$$CH_3Co(III)(chel) + Hg^{2+} \rightleftharpoons Co(III)(chel)^+ + CH_3Hg(II)^+ \qquad [8.2]$$

The reaction can be considered to occur by CH_3^- (carbanion) transfer. The second type of heterolytic mechanism, attack by a nucleophile, results in two-electron oxidation and reduction, and is considered to occur by CH_3^+ (carbocation) transfer. This process, which is rare for cobalt complexes in the context of this review, is exemplified by attack of $(CH_3)_3S^+$ on $As(OH)_3$ (equation 8.3):

$$(CH_3)_3S^+ + As(OH)_3 \rightarrow (CH_3)_2S + (CH_3)As(O)(OH)_2 + H^+ \qquad [8.3]$$

i.e. As(III) $\rightarrow$ As(V), and S(IV) $\rightarrow$ S(II)

The sulphur is formally reduced by two electrons, and arsenic is oxidized from arsenic(III) to arsenic(V).

Individual complexes show large differences in reactivity as methyl donors. The mechanism and rates of reactions depend not only on the nature of the methyl acceptor, but also on the nature of the chelating ligand (if any) and groups other than the methyl being transferred. As a consequence this chapter will describe a wide variety of methyl transfer reactions and their mechanisms.

8.2   Methyl transfer to mercury

8.2.1   Organic reagents and organometalloid compounds as methyl donors

Certain organosilicon and organoboron compounds and humic matter react in the dark with mercury(II). Thayer[8] noted that methylsiloxanes react rapidly with solid mercury(II) nitrate to form nitrogen dioxide (NO_2), but did not identify methylmercury, i.e. $CH_3Hg(NO_3)$. DeSimone[9] observed that two nuclear magnetic resonance (NMR) reference compounds, sodium salts of 3-trimethylsilylpropionate and 2,2-dimethyl-2-silapentane-5-sulphonate, react quantitatively with mercury(II) acetate in water to form methyl mercuric acetate which was identified by ^{1}H NMR. Honeycutt and Riddle[10] observed that triethylboron $(C_2H_5)_3B$ alkylates mercury(II) chloride and oxide in basic aqueous solution forming diethylmercury. Nagase[11] showed that humic substances methylate aqueous mercury(II) nitrate in the dark at 70 °C over a wide pH range. He determined methyl mercury, CH_3Hg^+, by gas chromatography (GC) with an electron capture detector (GC–ECD) and identified it by mass spectrometry (MS). In a later paper Nagase and co-workers[12] observed that methyl donors in humic acid include 2,6-di(t-butyl)-4-methylphenol, *p*-xylene and mesitylene.

Methyl transfer to mercury from organic compounds is quite common in photochemical processes. Methyl donors include acetate ion,[13,14] iodomethane (methyl iodide)[15] and amino acids.[16] Akagi and Takabatake[13] demonstrated that photolysis of aqueous mercury(II) acetate forms methylmercury and that the yield increases in the presence of mercuric oxide (equation 8.4).

$$Hg(CH_3CO_2)_2 \xrightarrow[Hg(II)]{h\nu} CH_3Hg^+ \qquad [8.4]$$

Addition of hydrochloric acid followed by MS determination of methylmercury chloride (CH_3HgCl) proves the methyl transfer process. Similar experiments[14] in the presence of red or black mercuric sulphide also yielded methylmercury as CH_3HgCl determined by GC. The authors contend that intermediate polymeric S_n species enhance the photosensitized methyl transfer process. Many years ago Maynard[15] showed that the photochemical reaction of mercury metal and iodomethane forms methylmercury (as CH_3HgI) after a 3 to 10 hour lag time (equation 8.5), and that addition of mercurous iodide (Hg_2I_2) considerably increases the reaction rate. Hayashi *et al.*[16] demonstrated the photomethylation of mercury(II) chloride in water by amino acids (equation 8.6). With glycine (R = H) as methyl donor or with alkylmercury(II) chlorides, $RHgCl(R = C_2H_5,$ n-C_3H_7 or i-C_3H_7) as methyl acceptors, no photochemically induced methyl transfer occurs.

$$Hg + CH_3I \xrightarrow{\text{sunlight}} CH_3HgI \qquad\qquad [8.5]$$

$$D,L\text{-}RCH(NH_2)CO_2H + HgCl_2 \rightarrow CH_3HgCl \qquad\qquad [8.6]$$

$$R = CH_3,\ i\text{-}C_3H_7,\ i\text{-}C_4H_9,\ s\text{-}C_4H_9$$

8.2.2 Reactions involving sulphide ions

Two 1978 papers[17,18] postulate the role of hydrogen sulphide in the environment transport of methylmercury (CH_3Hg^+) (equation 8.7).

$$2CH_3Hg^+ + H_2S \rightarrow (CH_3Hg)_2S + 2H^+ \qquad\qquad [8.7]$$

Craig and Bartlett [18] identified $(CH_3Hg)_2S$ by its 142 °C melting point, and MS. They concluded it decomposes forming dimethylmercury ($(CH_3)_2Hg$), which was identified by various mercury ($(CH_3)_2Hg^+$, CH_3Hg^+, and Hg^+) peaks in an MS experiment.

8.2.3 Methylcobalamin as methyl donor

Many groups[19-23] have studied methyl transfer reactions from methyl-cobalamin (CH_3CoB_{12}) to mercury(II) acetate (equation 8.8):

$$CH_3CoB_{12} + Hg(CH_3CO_2)_2 \xrightarrow{H_2O} H_2OCoB_{12}^+ + CH_3Hg(CH_3CO_2)$$
$$+ CH_3CO_2^- \qquad\qquad [8.8]$$

Certain features of this reaction are non-controversial: the reaction is first order in mercury(II) and CH_3CoB_{12}, methylmercury (CH_3Hg^+) and $H_2OCoB_{12}^+$ are the products: mercury(II) displaces cobalt from the benzimidazole (Bz) ligand, and methyl transfer occurs via electrophilic attack of a second mercuric ion on methyl bound to cobalt. The generally accepted mechanism is shown in Fig. 8.2. In rapid pre-equilibrium reactions Hg(II) (K) and $H^+(K_o)$ displace the benzimidazole (Bz) nitrogen from cobalt and bind to it. The constants for the pre-equilibria are: $K = 72\ dm^3\ mol^{-1}$, $k_1 = 4000\ dm^3\ mol^{-1}\ s^{-1}$,

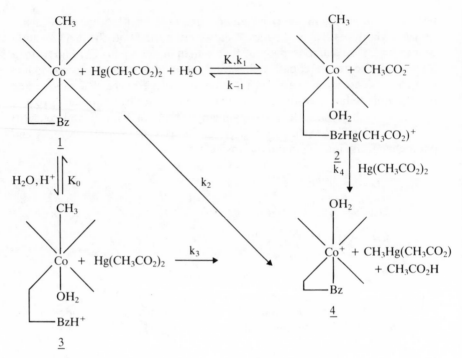

Fig. 8.2   Reaction of CH_3CoB_{12} with mercury(II) chloride

$k_{-1} = 54\,dm^3\,mol^{-1}\,s^{-1}$, $K_o = 0.0021\,dm^3\,mol^{-1}$.[19] The second-order rate constant (k_2) is $380\,dm^3\,mol^{-1}\,s^{-1}$. Other authors have also determined k_2 ($dm^3\,mol^{-1}\,s^{-1}$): 370,[20,23] 349,[21] and 300.[22] Slow demethylation of the base-off species is confirmed by slow reactions between mercury(II) acetate and methylcobinamide (Fig. 1.1) which has water coordinated *trans* to the cobalt methyl group. The cobalamin/cobinamide rate constant ratio is 3100[20] or 4700[22].

Kinetics of reactions in Fig. 8.2 are considerably complicated by the variety of mercury(II) species in equilibrium.[21,24] Presumably mercuric ion (Hg^{2+}) is the best electrophile, the various HgX^+ are poorer, and HgX_2 are poorest. Yamamoto *et al.*[24] observed that relative rates for anions X^- are: $CH_3CO_2^- \gg Cl^- \geqslant SCN^- \geqslant Br^- \gg CN^-$. The order as expected is inversely related to the stability constant of the reaction shown in equation 8.9:

$$HgX^+ + X^- \rightleftharpoons HgX_2 \qquad\qquad [8.9]$$

Despite high stability constant values of 10^4 to 10^{17} for this reaction (8.9), the authors concluded that HgX^+ is the attacking electrophile. Chu and Gruenwedel[21] in detailed studies also found that k_2 is 800 times faster in 0.1 mol dm^{-3} acetate solution rather than in 0.1 mol dm^{-3} chloride. The $CH_3CO_2^-/Cl^-$ stability constant ratio is ~ 0.01 for the reaction shown in equation 8.9.[24]

The anionic surfactant sodium dodecyl sulphate decreases the k_2 rate constant of Fig. 8.2 from 380 dm^3 mol^{-1} s^{-1} to 2.33 dm^3 mol^{-1} s^{-1}, a factor of 160.[19] The reason for the decrease is an enhanced equilibrium constant (K_o) for protonation of base-on form 1 (Fig. 8.2).

Completely unresolved is whether or not CH_3CoB_{12} can add a second methyl group to mercury(II) (equation 8.10). Two groups[25,26] reacted CH_3CoB_{12} with mercury(II):

$$CH_3CoB_{12} + CH_3Hg^+ \xrightarrow{H_2O} (CH_3)_2Hg + H_2OCoB_{12}^\dagger \qquad [8.10]$$

and detected dimethylmercury (($CH_3)_2Hg$), while Wood et al.[27] observed dimethylmercury in the presence of zinc metal and NH_4Cl. The thin layer chromatography and GC–ECD techniques used do not definitively demonstrate methyl transfer from CH_3CoB_{12} to monomethylmercury (CH_3Hg^+) because any dimethylmercury (($CH_3)_2Hg$) observed could originate from dismutation and disproportionation reactions (equations 8.11 and 8.12):

$$2CH_3HgCl \rightleftharpoons (CH_3)_2Hg + HgCl_2 \qquad [8.11]$$

$$2CH_3Hg^+ \xrightarrow{\text{reduction}} 2`CH_3Hg' \rightarrow (CH_3)_2Hg + Hg \qquad [8.12]$$

Finally Chu and Gruenwedel[28] reacted methylmercury as CH_3HgOH with CH_3CoB_{12} and obtained a k_2 value of 2.9 dm^3 mol^{-1} s^{-1} for the reaction of base-on species 1. However, they observed only the disappearance of CH_3CoB_{12}, not the formation of dimethylmercury.

8.2.4   Methylmetal complexes as methyl donors

Most of the methyl donors in this section are methylcobalt complexes. Generally those with a *trans*-dimethyl arrangement are more reactive toward electrophiles like mercury(II) ion than monomethyl complexes. Among a group of dimethyl or monomethyl complexes, reactivity decreases as the overall complex charge increases from zero to plus. All evidence strongly supports S_E2 mechanisms. Reactions discussed in this section are in water unless otherwise noted. The initial discussion will be of methylcobalt donors and that will be followed by miscellaneous methylmetal donors.

Two groups have studied reactions of dimethylcobalt complexes with mercury(II) and alkylmercury (RHg^+, R = CH_3, C_6H_5). Mestroni et al.[29] found that $(CH_3)_2Co((DO)(DOH)pn)$ (Fig. 8.1) reacted in acetonitrile (CH_3CN) with mercury(II) chloride (equations 8.13 and 8.14):

$$(CH_3)_2Co((DO)(DOH)pn) + excess \; HgCl_2 \rightarrow Co((DO)(DOH)pn)^{2+}$$
$$+ 2CH_3Hg^+ + 4Cl^- \qquad [8.13]$$

$$excess \; (CH_3)_2Co((DO)(DOH)pn) + HgCl_2 \rightarrow 2CH_3Co((DO)(DOH)pn)^+$$
$$+ (CH_3)_2Hg + 2Cl^- \qquad [8.14]$$

In these and the following reactions the solvent, which probably replaces transferred methyl groups on cobalt, is omitted. Reaction 8.13 demonstrates that the excellent electrophile, mercury(II) as Hg^{2+}, can remove both the first and second much less reactive methyl groups from $(CH_3)_2Co((DO)(DOH)pn)$. Reaction 8.14 suggests that $(CH_3)_2Co((DO)(DOH)pn)$ can transfer its more reactive first methyl group to methylmercury (CH_3Hg^+) to form dimethylmercury. The reaction in equation 8.15 demonstrates this point clearly:

$$(CH_3)_2Co((DO)(DOH)pn) + CH_3Hg^+ \rightarrow CH_3Co((DO)(DOH)pn)^+$$
$$+ (CH_3)_2Hg \qquad\qquad [8.15]$$

Espenson *et al.*[30] carried out kinetics studies of reactions between dimethylcobalt donors and phenylmercury $(c_6H_5Hg^+)$ in a 1:1 tetra-hydrofuran/water mixture (equation 8.16).

$$(CH_3)_2Co(chel) + C_6H_5Hg^+ \rightarrow CH_3Co(chel)^+ + (CH_3)(C_6H_5)Hg \qquad [8.16]$$

Reaction rates are $2 \times 10^6 \ dm^3 \ mol^{-1} \ s^{-1}$ for $(CH_3)_2Co((DO)(DOH)pn)$ and $1.3 \times 10^5 \ dm^3 \ mol^{-1} \ s^{-1}$ for $(CH_3)_2Co(tim)^+$. The cationic dimethylcobalt donor, as expected, reacts more slowly in the bimolecular electrophilic substitution reaction. The group also demonstrated by NMR that dimethylmercury $((CH_3)_2Hg)$ is the product when monomethylmercury (CH_3Hg^+) reacts with $(CH_3)_2Co((DO)(DOH)pn)$. Witman and Weber[31] have reviewed the chemistry of diorganocobalt complexes.

Several studies[32-35] demonstrate that $CH_3Co((DO)(DOH)pn)^+$ is a methyl donor to mercuric ion (equation 8.17):

$$CH_3Co((DO)(DOH)pn)^+ + Hg^{2+} \rightarrow Co((DO)(DOH)pn)^{2+}$$
$$+ CH_3Hg^+ \qquad\qquad [8.17]$$

The second-order rate constants vary from 1.6 to 5.9 $dm^3 \ mol^{-1} \ s^{-1}$ (23 to 25 °C). However, replacement of a H-bonded proton of the chelate with electron-withdrawing BF_2 to form $CH_3Co((DO)(DOBF_2)pn^+$ decreases k_2 of reaction 8.17 to $6.6 \times 10^{-4} \ dm^3 \ mol^{-1} \ s^{-1}$ at 40 °C[33] or $1.5 \times 10^{-4} \ dm^3 \ mol^{-1} \ s^{-1}$ at 23 °C.[34] The dicationic complex $CH_3Co(tim)(H_2O)^{2+}$ has a much smaller ($4.5 \times 10^{-4} \ dm^3 \ mol^{-1} \ s^{-1}$) k_2 value[32] than $CH_3Co((DO)(DOH)pn)^+$.

There are many studies[22,32,35-38] of the reaction of $CH_3Co(DH)_2(H_2O)$ with mercury(II). In analogy to protonation of benzimidazole of CH_3CoB_{12} (Fig. 8.2), $CH_3Co(DH)_2(H_2O)$ is protonated at an oxime oxygen before methyl transfer reactions occur. Mercuric ion attacks only the non-protonated complex in equilibrium with the protonated one (equations 8.18 and 8.19):[36,37]

$$CH_3Co(DH)_2 + H^+ \overset{K_B}{\rightleftharpoons} CH_3Co(DH_2)(DH)^+ \qquad [8.18]$$

$$CH_3Co(DH)_2 + Hg^{2+} \overset{k_2}{\rightarrow} Co(DH)_2^+ + CH_3Hg^+ \qquad [8.19]$$

K_B is 3.5 $dm^3 \ mol^{-1}$ and k_2 is $c.$ 60 $dm^3 \ mol^{-1} \ s^{-1}$, demonstrating that CH_3CoB_{12} transfers a methyl group to mercuric ion much more rapidly than $CH_3Co(DH)_2H_2O$. Complex $CH_3Co(DBF_2)_2(H_2O)$, a BF_2 analogue of

$CH_3Co(DH)_2(H_2O)$, reacts much more slowly, and the k_2 value of $< 3.5 \times 10^{-4}$ dm^3 mol^{-1} s^{-1} is independent of pH demonstrating that reaction 8.18 does not occur.[37] The anionic surfactant sodium lauryl sulphate[35] increases the rate of methyl transfer to mercuric ion from $CH_3Co(DH)_2(H_2O)$ by a factor of 19 and from $CH_3Co((DO)(DOH)pn)^+$ by a factor of 131. A possible explanation for this behaviour is the proximity of the methyl donor and mercury(II) at the surface of the anionic micelles.

Two groups[32,33] have studied reactions of $CH_3Co(chel)$ (chel = salen, saloph, and $7,7-(CH_3)_2salen$) with mercury(II). Espenson *et al.*[32] noted that K_B values (equation 8.18) ranged from 6 to 17 dm^3 mol^{-1}, and that k_2 (equation 8.19) is about 10^4 dm^3 mol^{-1} s^{-1}. Tauzher *et al.*[33] interpreted similar data for salen differently. They suggested a rapid pre-equilibrium association with mercuric ion (Hg^{2+}) (equation 8.20) and the adduct's reaction (equation 8.21) in addition to chelate protonation (equation 8.18):

$$CH_3Co(salen) + Hg^{2+} \overset{K_3}{\rightleftharpoons} CH_3Co(salen) \cdot Hg^{2+} \qquad [8.20]$$

$$CH_3Co(salen) \cdot Hg^{2+} \overset{k_4}{\rightarrow} Co(salen)^+ + CH_3Hg^+ \qquad [8.21]$$

For salen the value of K_3 is 376 dm^3 mol^{-1}, K_B is 21 dm^3 mol^{-1}, and K_3k_4 (equivalent to k_2 of equation 8.19) is about 10^4 dm^3 mol^{-1} s^{-1}.

Toscano and Marzilli[1] in their review of organocobalt chemistry reviewed the electrochemistry of alkylcobalt complexes. The usual order of difficulty of reduction of cobalt(III) as a function of equatorial ligand is: bae > salen > methylcobalamin > ((DO)(DOH)pn). Table 8.1 shows that the second-order rate constants for the reactions of $CH_3Co(chel)$ generally agree with the electrochemical data and suggest an S_E2 mechanism for mercury(II) ion attack. That is, the decreasing difficulty in reducing the complexes relates to decreased negative character of methyl bound to cobalt, and decreased S_E2 rates. In addition the *trans* effect of one methyl on the other has a tremendous effect on the rates. $(CH_3)_2Co((DO)(DOH)pn)$ and $(CH_3)_2Co(tim)^+$ have k_2 values of 2×10^6 dm^3 mol^{-1} s^{-1} and 1.3×10^5 dm^3 mol^{-1} s^{-1}, respectively, towards the relatively weak electrophile phenylmercury ($C_6H_5Hg^+$). The relative rate constants of these two dimethylcobalt complexes demonstrate the decrease in k_2 with increased positive charge. This effect is much greater in the monomethyl analogues in which k_2 for $CH_3Co((DO)(DOH)pn)$ $(H_2O)^+$ is 10^4 greater than that for the doubly charged cation $CH_3Co(tim)(H_2O)^{+2}$. Further evidence is the approximately 10^4 decrease in k_2 values when electron withdrawing BF_2 replaces the hydrogen bonded proton to form $CH_3Co((DO)(DOBF_2)pn)(H_2O)^+$ or two protons to form $CH_3Co(DBF_2)_2(H_2O)$. All available evidence demonstrates that methylcobalt complexes react with mercuric ion by an S_E2 mechanism. Rate constants are given in Table 8.1.

There are a few metal complexes not described above that transfer methyl to mercury compounds. Espenson and Chao[39] studied reactions of $(H_2O)(DH)_2Co(CH_2)_4Co(DH)_2(H_2O)$ with mercury(II) in methanol, and observed that two moles of the non-protonated complex react with

Table 8.1   Second-order rate constants (k_2) for reactions between methylcobalt complexes and mercury(II)

Complex	k_2 (dm³ mol⁻¹ s⁻¹)	Reference
$(CH_3)_2Co((DO)(DOH)pn)$	2×10^6*	30
$(CH_3)_2Co(tim)^+$	1.3×10^5*	30
$CH_3Co(salen)$	$2.6 \times 10^4, 3.2 \times 10^4$	32, 33
$CH_3Co(7,7^1-(CH_3)_2salen)$	2.1×10^4	32
CH_3CoB_{12}	300 to 380	19–23
$CH_3Co(DH)_2(H_2O)$	54, 65	36, 37
$CH_3Co((DO)(DOH)pn)(H_2O)^+$	1.6 to 5.9	32–35
$CH_3Co(tim)(H_2O)^{2+}$	4.5×10^{-4}	32
$CH_3Co(DBF_2)_2(H_2O)$	$<3.5 \times 10^{-4}$	37
$CH_3Co((DO)(DOBF_2)pn)^+$	6.6×10^{-4}, 1.5×10^{-4}	33, 34

* Towards $C_6H_5Hg^+$.

mercury(II) to form methylmercury in methanol (CH_3OH).[40] The chromium(III) complex $CH_3Cr(chel)(H_2O)^{2+}$ (chel is a 15-membered ring with four nitrogen donors) reacts very rapidly with mercuric ion (Hg^{2+}) ($k_2 = 3.1 \times 10^6$ dm³ mol⁻¹ s⁻¹) and methyl mercury (CH_3Hg^+) ($k_2 = 1.6 \times 10^3$ dm³ mol⁻¹ s⁻¹).[41] These large rate constants show that Cr(III) complexes react much more rapidly than similar Co(III) complexes. In addition, $(CH_3)_3Sn^+$ reacts with mercury(II) chloride ($HgCl_2$) in a second-order process (equation 8.22).[42] The $(CH_3)_2Sn^{2+}$ product does not react with mercury(II).

$$(CH_3)_3Sn^+ + Hg^{2+} \rightarrow (CH_2)_2Sn^{2+} + CH_3Hg^+ \qquad [8.22]$$

Finally, there have been discussions of reactions between methylcobalt complexes and mercury(I).[29,43] Dodd and Johnson[43] noted that the reaction between $CH_3Co(DH)_2(H_2O)$ was much slower for mercury(I) perchlorate ($Hg_2(ClO_4)_2$) than for mercury(II) perchlorate ($Hg(ClO_4)_2$). Their conclusion is that the methyl acceptor is 0.8 per cent mercuric ion (Hg^{2+}) in equilibrium with the mercurous ion (Hg_2^{2+}). Mestroni *et al.*[29] contend that $(CH_3)_2Co((DO)(DOH)pn)$, $CH_3Co(bae)$, and $CH_3Co(salen)$ react with mercury(I) nitrate ($Hg_2(NO_3)_2$) to form mercury(0), but few details are given. As with CH_3CoB_{12} there is no direct proof of methyl transfer to mercury(I).

8.3   Methyl transfer to tin

8.3.1   Aquatic stability of organotin compounds

A small number of reports about the stability of organotins in water exists.

Most papers describe photodegradation of butyl-, cyclohexyl- and phenyltin fungicides, bacteriocides and antifoulants. Reviews by Blunden and Chapman[44] and Klein and Woggon[45] summarize the fate of these compounds in the environment. Direct environmental studies are covered in Chapter 4.

Photolytic degradation studies of butyltin compounds in water investigated by Maguire *et al.*[46] showed that different compounds decomposed at different rates in the presence of 300 or 350 nm radiation. The half-lives are: $(n-C_4H_9)_3Sn^+$ (18 days) > $(n-C_4H_9)_2Sn^{2+}$ (9 days) > $(n-C_4H_9)Sn^{3+}$ (0.4 day). In sunlight the half-life of $(n-C_4H_9)_3Sn^+$ is greater than 89 days. Bond cleavage proceeds via consecutive butyl group abstractions ultimately resulting in inorganic tin.

Blunden[47] recently studied rates of UV (ultraviolet) decomposition of the methyltin compounds $((CH_3)_3Sn^+, (CH_3)_2Sn^{2+},$ and $CH_3Sn^{3+})$ in D_2O using ^1H NMR, but observed no monomethyltin intermediate, methane, or ethane. The two alkanes have too low a solubility in water for NMR determination. Paper chromatography revealed only the presence of dimethyltin $((CH_3)_2Sn^{2+})$ and inorganic tin degradation products. Mössbauer spectroscopy proved that the inorganic tin product is hydrated tin(IV) oxide (SnO_2).

Soderquist and Crosby[48] reported that aqueous triphenyltin ions readily degrade by carbon–tin bond cleavage in sunlight or UV light to form bis(phenyltin) oxide $((C_6H_5)_2SnO)$. They also observed a water-soluble organotin polymer, $((C_6H_5)SnO_xH_y)_n$, but no tetraphenyltin, monophenyltin, or inorganic tin products. Studies with di- and monophenyltin species in water yield a similar distribution of products.

The paucity of the above studies demonstrates that the stability of carbon–tin bonds in the aquatic environment needs further investigation. Particularly, little is known about organotin stability to sunlight and the effects of binding to particulate matter during their chemical and biological transformations in natural water systems. Some fundamental aspects of light and water stability are also discussed in Chapter 1.

8.3.2   Organic reagents as methyl donors

The formation of tetramethyltin $((CH_3)_4Sn)$ from reactions of tin(II) and iodomethane (CH_3I) is controversial. Chau *et al.*[49] reported that reactions of tin(II) or tin(0) with the naturally occurring metabolite iodomethane in aqueous media yield methyltin compounds (viz. $(CH_3)_4Sn$, $(CH_3)_3Sn^+$, $(CH_3)_2Sn^{2+}$, and CH_3Sn^{3+}). They suggested a mechanism of successive methyl addition to tin(0) or tin(II) because of the relative yields (i.e. CH_3Sn^{3+} > $(CH_3)_2Sn^{2+}$ > $(CH_3)_4Sn$). Craig and Rapsomanikis[50] also observed the three methyltin ions but not tetramethyltin $((CH_3)_4Sn)$, when iodomethane and tin(II) react. They detected tetramethyltin only in the presence of tin(0), or tin(II) and magnesium(0). Standard acidic reduction potentials (Mg^{2+}/Mg, -2.4 V; Sn^{2+}/Sn, -0.14 V) suggest that magnesium(0) reduces tin(II) to tin(0). Then iodomethane can oxidatively add to tin(0) presumably forming an intermediate (CH_3SnI), which further reacts forming the dimethyl species $(CH_3)_2SnI_2$ (equations 8.23 and 8.24):

$$Sn + CH_3I \rightarrow CH_3SnI \qquad\qquad [8.23]$$

$$CH_3SnI + CH_3I \rightarrow (CH_3)_2SnI_2 \qquad\qquad [8.24]$$

Dismutation of $(CH_3)_2SnI_2$ leads to $(CH_3)_3SnI$ (equation 8.25):

$$2(CH_3)_2SnI_2 \rightarrow (CH_3)_3SnI + CH_3SnI_3 \qquad\qquad [8.25]$$

which further dismutates to tetramethyltin $((CH_3)_4Sn)$ and $(CH_3)_2SnI_2$. Reaction 8.26

$$2(CH_3)_3SnI \rightarrow (CH_3)_4Sn + (CH_3)_2SnI_2 \qquad\qquad [8.26]$$

is irreversible because the product $(CH_3)_2SnI_2$ reacts via equation 8.25. Reactions of tin(II) with iodomethane would not produce tetramethyltin $((CH_3)_4Sn)$ because of diminishing yields from the dismutation reaction of the original monomethyl oxidative addition product (CH_3Sn^{3+}). Rapsomanikis[51] also noted that in the heterogeneous reaction of iodomethane (with Sn(II) penicillamine) in water only sulphur of the amino acid methylates.

Some of the above reactions with iodomethane in aqueous media are precedented in organometallic chemistry. Murphy and Poller[52] reported that the reaction of tin(0) with iodomethane in hydrocarbons yields $(CH_3)_2SnI_2$ as the only product. Also Lappert *et al.*[53] established that the products of the reaction of tin(II) (SnY_2; Y = $((CH_3)_2Si)_2CH^-$ or $((CH_3)_3Si)_2N^-$) with iodomethane in hexane are CH_3SnY_2I.

Craig and Rapsomanikis[54] reported reactions of tin(0) and tin(II) salts with $(CH_3)_3S^+I^-$ and naturally occurring methylating agents dimethyl sulphide $((CH_3)_2S)$ and $(CH_3)_3\overset{+}{N}CH_2CO_2^-\cdot H_2O$. They did not find tetramethyltin, and noted that more $(CH_3)_2Sn^{2+}$ than $(CH_3)_3Sn^+$ occurs in the oxidative addition reactions.

Certain compounds promote dismutation of methyltin compounds. Craig and Rapsomanikis[50] proved that $(CH_3)_3Sn^+$ forms tetramethyltin in the presence of sulphide ion in light (equations 8.27 and 8.28):

$$2(CH_3)_3Sn^+ + S^{2-} \rightarrow ((CH_3)_3Sn)_2S \qquad\qquad [8.27]$$

$$3((CH_3)_3Sn)_2S \overset{h\nu}{\rightarrow} ((CH_3)_2SnS)_3 + 3(CH_3)_4Sn \qquad\qquad [8.28]$$

The photochemical dismutation step occurs with simulated daylight radiation (>240 nm). Reactions 8.27 and 8.28 might occur between sulphide ion and $(CH_3)_3Sn^+$ in natural waters. In a similar process[55] tropolone (T), a naturally occurring complexing agent[56] which has been used to extract methyl tin compounds, causes $(CH_3)_3Sn^+$ to dismutate in the presence of radiation

(>240 nm) or heat (>54 °C) (equation 8.29):

$$2(CH_3)_3SnT \xrightarrow[\text{heat}]{hv \text{ or}} (CH_3)_2SnT_2 + (CH_3)_4Sn \qquad [8.29]$$

Reactions 8.27 to 8.29 have important implications for the aqueous chemistry of $(CH_3)_3Sn^+$ and its environmental cycling via formation of air-transportable tetramethyltin.

8.3.3 Methylcobalamin as a methyl donor

Several studies[57-59] of reactions between tin(II) and CH_3CoB_{12} reveal that methyl transfer does not occur in the absence of dioxygen or other oxidizing agents. Dizikes *et al.*[58] found that $H_2OCoB_{12}^+$ or iron(III) is an appropriate oxidizing agent for initiation of methyl transfer, and formation of CH_3Sn^{3+}. In chloride media with excess tin(II) and iron(III) the second-order rate constant is 1.4 dm^3 mol^{-1} s^{-1}. Fanchiang and Wood[59] studied reactions between CH_3CoB_{12} and tin(II) chloride in 1 mol dm^{-3} chloride at pH 0 to 1 under aerobic or anaerobic conditions. No reaction occurs in the absence of dioxygen or a stoichiometric amount of the oxidizing agent $H_2OCoB_{12}^+$ equations 8.30 and 8.31):

$$CH_3CoB_{12} + Sn(II) \rightleftharpoons CoB_{12}r + CH_3Sn(III) \qquad [8.30]$$

$$CH_3Sn(III) + H_2OCoB_{12}^+ \rightleftharpoons CoB_{12}r + CH_3Sn(IV) + H_2O \qquad [8.31]$$

The function of $H_2OCoB_{12}^+$ is to oxidize the methyltin(III) ($CH_3Sn(III)$) intermediate. Dioxygen also fulfils the function of oxidizing agent (equations 8.32 and 8.33)

$$CH_3Sn(III) \stackrel{O_2}{\rightleftharpoons} CH_3Sn(IV) \qquad [8.32]$$

$$CoB_{12}r + H_2O \stackrel{O_2}{\rightleftharpoons} H_2OCoB_{12}^+ \qquad CoB_{12}r = Co(II)B_{12} \qquad [8.33]$$

and the second-order rate constant for reactions in 1 mol dm^{-3} HCl is 1.0 dm^3 mol^{-1} s^{-1}. The mechanism is an S_H2 attack of tin(II) on carbon. There is some controversy about the nature of the methyltin(IV) product. The Fanchiang–Wood paper on the basis of ^{13}C NMR in 1.0 mol dm^{-3} DCl–D$_2$O using 3 nmol dm^{-3} CH_3CoB_{12} and excess tin(II) chloride contend the product is a methylchlorotin anion with an unknown number of chloride ions. However, Craig and Rapsomanikis[60] using the ^1H NMR technique of Blunden[61] found different products for less than 5 mmol dm^{-3} concentrations of CH_3Sn^{3+}. The product distribution is: $CH_3Sn(OH)Cl_2(H_2O)_2$ (64 per cent), $(CH_3Sn(OH)(H_2O)_4)^{2+}$ (17 per cent) and $CH_3Sn(OH)_2Cl.nH_2O$ (19 per cent).

Reactions of CH_3CoB_{12} with methyltin(IV) compounds gave variable results with different researchers. Thayer[62] reported the slow disappearance of

CH_3CoB_{12} in the presence of $(CH_3)_3Sn(CH_3CO_2)$, but did not detect tetramethyltin. The above result does not agree with the results of Wood *et al.*[57] and Craig and Rapsomanikis.[60] The latter workers established that 5 to 15 mmol dm^{-3} solutions of $(CH_3)_3SnCl$ or $(CH_3)_2SnCl_2$ and 0.05 mmol dm^{-3} CH_3CoB_{12} do not react in 60 days. Reactions of 8.3 mmol dm^{-3} $(CH_3)_3SnCl$ or $(CH_3)_2SnCl_2$ with more concentrated 4.3 mmol dm^{-3} CH_3CoB_{12} produce less tetramethyltin than $(CH_3)_3SnCl$ or $(CH_3)_2SnCl_2$ blanks.

8.3.4   Cobalt complexes as methyl donors

There are few reports on reactions of tin compounds with metal complex methyl donors. Schrauzer *et al.*[63] studied demethylation of a large number of methylcobalt chelates. Methyl donors included methylcobinamide, CH_3CoB_{12}, $CH_3Co(DH)_2L$, $CH_3Co(DpH)_2(H_2O)$, $CH_3Co(DcH)_2Py$, $CH_3Co(GH)_2Py$, $CH_3Co(DBF_2)_2Py$, $CH_3Co((DO)(DOH)pn)(H_2O)^+$ and $CH_3Co(salen)(H_2O)$ (Fig. 8.1). The *trans* ligand L of $CH_3Co(DH)_2L$ is H_2O, $C_5H_5N(Py)$, $C_6H_{11}NC$, $P(n\text{-}C_4H_9)_3$, etc. Reactions of the methylcobalt donors with 0.22 mol dm^{-3} SnO_2^{2-} (stannite) in 3 mol dm^{-3} aqueous sodium hydroxide at 27°C produce Co(1)(chel) and sometimes methane as products, but no tetramethyltin. Under pseudo-first-order conditions relative rates are: methylcobinamide(1), $CH_3CoB_{12}(0.2)$, $CH_3Co(DH)_2(H_2O)(0.7)$, $CH_3Co((DO)(DOH)pn)(H_2O)^+$ (0.8), and $CH_3Co(salen)(H_2O)$ (very slow). The proposed mechanism has SnO_2^{2-} (stannite) attacking cobalt *trans* to the leaving methyl group. Cros *et al.*[64] studied reactions between $CH_3Co(DH)_2(S(CH_3)_2)$ and CH_3SnCl_3 in dichloromethane (CH_2Cl_2) and suggested formation of an adduct between CH_3SnCl_3 and the axial methyl group of the cobalt complex based on evidence of a large downfield shift of CH_3—Co in ^1H NMR spectra. In similar studies Darbieu and Cros[65] studied reactions in dichloromethane between CH_3SnCl_3 or $(CH_3)_2SnCl_2$ and several methylcobalt chelates (CH_3Co(chel), including $CH_3Co(DH)_2L$ (where L = H_2O, $S(CH_3)_2$), $CH_3Co(salen)$, $CH_3Co(bae)$, $CH_3Co((DO)(DOH)pn)(H_2O)^+$, and $CH_3Co(tim)Cl^+$). As noted above the authors observed upfield ^1H NMR shifts in CH_3—Co on addition of an organotin compound. CH_3SnCl_3 accepts a methyl group only from $CH_3Co(tim)Cl^+$ (equation 8.34):

$$CH_3Co(tim)Cl^+ + CH_3SnCl_3 \rightarrow (CH_3)_2SnCl_2 + Co(tim)Cl_2^+ \qquad [8.34]$$

8.4   Methyl transfer to lead

8.4.1   Aquatic stability of organolead compounds

Jarvie *et al.*[66] investigated the degradation of tetraalkyllead compounds in aqueous media. Sunlight, particulate surface area, and certain anions and cations accelerate the decomposition rate. Suspensions of tetraalkyllead compounds (R_4Pb, $R = C_2H_5$, CH_3) are quite stable in water in the dark. The ethyl compound decomposes 2 per cent over 77 days, and the methyl one 59 per cent in 22 days. Decomposition processes, which form R_3Pb^+ and traces of $(C_2H_5)_2Pb^{2+}$, accelerate in the presence of copper(II) or iron(II). Tetraalkyllead compounds rapidly adsorb on silica and decomposition to R_3Pb^+ occurs more rapidly for both $R = C_2H_5$ (97 per cent in 30 days) and $R = CH_3$ (55 per cent in 30 days).

The same paper[66] describes decomposition of the tri- and dialkyllead ions. Solutions of $R^1_3PbCl(R^1 = CH_3$, C_2H_5, n-C_4H_9) are stable in the dark; the only decomposition observed was 1 per cent of $(CH_3)_3PbCl$ in 220 days. Copper(II) and iron(II) do not increase the rate of R^1_3PbCl decomposition, but sulphide ion promotes formation of tetraalkyllead compounds. Sunlight increases the breakdown rate of all three trialkyllead ions. In 15 days there is 4 per cent loss of $(CH_3)_3Pb^+$, 99 per cent loss of $(C_2H_5)_3Pb^+$ and 25 per cent loss of (n-$C_4H_9)_3Pb^+$. Inorganic lead forms, and the ethyl compound yields a trace of $(C_2H_5)_2Pb^{2+}$. Silica adsorbs R^1Pb^+ and slightly increases their decomposition rates. $R^1_2PbCl_2$ compounds ($R^1 = CH_3$, C_2H_5, n-C_4H_9) also react slowly in the dark in water. Over 30 days there is 10 per cent decomposition for $R^1 = CH_3$, 6 per cent for $R^1 = C_2H_5$, and 4 per cent for $R^1 = $ n-C_4H_9. The products are $R^1_3Pb^+$, lead(II), and presumably R^1Cl. Decomposition reactions are much faster in sunlight. Within 40 days 5 per cent of $(CH_3)_2PbCl_2$, 25 per cent of $(C_2H_5)_2PbCl_2$, and 70 per cent of (n-$C_4H_9)_2PbCl_2$ react. The overall conclusion from this work is that alkyllead compounds, e.g. emitted from vehicle exhausts and deposited in waterways, undergo fairly rapid decomposition in the presence of light, various particulate surfaces, and other cations or anions. Direct environmental details are given in Chapter 4.

A study of degradation of pure tetraallkyllead compounds $(CH_3)_4Pb$ and $(C_2H_5)_4Pb$ and their commercial mixtures containing bromo- and chloroalkanes by Charlou et al.[67] presents more information about degradation processes of tetraalkyllead in pure water and seawater. Daylight radiation decomposes pure tetraethyllead in seawater at pH 8.1 forming $(C_2H_5)_3Pb^+$, $(C_2H_5)_2Pb^{2+}$, and eventually lead(II). The same process appears to take place with commercial tetraethyllead, but pH decreases, perhaps due to the formation of hydrochloric and -bromic acids (HCl and HBr) from the photolysis of dibromomethane and dichloroethane scavengers. Degradation accelerates with stirring, and the total lead in seawater increases due to greater solubility of $(C_2H_5)_3Pb^+$ and $(C_2H_5)_2Pb^{2+}$ than tetraethyllead (2 mg dm^{-3}). Inorganic lead(II) precipitates (see below) as brilliant white flakes. In 4 hours tetraethyllead decomposes 1 per cent in the dark, 20 per cent in daylight, and 97 to 100 per cent in artificial UV light. The presence of carbon dioxide (CO_2) in the aqueous medium of the last two experiments results in precipitation of $Pb_2(CO_3)_2(OH)_2$, and commercial tetraethyllead forms $Pb(OH)Cl$. Analysis of head space gases from degradation of pure and commercial tetraalkyleads ($(C_2H_5)_4Pb$ and $(CH_3)_4Pb$) by daylight radiation over six months in distilled water or seawater revealed methane, ethane, propane, butane, ethylene, acetylene, propylene, acetone, and ethanol. The only unusual organolead compound detected was $(n\text{-}C_4H_9)(C_2H_5)_3Pb$.

Hitchen et al.[68] described aqueous reactions of $(CH_3)_3Pb^+$ and $(CH_3)_2Pb^{2+}$ with zinc(0) (equations 8.35 and 8.36):

$$2(CH_3)_3PbCl + 4Zn + 3H_2O \rightarrow 6CH_4 + 2Pb + 3ZnO + ZnCl_2 \qquad [8.35]$$

$$(CH_3)_2PbCl_2 + 2Zn + H_2O \rightarrow 2CH_4 + Pb + ZnO + ZnCl_2 \qquad [8.36]$$

The authors expect the reactions to have applications for removal of organolead compounds from effluents resulting from the manufacture of tetraalkyllead compounds.

8.4.2   Organic reagents as methyl donors

The products formed from reactions between lead(II) and carbocation donors such as iodomethane are a source of controversy. Ahmad et al.[69] first reported aqueous organometallic reactions of lead(II) salts with iodomethane. The authors observed larger amounts of tetramethyllead formed at pH 13 (10^{-4} per cent yield) than at pH 5 to 6. Also $(CH_3)_3S^+I^-$, $(CH_3)_3SO^+I^-$, and $(CH_3)_3O^+SbCl_6^-$ appeared to produce tetramethyllead but in smaller overall yields. The authors explained tetramethyllead production in the presence of sodium hydroxide by postulating that it would remove iodide ion from the methyllead iodide intermediates. In similar experiments under closely controlled conditions Jarvie et al.[70] detected no tetramethyllead but observed $(CH_3)_2Pb^{2+}$ and $(CH_2)_3Pb^+$ by differential pulse anodic stripping voltametry (DPASV). Craig and Rapsomanikis[50,54] agree with the results of Jarvie and Whitmore.[70] They detected only $(CH_3)_3Pb^+$ and $(CH_3)_2Pb^{2+}$ by UV spectrophotometry and 400 MHz ^1H NMR after reactions of lead(II) and iodomethane. Addition of magnesium(0) to reactions resulted in enhanced $(CH_3)_3Pb^+$ and $(CH_3)_2Pb^{2+}$ yields but no tetramethyllead. Contrary to Ahmad et al.'s[69] results $(CH_3)_3S^+I^-$ did not form tetramethyllead from lead(II). Snyder and Bentz[71] agree with the Craig and Jarvie groups that reactions of lead(II) and iodomethane in water do not yield $(CH_3)_4Pb$. They observed that aluminium foil reduces lead(II) to lead(0) which is methylated with iodomethane to tetramethyllead. They concluded that aluminium foil used by earlier workers[69] caused tetramethyllead formation via lead(0) formed by reduction of lead(II). Addition reactions of lead(II) and $(CH_3)_3{}^+NCH_2COO^- \cdot H_2O$ produced no methyllead products,[54] and attempts to synthesize a monomethyllead(IV) complex by reactions of iodomethane and Pb(II) penicillamine[51] resulted only in methylation of sulphur of the amino acid.

There is no disagreement about the formation of methyllead compounds by oxidative addition reactions of lead(0). Jarvie and Whitmore[70] showed that reactions of lead(0) with iodomethane produced tetramethyllead as well as $(CH_3)_3Pb^+$ and $(CH_3)_2Pb^{2+}$. They also observed that $(CH_3)_3Pb^+$ dismutates in the presence of lead(0) (equation 8.37):

$$2(CH_3)_3Pb^+ + Pb \rightarrow (CH_3)_2Pb^{2+} + (CH_3)_4Pb \qquad [8.37]$$

Craig and Rapsomanikis[54] noted that $(CH_3)_3S^+I^-$ reacts with lead(0) to form $(CH_3)_2Pb^{2+}$.

In summary it appears that lead(0) undergoes oxidative addition by carbocation donors to form CH_3Pb^+ which reacts quickly forming $(CH_3)_2Pb^{2+}$ (equations 8.38 and 8.39):

$$Pb + CH_3I \rightarrow CH_3PbI \text{ (unstable, lead(II) compound)} \qquad [8.38]$$

$$CH_3PbI + CH_3I \rightarrow (CH_3)_2PbI_2 \qquad [8.39]$$

Then $(CH_3)_2PbI_2$ decomposes and the resulting $(CH_3)_3PbI$ dismutates (equations 8.40 and 8.41):

$$2(CH_3)_2PbI_2 \rightarrow (CH_3)_3PbI + PbI_2 + CH_3I \qquad [8.40]$$

$$4(CH_3)_3PbI \rightleftharpoons 2(CH_3)_4Pb + 2(CH_3)_2PbI_2 \qquad [8.41]$$

Lead(0) increases the rate of reaction 8.40 by removing iodomethane via reaction 8.38. The reaction of Pb(II) with CH_3I probably proceeds via oxidative addition followed by dismutation of CH_3Pb^{3+} (equations 8.42 and 8.43):

$$PbX_2 + CH_3I \rightleftharpoons CH_3PbIX_2 \qquad [8.42]$$

$$2CH_3PbIX_2 \rightarrow (CH_3)_2PbX_2 + PbI_2X_2 \qquad [8.43]$$

Reaction 8.42 is not an entirely satisfactory explanation for the observed $(CH_3)_2Pb^{2+}$ products, because CH_3PbIX_2 might not be sufficiently stable to dismutate. Formation of tetramethyllead via equations 8.40–8.43 is not observed because of a low yield from four consecutive reactions. However, Rapsomanikis[51] identified lead(II) iodide as a product by lead analysis and Fourier transform far infrared spectroscopy in reactions between lead(II) nitrate and iodomethane. Lead(II) iodide could originate from decomposition of the PbI_2X_2 product of reaction 8.43.

Jarvie *et al.*[72] first reported that methylation of $(CH_3)_3Pb^+$ to tetramethyllead was not the biological process first reported.[73] Instead all methyl groups in tetramethyllead originated from $(CH_3)_3Pb^+$ via reaction with sulphide ion in the sediment (equations 8.44 and 8.45):

$$2(CH_3)_3Pb^+ + S^{2-} \rightarrow ((CH_3)_3Pb)_2S \qquad [8.44]$$

$$((CH_3)_3Pb)_2S \rightarrow (CH_3)_4Pb + (CH_3)_2PbS \qquad [8.45]$$

The conclusion is strengthened by the fact that $(C_2H_5)_3Pb^+$ produces tetraethyllead not $(C_2H_5)_3CH_3Pb$. Jarvie *et al.*[74] reaffirmed the sulphide-mediated pathway for tetramethyllead formation by studies of $(R_3Pb)_2S$ (R is CH_3 and C_2H_5). Both compounds decompose over 19 hours in aqueous solution at pH 7 to form R_4Pb. Under acidic conditions only dimethyl sulphide is produced. Aqueous reactions of $(CH_3)_3PbCl$ and sodium sulphide (2/1 Pb/S molar ratio) produce maximum amounts of tetramethyllead, but excess sulphide ion diminishes the yield. Anions such as hydroxide, carbonate or selenide, do not promote tetramethyllead formation. In agreement with the above results Reisinger et al.[75] found that the presence of sodium sulphide, cysteine, or methionine enhances yields of tetraalkyllead (R = CH_3, C_2H_5) from R_3Pb^+ in aqueous media. Craig[76] also observed that production of tetramethyllead from $(CH_3)_3Pb^2$ in aqueous sterile sediments is due to reactions 8.44 and 8.45.

8.4.3   Methylcobalamin as a methyl donor

There are some studies of reactions between CH_3CoB_{12} and inorganic lead salts. No publications contend that CH_3CoB_{12} and lead(II) salts react in water to form organolead compounds. Two publications,[60,77] describing research done over a wide variety of experimental conditions, report negative results. In contrast Taylor and Hanna[78] reported demethylation of $^{14}CH_3CoB_{12}$ by Pb_3O_4, lead(IV) oxide and lead(IV) acetate at pH 2 in the dark, but detected no organolead products. Thayer[79] observed that lead(IV) oxide and CH_3CoB_{12} in the presence of acetic acid produce traces of tetramethyllead, methane, methanol and acetone. This means that lead(IV) oxides in sediments would form tetramethyllead by reacting with CH_3CoB_{12}.

Methylation of organolead ions by CH_3CoB_{12} is more controversial than methylation of inorganic lead. Jarvie et al.[72] found no tetramethyllead after attempted methylation of $(CH_3)_3Pb^+$ or $(CH_3)_2Pb^{2+}$ by CH_3CoB_{12} in aqueous solution under conditions that methylate mercury(II). Craig and Rapsomanikis[60] made similar observations and noted that blanks produce more tetramethyllead than the reactions studied. Rhode and Weber[80] also obtained negative results. In contrast Wood et al.[57] noted that CH_3CoB_{12} and $(C_2H_5)_2PbCl_2$ reacted very slowly to form $(CH_3)(C_2H_5)_2PbCl$. Thayer[62,81] reported slow demethylation of CH_3CoB_{12} by $(CH_3)_3Pb(CH_3CO_2)$, but found no organolead products.

8.4.4   Metal complexes as methyl donors

Several papers describe methylation of lead(II) by $(CH_3)_2Co((DO)(DOH)pn)$ or $(CH_3)_2Co(tim)^+$ in non-aqueous solvents. Witman and Weber[82,83] reacted lead(II) and $(CH_3)_2Co((DO)(DOH)pn)$ (1/1 molar ratio) in 2-propanol, and postulated $CH_3Pb(II)^+$ as an intermediate with a 60 hour half-life. Reactions with excess methyl donor have a 2 methyl donor/1 Pb^{2+} stoichiometry. The methyllead product evolves 1 mole of methane over about 1 hour (equations 8.46–8.48):

$$(CH_3)_2Co((DO)(DOH)pn) + Pb^{2+} \xrightarrow{\text{fast}} CH_3Co((DO)(DOH)pn)^+$$
$$+ CH_3Pb^+ \hspace{4cm} [8.46]$$

$$(CH_3)_2Co((DO)(DOH)pn) + CH_3Pb^+ \xrightarrow{slow} CH_3Co((DO)(DOH)pn)^+$$
$$\text{'(CH}_3)_2Pb\text{'} \qquad\qquad\qquad\qquad\qquad\qquad\qquad\qquad [8.47]$$

$$\text{'(CH}_3)_2Pb\text{'} + CH_3CHOHCH_3 \rightarrow CH_4 + \text{Unknown methyllead product} [8.48]$$

The researchers did not characterize the methyllead product of reaction 8.48. Reactions between lead(II) and $(CH_3)_2Co((DO)(DOH)pn)$ in acetonitrile[84] produce tetramethyllead when the methyl donor/Pb^{2+} molar ratio is 2/1 or greater (equations 8.49 and 8.50). The authors identified lead(0) by X-ray powder diffraction and tetramethyllead by GC.

$$2(CH_3)_2Co((DO)(DOH)pn) + Pb^{2+} \rightarrow 2CH_3Co((DO)(DOH)pn)^+$$
$$+ \text{'(CH}_3)_2Pb\text{'} \qquad\qquad\qquad\qquad\qquad\qquad\qquad [8.49]$$

$$\text{'(CH}_3)_2Pb\text{'} \xrightarrow{fast} 0.5\,Pb + 0.5(CH_3)_4Pb \qquad\qquad [8.50]$$

The lead is quantitatively 50 per cent each of lead(0) and tetramethyllead.

Aqueous reactions between $(CH_3)_2Co(tim)^+$ and lead(II) produce low yields of tetramethyllead.[80,85,86] Rhode and Weber[80] studied the separate and combined effects of iodomethane, $(CH_3)_2Co(tim)^+$, and manganese(IV) oxide on methyl transfer to lead(II). They observed that tetramethyllead formed in anaerobic reactions in the dark only when $(CH_3)_2Co(tim)^+$ is present. Rapsomanikis *et al.*[86] quantitatively determined low (0.04 to 0.18 per cent) yields for the reactions.

Methyl transfer from methylcobalt donors to methyllead(IV) ions also occurs. Dimmit and Weber[84] proved that in acetonitrile $(CH_3)_2Co(chel)$ (chel = $((DO)(DOH)pn)$, tim) quantitatively forms tetramethyllead (equations 8.51 and 8.52):

$$(CH_3)_2Co(chel) + (CH_3)_3Pb^+ \rightarrow CH_3Co(chel)^+ + (CH_3)_4Pb \qquad [8.51]$$

$$2(CH_3)_2Co(chel) + (CH_3)_2Pb^{2+} \rightarrow 2CH_3Co(chel)^+ + (CH_3)_4Pb \qquad [8.52]$$

Reactions in water[86] between $(CH_3)_2Co(tim)^+$ and methyllead(IV) ions result in high tetramethyllead yields of 33–61 per cent $(CH_3)_3Pb^+$ and 41–65 per cent for $(CH_3)_2Pb^{2+}$. The use of $(CD_3)_2Co(tim)^+$ as methyl donor followed by MS analysis demonstrates that CD_3 is transferred from cobalt to methyllead(IV) ions.

Ethylation of lead(II) compounds by triethylboron $((C_2H_5)_3B)$ occurs in aqueous and organic media. This reaction, reported by Honeycutt and Riddle[10] produces tetraethyllead in 18–42 per cent yields. $Na^+B(C_2H_5)_4^-$ [87] produces yields of 14–98 per cent depending on the reaction conditions and lead(II) species. They did not report analogous methylation reactions.

8.5   Methyl transfer to other metals and metalloids

8.5.1   Platinum(II/IV) – a redox mechanism

Many groups[77,81,88–94] have studied reactions of platinum compounds with

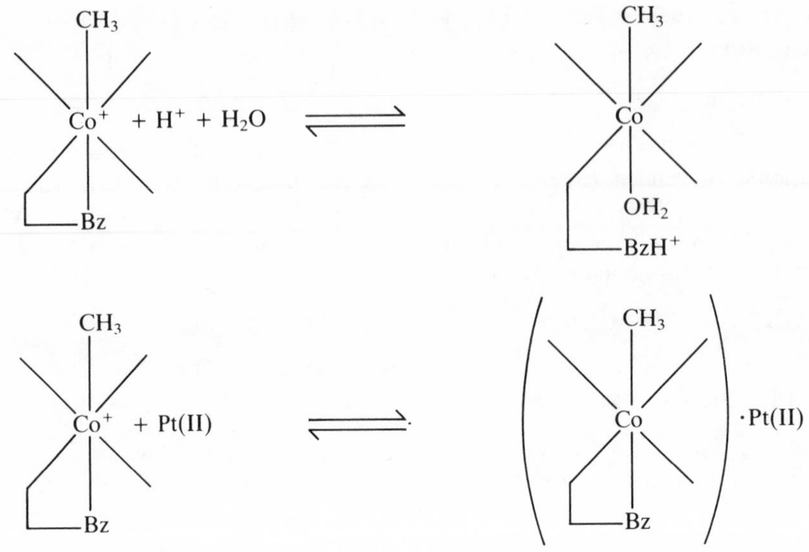

$$CH_3\text{-}Co^+\text{-}Bz + H^+ + H_2O \rightleftharpoons CH_3\text{-}Co\text{-}OH_2\text{-}BzH^+$$

$$CH_3\text{-}Co^+\text{-}Bz + Pt(II) \rightleftharpoons \left(CH_3\text{-}Co\text{-}Bz\right)\cdot Pt(II)$$

$$\underline{1}$$

$$\underline{1} + Pt(IV)^* \rightleftharpoons \underline{1}\cdot Pt(II)\cdot Pt(IV)^* \text{ (Species } \underline{2})$$

$$\underline{2} \xrightarrow{\text{fast}} H_2OCoB_{12}^+ + CH_3Pt(IV) + Pt(II)^*$$

Possible intermediate steps between $\underline{2}$ and products (assuming a Pt(IV)-Cl-Pt(II)bridge):

$$Pt(IV)^*\text{-}Cl\text{-}Pt(II)\cdot CH_3CoB_{12} \xrightarrow[\text{transfer}]{Cl^+} Pt(II)^* + Cl\text{-}Pt(IV)\cdot CH_3CoB_{12}$$
$$\underline{2}$$

$$\xrightarrow[\text{transfer}]{CH_3^-} ClPt(IV)CH_3 + H_2OCoB_{12}^+$$

Fig. 8.3 Reaction of CH_3CoB_{12} with platinum(II) and (IV)

CH_3CoB_{12}. The more recent detailed studies agree on the following points (Fig. 8.3): (1) platinum(II) and platinum(IV) are both necessary for methyl transfer reactions; (2) platinum(II) is recycled and platinum(IV) is consumed, and $H_2OCoB_{12}^+$ is the cobalt product; (3) platinum(II) and platinum(IV) associate with CH_3CoB_{12} before methyl transfer, i.e., there is a trinuclear intermediate (species $\underline{2}$) containing platinum(II), platinum(IV), and CH_3CoB_{12}; (4) labelled reactant platinum(IV) forms a platinum(II) product, and reactant platinum(II) forms, at least initially, a methylplatinum(IV) product. That is, platinum(II) is oxidized and platinum(IV) reduced. Fanchiang et al.[93] contend the redox process occurs via transfer of a bridging chloride from platinum(IV) to platinum(II) as positive chlorine (Cl^+). A reasonable alternative might involve carbocation transfer after oxidation as observed in electrochemical redox studies of CH_3CoB_{12}.[95]

The major disagreement between the Fanchiang[92] and Taylor[90] groups is the nature of the methylplatinum product. Taylor *et al.*[90] contend that $CH_3Pt(II)Cl_3^{2-}$ is the product despite their agreement that a methylplatinum(IV) ($CH_3Pt(IV)$) product is initially formed. They offer no plausible explanation of what species donates electrons for the reduction of methylplatinum(IV) to methylplatinum(II). Their evidence for $CH_3Pt(II)Cl_3^{2-}$ includes finding 3.1 mol Cl^-/mol Pt complex by ^{35}Cl NMR, and identification of platinum(II) by X-ray photoelectron spectroscopy (ESCA) measurements. Fanchiang *et al.*[92] persuasively contend that the product is $CH_3Pt(IV)Cl_5^{2-}$. Their major evidence is that ^{195}Pt NMR of the product has a chemical shift of -750 p.p.m. compared to -1619 p.p.m. for $PtCl_4^{2-}$ and 0.0 p.p.m. for $PtCl_6^{2-}$ (reference compound). Since methylation of *trans*-$PtCl_2(P(C_2H_5)_3)_2$ (-3938 p.p.m.) to *trans*-$CH_3PtCl(P(C_2H_5)_3)_2$ (-4509 p.p.m.) causes an upfield shift, it is reasonable that $CH_3PtCl_5^{2-}$ may be 750 p.p.m. upfield of $PtCl_6^{2-}$. Furthermore they contend that Taylor *et al.*'s ESCA results are ambiguous. Figure 8.3 shows the overall mechanism and a reasonable choice of a trinuclear intermediate (some Cl^- groups are omitted).

Finally $(CH_3)_3Sn^+$ reacts with $PtCl_4^{2-}$ in a second-order process to form platinum(0) and chloromethane (equations 8.53 and 8.54):[42]

$$(CH_3)_3Sn^+ + PtCl_4^{2-} \rightarrow (CH_3)_2Sn^{2+} + \text{'}CH_3PtCl_3^{2-}\text{'} + Cl^- \qquad [8.53]$$

$$\text{'}CH_3PtCl_3^{2-}\text{'} \rightarrow CH_3Cl + Pt + 2Cl^- \qquad [8.54]$$

A proposed intermediate of Rapsomanikis and Weber, '$CH_3PtCl_3^{2-}$' undergoes reductive elimination to the observed final products in reaction 8.54.

8.5.2 Gold(III) – a redox mechanism

Several groups[77,81,96] have studied reactions of CH_3CoB_{12} with gold(III) compounds. Fanchiang[96] has done the most detailed study using $Au(III)X_4^-$ ($X = Cl^-$, Br^-). Under anaerobic conditions 2 mol of $Au(III)X_4^-$ per mol of CH_3CoB_{12} effect a reaction (equation 8.55). According to this process CH_3CoB_{12} is

$$CH_3CoB_{12} + 2Au(III)X_4^- \rightarrow CH_3CoB_{12}(OX) + 2Au(0) + CH_3X \qquad [8.55]$$

oxidized by four electrons. In the presence of air, the reduced gold species are reoxidized to gold(III) and the $Au(III)X_4^-/CH_3CoB_{12}$ ratio is much less than one. The reaction is pseudo-first-order in $Au(III)X_4^-$ in the presence or absence of air, and the presence of gold(I) (AuX_2^-) does not affect the kinetics. The latter observation demonstrates that in contrast to platinum(II) and platinum(IV) gold compound reactions with CH_3CoB_{12} do not require gold(I) with gold(III). Previous work[77] shows that gold(I) alone does not react with CH_3CoB_{12}. The reaction mechanism[96] is shown in equations 8.56 to 8.60:

$$CH_3CoB_{12} + Au(III)X_4^- \overset{K}{\rightleftharpoons} CH_3CoB_{12} \cdot Au(III)X_4^- \qquad [8.56]$$

$$CH_3CoB_{12} \cdot Au(III)X_4^- \xrightarrow[\text{transfer}]{\text{electron}} CH_3CoB_{12}^+ \cdot Au(II)X_4^{2-} \qquad [8.57]$$

$$CH_3CoB_{12}^+ \cdot Au(II)X_4^{2-} + H_2O \overset{\text{fast}}{\rightleftharpoons} H_2OCoB_{12} + \text{'}Au(II)\text{'} + CH_3^\bullet \quad [8.58]$$

$$CH_3^\bullet + Au(III)X_4^- \rightarrow CH_3X + \text{'}Au(II)\text{'} \qquad [8.59]$$

$$2\text{'}Au(II)\text{'} + H_2OCoB_{12} \rightarrow 2Au(0) + H_2OCoB_{12}(OX) \qquad [8.60]$$

The one-electron oxidation and methyl radical transfer are analogous to the proposed electrochemical one-electron oxidation and carbocation (CH_3^+) transfer proposed for CH_3CoB_{12}.[95] Direct carbocation (CH_3^+) transfer is an alternative possibility.

The reaction between $(CH_3)_3Sn^+$ and $Au(III)Cl_4^{-}$[42] occurs by second-order kinetics (equation 8.61). The products given for the reaction

$$(CH_3)_3Sn^+ + Au(III)Cl_4^- \rightarrow (CH_3)_2Sn^{2+} + CH_3Cl + Au^0 \qquad [8.61]$$

are somewhat puzzling because reductive elimination from methylgold(III) would form gold(I), not gold(0), and chloromethane.

8.5.3   Chromium(II) − a methyl radical acceptor

Espenson and Shveima[97] studied reactions of $CH_3Co(DH)_2(H_2O)$ with chromium(II) ion in acidic media. The reaction goes by 1:1 stoichiometry (equation 8.62):

$$CH_3Co(DH)_2(H_2O) + Cr^{2+} + 2H^+ \rightarrow Co(H_2O)_6^{2+} + 2DH_2$$
$$+ CH_3Cr(H_2O)_5^{2+} \qquad [8.62]$$

Unlike the analogous mercury(II) reaction (equations 8.18 and 8.19), methyl transfer occurs from non-protonated $CH_3Co(DH)_2(H_2O)$ ($k_2 = 14$ $dm^3 mol^{-1} s^{-1}$) *and* protonated $CH_3Co(DH)$ $(DH)(H_2O)^+$ ($k_2 = 23$ dm^3 $mol^{-1} s^{-1}$). These results strongly dispute an S_E2 mechanism such as that found for mercury(II). The most reasonable mechanism based on the observations is a methyl radical transfer and S_H2 mechanism. Such a process explains the methylchromium(III) product; $Co(H_2O)_6^{2+}$ and protonated ligand (DH_2) occur because the initial product $Co(II)(DH)_2$ decomposes quickly in acidic aqueous media. This paper gives similar results for $CH_3Co(salen)(H_2O)$ ($k_2 = 76$ $dm^3 mol^{-1} s^{-1}$) and $CH_3Co(bae)$ (H_2O) ($k_2 = 57$ $dm^3 mol^{-1} s^{-1}$). In contrast the S_E2 reactions with mercury(II) discussed above (Table 8.1) have much faster rates for salen than for $(DH)_2$.

Espenson and Sellers[98] made similar observations in reactions of CH_3CoB_{12} and chromium(II) (equation 8.62). The reaction is pseudo-first-order in chromium(II)

$$CH_3CoB_{12} + Cr^{2+} \rightarrow CoB_{12}r + CH_3Cr(III) (H_2O)_5^{2+}$$
$$(NB\, CoB_{12}r = Co(II)B_{12} = CoB_{12}^\bullet) \qquad [8.63]$$

The rate constant of 360 dm^3 mol^{-1} s^{-1} is independent of pH so the base-on and base-off forms react at similar rates. As in the preceding discussion of $CH_3Co(chel)$, the kinetics and products suggest an S_H2 mechanism.

8.5.4 Palladium(II) – a carbanion acceptor

Scovell[99] and a Russian group[100-102] studied reactions of CH_3CoB_{12} and palladium(II). The Russian group[100-102] noted that $PdCl_4^{2-}$ and CH_3CoB_{12} rapidly form an adduct of 1:1 stoichiometry in dimethylsulphoxide and conclude that palladium(II) is bonded to the benzimidazole group. Scovell[99] observed the products of reaction (8.64). The chloromethane and palladium(0) products

$$CH_3CoB_{12} + PdCl_4^{2-} \rightarrow H_2OCoB_{12}^+ + CH_3Cl + Pd + 3Cl^- \qquad [8.64]$$

presumably originate from the reductive elimination of chloromethane from $CH_3PdCl_3^{2-}$. The mechanism is like that shown for mercury(II) in Fig. 8.2. $PdCl_4^{2-}$ binds to benzimidazole of CH_3CoB_{12} (K = 150 dm^3 mol^{-1}) and the remaining equilibrium concentration of base-on CH_3CoB_{12} reacts with $PdCl_4^{2+}$ (k_2 = 7.7 × 10^{-3} dm^3 mol^{-1} s^{-1}). However, the low k_2 value shows that palladium(II) is a much poorer electrophile towards CH_3CoB_{12} than mercury(II).

8.5.5 Thallium(III) – a carbanion acceptor

Abley *et al.*[36] noted that thallium(III) perchlorate ($Tl(ClO_4)_3$) and $CH_3Co(DH)_2(H_2O)$ react with 1:1 stoichiometry, but detected no methylthallium(III) (CH_3Tl^{2+}). This result is not surprising due to the instability of $CH_3Tl(CH_3CO_2)_2$ in H_2O.[103] The non-tin products of the reaction between $(CH_3)_3Sn^+$ and thallium(III) chloride are chloromethane and thallium(I), suggesting that intermediate CH_3TlCl_2 decomposes by reductive elimination of chloromethane.[42]

Studies of thallium(III) reactivity towards CH_3CoB_{12} are numerous, but not very detailed.[62,77,81] Agnes *et al.*[77] observed that thallium(III), but not thallium(I), reacts with CH_3CoB_{12}. The reaction shows second-order kinetics over a wide pH range and added chloride ion yields chloromethane as a product. Thayer[81] noted that TlX_4^- (X = $CH_3CO_2^-$, Cl^-, Br^-) and CH_3CoB_{12} react with second-order kinetics, and that thallium(I) does not affect the kinetics. Thayer[62] also showed that $(CH_3)_2Tl(CH_3CO_2)$ formed $H_2OCoB_{12}^+$ from CH_3CoB_{12} at pH 4.55 by demethylation.

8.5.6 Arsenic(III) – a carbocation acceptor

Antonio *et al.*[104] demonstrated that several carbocation-donating reagents react with arsenic(III) species to form methylarsenic(V) compounds. For example, $Na^+ CH_3SO_4^-$ partially reacts with AsO_3^{3-} and $C_6H_5AsO_2^{2-}$ at 25 °C to form $CH_3AsO_3^{2-}$ and $(C_6H_5)(CH_3)AsO_2^-$, respectively. S-methylmethionine sulphonium ion $((CH_3)_2SCH_2CHNH_2CO_2H)$ partially transfers methylcarbocation (CH_3^+) to $CH_3AsO_2^{2-}$. Trimethysulphonium ion $((CH_3)_3S^+)$ reacts at pH 12 and 80 °C with arsenous acid (equation 8.65), $CH_3As(OH)_2$ (equation 8.66), and $(CH_3)_2As(OH)$ (equation 8.67):

$$As(OH)_3 + (CH_3)_3S^+ \rightarrow CH_3AsO(OH)_2 + (CH_3)_2S + H^+ \qquad [8.65]$$

$$CH_3As(OH)_2 + (CH_3)_3S^+ \rightarrow (CH_3)_2AsO(OH) + (CH_3)_2S + H^+ \qquad [8.66]$$

$$(CH_3)_2As(OH) + (CH_3)_3S^+ \rightarrow (CH_3)_3AsO + (CH_3)_2S + H^+ \qquad [8.67]$$

In the above examples the exact nature of the arsenic(III) and arsenic(V) compounds depends on pH. However, attempted reactions between CH_3CoB_{12} and the arsenic(V) compounds $CH_3AsO_3^{2-}$ and $(CH_3)_2AsO_2^-$ show no formation of $H_2OCoB_{12}^+$ from CH_3CoB_{12}.[62] For further details on the environmental aspects of organarsenic chemistry see Chapter 5.

8.5.7   Miscellaneous metals as methyl acceptors

There are in the literature methyl-accepting reactions of several other metals. Robinson and Kiesel,[105] on the basis of slight evidence and non-reproducible experiments, say that CH_3CoB_{12} methylates cadmium(II), but another group[77] disagrees. Not even $(CH_3)_3Sn^+$ transfers methyl to cadmium(II).[42] Two groups[21,23] noted that CH_3CoB_{12} does not transfer a methyl group to silver(I). Chu and Gruenwedel[21] observed that copper(II) does not demethylate CH_3CoB_{12}, but Thayer[81] using copper(II) bromide found bromomethane as a product. It probably occurs after methyl transfer followed by reductive elimination. Agnes et al.[106] observed a free radical demethylation process when acidic iron(III) was added to CH_3CoB_{12} but no stable methylation product formed.

8.6   Conclusions

This chapter has reviewed environmentally significant reactions of metals and metalloids. Emphasis was on methyl transfer reactions because methyl is the only organic group known to be transferred to a metal or metalloid in the environment. Non-methyl organometallic compounds in the environment have an anthropogenic origin. Aquatic methylating agents include naturally occurring iodomethane, CH_3CoB_{12}, humic acids, etc., and synthetic compounds such as $(CH_3)_2Co(tim)^+$, $(CH_3)_3S^+I^-$ and $(CH_3)_3Sn^+$. This variety of methylating agents reveals that methyl transfer reactions encompass electrophilic, nucleophilic, or free radical attack by the methyl acceptor on the donor. All three mechanisms occur in aqueous media depending on the oxidation state, charge, and other ligands of the methyl donors and acceptors. Naturally occurring ligands like sulphide ion, cysteine, methionine, and tropolone are probably involved in converting methylmetal ions into fully methylated, volatile molecules via dismutation processes. The review also critically evaluated controversies over the mechanisms of and products from several reactions.

Historically, most researchers have studied organometallic reactions in non-aqueous media because of imagined or real reactions of water with reactants, intermediates, or products. It is true that water as solvent often decreases yields of organometallic reactions, but bioaccumulation makes processes with low yields important. It is emphasized that alkyl transfer reactions do occur in aqueous media, and in some cases, such as methylation of $(CH_3)_3Pb^+$ and $(CH_3)_2Pb^{2+}$ by $(CH_3)_2Co(tim)^+$ and ethylation of Pb^{2+} by $(C_2H_5)_4B^-$, are quantitative. The above examples present synthetically and environmentally significant results, and it is hoped they will trigger further interest in aqueous organometallic chemistry.

Laboratory experiments described in this chapter are a secure foundation for more difficult environmental experiments. This is true because the use of synthesized, well-characterized reagents and identification of pure products improve understanding of more complex processes. Also, such reactions help distinguish between chemical and enzymatic (within a living cell) methylation of metals and metalloids. However, because of extremely low yields of organometallic products, they are often difficult to identify and quantify. The use of the most sensitive analytical techniques, knowledge of potential sources of contamination, and isotopic labelling considerably increase the probability of obtaining unambiguous results. Furthermore, studies of naturally occurring methyl donors such as choline, *S*-adenosylmethionine, as well as a quest for identifying new ones, would enhance knowledge of environmental processes. The final result would be a better understanding of the fate and transport of anthropogenic, organometallic compounds in the environment.

8.7   Acknowledgements

The authors thank the National Science Foundation for partial support of this work through Grant NSF CEE 81-16960.

References

1. Toscano P J, Marzilli L G 1984 *Prog Inorg Chem* **31**: 105–204.
2. Craig P J 1982 *Comp Organomet Chem*, Vol 2 (eds Abel E W, Stone E G A, Wilkinson G). Pergamon, Oxford, pp 979–1020

3. Saxena J, Howard P H 1977 *Adv Appl Microbiol* **21**: 185–226
4. Davies A G, Smith P J 1980 *Adv Inorg Radiochem* **23**: 1–77
5. Thayer J S, Brinckman F E 1982 *Adv Organomet Chem* **20**: 314–56
6. Fanchiang Y-T, Ridley W P, Wood J M 1979 *Adv Inorg Biochem* **1**: 147–62
7. Kochi J K 1978, *Organometallic Mechanisms and Catalysis*. Academic Press, New York, chs 13, 16, 18
8. Thayer J S 1978 *Synth React Inorg Metal-Org Chem* **8**: 371–9
9. DeSimone R E 1972 *J Chem Soc Chem Commun*: 780–1
10. Honeycutt J B, Jr, Riddle J M 1960 *J Amer Chem Soc* **82** : 3051–2
11. Nagase H 1982 *Sci Total Environ* **24**: 133–42
12. Nagase H, Ose Y, Sato T, Ishikawa T 1984 *Sci Total Environ* **32**: 147–56
13. Akagi H, Takabatake E 1973 *Chemosphere* **2**: 131–3
14. Akagi H, Fujita Y, Takabatake E 1975 *Chem Lett*: 171–6
15. Maynard J L 1932 *J Amer Chem Soc* **54**: 2108–12
16. Hayashi K, Kawai S, Ohno T, Maki Y 1977 *J Chem Soc Chem Commun* 158–9
17. Deacon G B 1978 *Nature (London)* **275**: 344
18. Craig P J, Bartlett P D 1978, *Nature (London)* **275**: 635–7
19. Robinson G C, Nome F, Fendler J H 1977 *J Amer Chem Soc* **99**: 4969–76
20. DeSimone R E, Penley M W, Charbonneau L, Smith S G, Wood J M, Hill H A O, Pratt J M, Ridsdale S, Williams R J P 1973 *Biochim Biophys Acta* **304**: 851–63
21. Chu V C W, Gruenwedel D W 1977 *Bioinorg Chem* **7**: 169–86
22. Schrauzer G N, Weber J H, Beckham T M, Ho R K Y 1971 *Tetrahedron Lett*: 275–7
23. Hill H A O, Pratt J M, Ridsdale S, Williams F R, Williams R J P 1970 *J Chem Soc Chem Commun*: 341
24. Yamamoto H, Yokoyama T, Chen J-L, Kwan T 1975 *Bull Chem Soc Japan* **48**: 844–7
25. Craig P J, Morton S F 1978 *J Organometal Chem* **145**: 79–89.
26. Imura N, Sukegawa E, Pan S-K, Nagao K, Kim J-Y, Kwan T, Ukita T 1971 *Science* **172**: 1248–9
27. Wood J M, Kennedy F S, Rosen C G 1968 *Nature* **220**: 173–4
28. Chu V C W, Gruenwedel D W 1976 *Z Naturforsch* **31C**: 753–5
29. Mestroni G, Zassinovich G, Camus A, Costa G 1975/76 *Transition Metal Chem* **1**: 32–6
30. Espenson J H, Fritz H L, Heckman R A, Nicolini C 1976 *Inorg Chem* **15**: 906–8
31. Witman M W, Weber J H 1977 *Inorg Chim Acta* **23**: 263–75
32. Espenson J H, Bushey W R, Chmielewski M E 1975 *Inorg Chem* **14**: 1302–5
33. Tauzher G, Dreos R, Costa G, Green M 1974 *J Organometal Chem* **81**: 107–10
34. Magnuson V E, Weber J H 1974 *J Organometal Chem* **74**: 135–41

35. Allen R J, Bunton C A 1976 *Bioinorg Chem* **5**: 311−24
36. Abley P, Dockal R, Halpern J 1973 *J Amer Chem Soc* **95**: 3166−70
37. Adin A, Espenson J H 1971 *J Chem Soc Chem Commun*: 653−4
38. Kim J-Y, Imura N, Ukita T, Kwan T 1970 *Bull Chem Soc Japan* **44**: 300
39. Espenson J H, Chao T-H 1977 *Inorg Chem* **16**: 2553−7
40. Espenson J H, Kirker G W 1980 *Inorg Chim Acta* **40**: 105−10
41. Samuels G J, Espenson J H 1980 *Inorg Chem* **19**: 233−5
42. Brinckman F E 1981 *J Organometal Chem Libr* **12**: 343−76
43. Dodd D, Johnson M D 1974 *J Chem Soc Perkin Trans* **II**: 219−23
44. Blunden S J, Chapman A H 1982 *Environ Tech Lett* **3**: 267−72
45. Klein S, Woggon H 1983 *Z Gesamte Hyg Ihre Grenzgeb* **29**: 246−9
46. Maguire R J, Carey J H, Hale E J 1983 *J Agric Food Chem* **31**: 1060−5
47. Blunden S J 1983 *J Organometal Chem* **248**: 149−60
48. Soderquist C J, Crosby D G 1980 *J Agric Food Chem* **28**: 111−7
49. Chau Y K, Wong P T S, Kramar O, Bengert G A 1981 *Abstracts of the International Conference on Heavy Metals in the Environment, Amsterdam.* CEP, Edinburgh, pp 641−3
50. Craig P J, Rapsomanikis S 1982 *J Chem Soc Chem Commun*: 114−16
51. Rapsomanikis S 1983 PhD Thesis, Leicester Polytechnic, UK
52. Murphy T, Poller R C 1979 *J Organometal Chem Libr* **9**: 189−222
53. Gynane M J S, Lappert M F, Miles S J, Power P P 1976 *J Chem Soc Chem Commun*: 256−7
54. Craig P J, Moreton P A, Rapsomanikis S 1983 *International Conference on Heavy Metals in the Environment, Heidelberg.* CEP, Edinburgh, pp 788−92
55. Craig P J, Rapsomanikis S 1983 *Inorg Chim Acta* **80**: L19−L21
56. Nozoe T 1971 *Pure Appl Chem* **28**: 239−80
57. Wood J M, Cheh A, Dizikes L J, Ridley W P, Rakow S, Lakowicz J R 1978 *Fed Proc* **37**: 16−21
58. Dizikes L J, Ridley W P, Wood J M 1978 *J Amer Chem Soc* **100**: 1010−2
59. Fanchiang Y-T, Wood J M 1981 *J Amer Chem Soc* **103**: 5100−3
60. Craig P J, Rapsomanikis S 1985 *Inorg Chim Acta* **107**: 39−43
61. Blunden S J, Smith P J, Gillies D G 1982 *Inorg Chim Acta* **60**: 105−9
62. Thayer J S 1979 *Inorg Chem* **18**: 1171−2
63. Schrauzer G N, Seck J A, Beckham T M 1973 *Bioinorg Chem* **2**: 211−29
64. Cros G, Darbieu M H, Laurent J P 1980 *Inorg Nucl Chem Lett* **16**: 349−53
65. Darbieu M H, Cros G 1983 *J Organometal Chem* **252**: 327−40
66. Jarvie A W P, Markall R N, Potter H R 1981 *Environ Res* **25**: 241−9
67. Charlou J L, Caprais M P, Blanchard G, Martin G 1982 *Environ Tech Lett* **3**: 415−24
68. Hitchen M H, Holliday A K, Puddephatt R J 1979 *J Organometal Chem* **172**: 427−44
69. Ahmad I, Chau Y K, Wong P T S, Carty A J, Taylor L 1980 *Nature (London)* **287**: 716−17

70. Jarvie A W P, Whitmore A P 1981 *Environ Tech Lett* **2**: 197–204
71. Snyder L J, Bentz J M 1982 *Nature (London)* **296**: 228–9
72. Jarvie A W P, Markall R N, Potter H R 1975 *Nature (London)* **255**: 217–18
73. Wong P T S, Chau Y K, Luxon P L 1975 *Nature (London)* **255**: 263–4
74. Jarvie A W P, Whitmore A P, Markall R N, Potter H R 1983 *Environ Poll (Ser B)* **6**: 69–79
75. Reisinger K, Stoeppler M, Nürnberg H W 1981 *Nature (London)* **291**: 228–30
76. Craig P J 1980 *Environ Tech Lett* **1**: 17–20
77. Agnes G, Bendle S, Hill H A O, Williams F R, Williams R J P 1971 *J Chem Soc Chem Commun*: 850–1
78. Taylor R T, Hanna M L 1976 *J Environ Sci Health* **A11**: 201–11
79. Thayer J S 1983 *J Environ Sci Health* **A18**: 471–81
80. Rhode S F, Weber J H 1984 *Environ Tech Lett* **5**: 63–8
81. Thayer J S 1981 *Inorg Chem* **20**: 3573–6
82. Witman M W, Weber J H 1976 *Inorg Chem* **15**: 2375–8
83. Witman M W, Weber J H 1977 *Inorg Chem* **16**: 2512–5
84. Dimmit J H, Weber J H 1982 *Inorg Chem* **21**: 1554–7
85. Rhode S F 1983 MS Thesis, University of New Hampshire, USA
86. Rapsomanikis S, Ciejka J J, Weber J H 1984 *Inorg Chim Acta* **89**: 179–83
87. Honeycutt J B, Riddle J M 1961 *J Amer Chem Soc* **83**: 369–73
88. Taylor R T, Hanna M L 1976 *Bioinorg Chem* **6**: 281–93
89. Hogenkamp H P C, Kohlmiller N A, Howsinger R, Walker T E, Matwiyoff N A 1980 *J Chem Soc Dalton*: 1668–73
90. Taylor R T, Happe J A, Wu R 1978 *J Environ Sci Health* **A13**: 707–23
91. Taylor R T, Happe J A, Hanna M L, Wu R 1979 *J Environ Sci Health* **A14**: 87–109
92. Fanchiang Y-T, Pignatello J J, Wood J M 1983 *Organometallics* **2**: 1748–51
93. Fanchiang Y-T, Pignatello J J, Wood J M 1983 *Organometallics* **2**: 1752–8
94. Fanchiang Y-T, Ridley W P, Wood J M 1979, *J Amer Chem Soc* **101**: 1442–7
95. Rubinson K A, Itabashi E, Mark H B Jr 1982 *Inorg Chem* **21**: 3571–3
96. Fanchiang Y-T 1982 *Inorg Chem* **21**: 2344–8
97. Espenson J H, Shveima J S 1973 *J Amer Chem Soc* **95**: 4468–9
98. Espenson J H, Sellers T D Jr 1974 *J Amer Chem Soc* **96**: 94–7
99. Scovell W M 1974 *J Amer Chem Soc* **96**: 3451–6
100. Yurkevich A M, Chauser E G, Rudakova I P 1977 *Bioinorg Chem* **7**: 315–24
101. Chauser E G, Rudakova I P, Yurkevich A M 1975 *J Gen Chem USSR* (Eng) **45**: 1181–2
102. Chauser E G, Rudakova I P, Yurkevich A M 1976 *J Gen Chem USSR*

(Eng) **46**: 356-9
103. Pohl U, Huber F 1976 *J Organometal Chem* **116**: 141-51
104. Antonio T, Chopoa A K, Cullen W R, Dolphin D 1979 *J Inorg Nucl Chem* **41**: 1220-1
105. Robinson J W, Kiesel E L 1981 *J Environ Sci Health* **A16**: 341-52
106. Agnes G, Hill H A O, Pratt J M, Ridsdale S C, Kennedy F S, Williams R J P 1971 *Biochim Biophys Acta* **252**: 207-11

Chapter 9

Organometallic compounds in polymers – their interactions with the environment

9.1 Introduction

Organometallic compounds do not occur in polymer systems in either a unique or systematic fashion. There is no single range of compound types or elemental species which can be singled out and used as a central theme in a chapter of this nature.

Organometallic compounds are present in polymer systems by accident, by necessity, by design (because they confer some cost-effective benefit) or because they are part of the polymer backbone. The concentration of the organometallic species in the polymer depends on the reason for its incorporation. Species present by accident, or perhaps because it is cost effective not to remove them, would occur at levels probably less than parts per million by weight. Reaction catalysts or antidegradants, etc. are present over the range 0.01–10 per cent while higher percentage levels occur for the metal-containing polymers.

It is convenient to discuss the question of organometallics in polymers organized on the basis of starting with low level (< 1 μg g^{-1}) impurities and working towards the metal-containing polymers as indicated in Table 9.1. Inevitably there are areas for which little information concerning environmental impact is as yet available. In some instances this is because no effect has been observed, or other species in the polymer system are far more active. Concern over toluene di-isocyanate (TDI) in polyurethane processing for example, currently is far more important than concern over low concentrations of organotin-based catalysts in the polymer. Any precautions taken to reduce exposure to isocyanates will automatically also reduce the exposure to the organometallic catalyst system.

For some of the organometallic systems to be discussed the potential for

Table 9.1 Organometallic compounds in polymer systems

Application	Level of organometallic compound (%)	Metal	Polymer system
Polymer production	Sub μg g^{-1}	Sn, Zn, Li, W, Mg, Al, Fe, Ru	Olefinic polymers
Coupling agents	0.001−0.1	Si	Glass reinforced materials Primer coats
Reaction catalysts	0.01−4.0	Sn, Mg, Pb Sn Sn	Polyurethanes Silicones Epoxy resins
Cross-linking agents	0.01−1.0	Si	Rubbers Polyethylene
Stabilizers	1.0−10	Sn	Polyvinyl chloride
Antifouling paints	1.0−10	Sn, Cu	Various acrylate polymers
Biocide polymer systems	1.0−10	Sn, Ti, Mn, As, Sb, U	Various
New organometallic polymer systems	Incorporated into the polymer backbone	Si, Ge, Sn, Os, Ti, Fe, Pt, Li, Pd, Tb, Eu, Cr, W, Mo, Mn, Re etc.	Various

changing the environment is dramatic and indeed the polymer systems are tailored to enhance this effect. Two areas are of particular concern here − the biocidal properties of many of the polymer systems to be discussed and also the pharmacological properties which are currently being investigated. It is to be hoped that none of the organometallic species, presently incorporated into polymers, will have unexpected toxicological properties which will merit treatment in a manner similar to vinyl chloride monomer (VCM) or other toxic species.

9.2 Polymer formation

A wide range of organometallic compounds has been used as one of the constituents of the Ziegler−Natta catalysts used in the polymerization of polyolefins. The choice of the catalyst components frequently defines the stereo regularity of the resultant polymer. The polymerization of olefins is generally thought to involve the adsorption and complexing of the monomers. The resultant stereoregularity produced by these heterogeneous catalysts may result from steric hindrance, at the catalyst site, which allows adsorption to occur in one orientation only.

9.2.1 Organolithium catalysts

Butyllithium is the most commonly used organoalkali metal catalyst for the production of synthetic elastomers. Approximately 100 tonnes per annum are used in the western world in the production of materials for tyres, hoses and footwear. Butadiene polymers prepared using n-butyllithium in heptane posses 40−50 per cent *trans*, 40−50 per cent *cis* and 5−10 per cent vinyl structures.[1]

9.2.2 Organosodium catalysts

Sodium based Alfin catalysts[2,3] are a complex mixture which includes sodium chloride, sodium alkoxide and allylsodium. Polymer microstructure using such a catalyst system is 67 per cent *trans*, 17 per cent *cis* and 16 per cent vinyl. One major problem with polymers produced using this catalyst system, e.g. polybutadiene, is the very high molecular weight. The polymers are virtually non-processable by conventional techniques. The Alfin process is probably now obsolete.

9.2.3 Organomagnesium catalysts (alkylmagnesium)

The most important use of alkylmagnesium compounds is as second generation Ziegler−Natta catalysts for polyolefins. Most if not all of these compounds are used in the production of polyethylene, but the application to polypropylene production is currently being considered. The magnesium based catalysts are often active enough to obviate costly steps to remove catalyst residues from the resultant polymer system. The commercially available dialkylmagnesium compounds are: butylethyl, di-n-butyl, di-n-hexyl, n-butyl and n-octyl.[4,5]

9.2.4 Organoaluminium catalysts

Organoaluminium compounds are used as part of the catalyst system for the production of both olefins and dienes.[6,7] The world production of alkylaluminium compounds is in the range $1−2 \times 10^5$ tonnes per year, the

Table 9.2 Organoaluminium compounds used in Ziegler–Natta catalysis

Trialkylaluminium	R_3Al	alkyl = ethyl, isobutyl
Dialkylaluminium halides	R_2AlX	alkyl = ethyl, X = chloride
Alkylaluminium sesquihalides	$R_3Al_2X_3$	alkyl = ethyl, X = chloride
Dialkylaluminium hydrides	R_2AlH	alkyl = ethyl
Alkylaluminium dihalides	$RAlX_2$	alkyl = ethyl, X = chloride
Alkylaluminium alkoxides	R_2AlOR^1	alkyl = ethyl, X = chloride

major proportion of this quantity being used in the manufacture of synthetic polymers. Mono- di- and trialkyl-substituted aluminium compounds of the types shown in Table 9.2 are commercially available. These materials are generally thermally stable provided they are maintained below 50 °C under an inert atmosphere. Some idea of the range of polymers which can be prepared using these catalysts can be obtained from Table 9.3.[8]

Table 9.3 Some olefin polymers prepared using Ziegler–Natta catalysts*

Polymer	Type
Ethylene	Linear low density
Propylene	Isotactic/syndiotactic
1-Butene	Isotactic
3-Methyl-1-butene	Isotactic polymer
Styrene and substituted styrenes	Isotactic styrene and ring-substituted species
Acetylenes	Essentially completely conjugated polymers

* See Ref. 8

9.2.5 Organonickel catalysts

The nickel based catalysis of polymerization (butadiene) is of considerable industrial interest since the catalytic efficiency of nickel is high: $10^3 - 10^4$ g of polybutadiene are produced per gram of nickel.[9]

Of the two major groups of nickel catalysts only the π-allylnickel complexes fall within the scope of this chapter, i.e. organic. The π-allylnickel halides show catalytic activity with butadiene; a *cis* polymer is obtained from the chloride whereas the bromide and iodide produce a *trans* configuration.[10]

Many examples of polymerization using nickel in binary catalyst systems have been reported. Complexes of cyclo-octadienyl-, bis(cyclo-octadienyl)- and bis(cyclopentadienyl) nickel(0) with metallic halides yield highly stereospecific[11,12] 1,4-polybutadienes.

9.2.6   Tungsten and molybdenum catalysts

Efficient polymerization of low strain cyclo-olefins (cyclopentene) was first achieved in 1963 using a combination of tungsten, molybdenum and organoaluminium catalysts.[13] The transition metals have been employed in a variety of forms in mixed metal catalyst systems, but those using carbene compounds of tungsten are particularly significant.[14] Some of these compounds are able to catalyse the polymerization of cyclopentene at remarkably low concentrations. For example $(CO)_5W{=}C(OC_2H_5)R$ is effective at a cyclopentene–tungsten ratio of 10^6 while $Br(CO)_4W \equiv CCH_3$ (a carbyne), is active at ratios of 10^9.[15]

9.3   Organosilicon compounds as bonding promoters

Organofunctional silanes can couple inert, inorganic fillers to an organic polymer thereby allowing the use of low cost mineral fillers as extenders. They facilitate the use of glass reinforcement of cheaper resins to be used in place of engineering plastics. Organofunctional silanes can cross-link polymers to improve physical properties, and for the manufacture of insulated glass windows (for windscreens) and for other products organosilanes can add to the breadth of materials available.

With one exception (cross-linking of polymers) the organosilicon chemicals are used to modify the surface characteristics of many materials; glass fibres can be given a reactive sheath which will combine with unactivated polyesters, epoxy or melamine resins. Paper can be modified to retain oil and pass water and titanium dioxide can be made to disperse better in polystyrene. The potentialities of organosilane compounds are unusually broad and much interest in their use is anticipated.

Non-elastomeric aspects of organosilicon compounds are discussed in Chapter 6.

9.3.1   Compound types

At present the organosilicon compounds can be divided into the three major groups summarized in Table 9.4.

Since all these compounds are available off-the-shelf, the preparation, properties and typical non-environmental reactions are not discussed here

Table 9.4 Organosilane coupling agents (bond promoters)

Organochlorosilanes	R_nSiCl_{4-n}; where $R = CH_2{=}CH$, CH_3, $CH_3CH{=}CH$
Organosilane esters	$CH_3Si(OC_2H_5)_3$ and $CH_3Si(OCH_3)_3$
Organofunction silanes	
Vinyl	$CH_2{=}CHSiR_3$; where $R{=}OC_2H_5, OCH_3, OCOCH_3$ etc.
Methacrylate	$CH_2 = CH(CH_3)CO_2(CH_2)_3SiR_3$; where $R = OC(CH_3)_3$ and $(OC_2H_4OCH_3)$
Epoxy	$\overset{\displaystyle O}{CH_2{-}CH}CH_2O(CH_2)_3Si(OCH_3)_3$
Mercapto	$HS(CH_2)_3SiR_3$; where $R = OCH_3$ and OC_2H_5

since these data are readily available in the literature.[16-19] Perhaps the important property to note here is that the effectiveness of the organosilicone material parallels the reactivity between the organofunctional group and the resin material itself.[20]

9.3.2 Applications

The following information provides a brief survey of the common areas of application of organosilicon compounds and provides some notional values for the concentrations of material utilized.

Organosilicones are used at the 0.5–1 per cent level by weight to provide improved dry adhesion of thermoset and thermoplastic adhesives to glass and metal substrates.[21] Typically, improvement in the strength of structural adhesive joints is of the order of 50–100 per cent.

Suitable organosilicon compounds can be chosen for use with the following range of adhesive systems, nitrile/phenolic, vinyl butyral/phenolics epoxy systems, polyurethane and hot melt materials. The areas covered by the use of these different adhesives include the bonding of brake shoes, printed circuit manufacture and the application of labels to glass bottles.

There are two major areas in which sealants are used in large volumes, the automobile industry and the building industry. In both cases maximum bond strength is required and the combination of a suitable organosilicon compound with any of the sealants (polysulphide, polyurethane, elastomeric or plastisol) will improve the metal or glass to sealant bond.

In the printing and coating industries, silanes are used either to aid dispersion of fillers or pigments or as primers which will produce a better bond between the substrate and the final coat. The addition of approximately 1 per cent organosilane to acrylic paints improves the adhesion to glass to a level that can withstand the action of boiling detergents. The loss of adhesion between urethane coatings and substrates as a result of water attack can be eliminated by the use of silane based primers.[21]

Particulate mineral fillers are added to many polymers in order to change and hopefully improve some property of the polymer. A high filler level required to achieve one desired property may well seriously degrade the polymer in another direction. The success of the silane coupling agents with fibre glass has suggested their use with other fillers.[22]

Significant improvement in the flexural strength of polyesters filled with a range of treated fillers which included amorphous silica, Kaolin Clay, Mica, Talc, Wollastonite and alumina trihydrate, has been demonstrated. Similar improvements in polymer to filler bond strength have led to the use of organosilane coupling agents in a wide range of wire and cable and mechanical goods applications; for example, cable insulation, hoses to improve solvent resistance, drive belts, shoe soles and tyres, to improve abrasion resistance and even in pharmaceuticals to reduce moisture permeability.

The ability of the organosilane to hydrolyse and form a polymer is used to cross-link polyethylene cable insulation. A peroxide-induced graft of vinyl silane on to polyethylene can be cross-linked, after cable fabrication, by the action of moisture thus producing a high quality product without the use of a high temperature salt bath cure.[23]

Block[24] reports improved cross-linking and adhesion to carbon black fillers in acrylate polymers resulting from the use of amino propyltriethoxysilane. This interaction speeds up the cure and improves the strength and dynamic properties of the rubber.

Chlorosulphonated polyethylene compositions containing 5–15 parts per hundred of polymethylsilazanes showed improved resistance to heat, ageing and the action of corrosive media (acids and caustic potash (KOH)). Organosilicon heterocyclic diamine compounds produced the best properties in the polymer, for example 30-day exposure to either 40 per cent alkali or 10 per cent nitric acid produced little change in the material.

Block also indicated the advantages of using silicon based oligodiols for castable polyurethane elastomers. Better heat stability (220 °C for six hours) and a relatively high resistance to the action of both stress and temperature were reported.

Silicone coupling agents, in conjunction with various latex polymers, have been investigated for water harvesting in areas where rainfall is lost by rapid evaporation. Acrylic and styrene–butadiene rubber (SBR) latices modified with silicone coupling agents were effective in this respect. Topsoil treated in this way allowed maximum water intake, but constituted a significant barrier to rapid evaporation. Leaching of minerals from soils was similarly reduced and the reclamation of land covered with mine spoil waste was enhanced by the use of a 0.6 per cent silicone modified latex emulsion.[25] In further tests, the yield of sweet corn from a plot of soil protected by a silicone latex was similar to that from a plot protected by conventional mulching techniques.[26]

9.4 Reaction Catalysts

Many organometallic compounds are used to catalyse the curing of polymer systems, notably in urethane and silicone polymer systems. The principal metal involved in these catalysts is tin, although specialized polymer systems use other metals, for example, organomercury compounds are used in cast elastomers and Reaction Injection Moulding (RIM) systems.[27]

The organometallic catalyst may be present in the polymer at levels between 0.001 and 1 per cent depending on the particular application. World total use of tin compounds in urethane compounds is estimated to be 1000 tonnes for 1980.[28] Other aspects of organotin chemistry are covered in Chapter 3.

9.4.1 Catalysts in urethane chemistry

Tin catalysts are preferred for use with urethane systems because of their low odour and the small amounts required to achieve high reaction rates, for example di-n-butyltin(IV) diacetate was found to be 2400 times more reactive than the comparitive amine catalyst. Synergistic effects between tin compounds and amines to produce even better catalyst systems[29] are shown in Table 9.5. In standard or hot-cure flexible foam production, great care must be taken to ensure the correct proportions of tin and amine catalyst.

Table 9.5 Synergistic effects in polyurethane catalysis

Catalyst	Mol %	$K \times 10^4$ $dm^{-1} mol^{-1} s^{-1}$*
Triethylamine	0.88	2.4
Di-n-butyltin diacetate	0.00105	20
Triethylamine/di-n-butyltin diacetate	0.99/0.00098	88

* Rate of cure.

If the tin catalysis is too rapid a tight foam is obtained because a large number of closed cells are produced by fast gelation. A high amine concentration may lead to collapse of the foam from lack of gel strength. This balance has led to the use of a range of tin compounds tailored to different end-products. Dibutyltin(IV) dilaurate for example, is satisfactory for systems with limited storage expectancy. Where longer storage is required, dialkyltin(IV) salts with better resistance to hydrolysis are utilized. Delayed action catalysts for use in RIM installations use either dialkyltin(IV) dimaleate or the dimercaptides. Organolead compounds catalyse urethane reactions readily, for example phenyllead(IV) triacetate ($C_6H_5Pb(OAc)_3$) compared well with tin(II) dioctate. Delayed-action catalysts can also be obtained by using sparingly soluble catalysts such as phenyl mercuric propionate which also has been cited as having biocidal value when used in underground installations.[27]

Some of the more common uses of these catalysts in the urethane industry are cited in Table 9.6.[30-38] Some details of organolead catalysts are also available.[39-41]

Table 9.6   Organometallic catalysts used in the urethane industry

Catalyst*	Polymer system	Reference
Dibutyltin dilaurate	High resilience foams to improve compression set	30
	Moulded and powder coated sheets	31,32
	Preparation of *in vitro* test membranes	33
	Rigid and cold-cure foams	34
	Reaction injection moulding systems	27
Dibutyltin diacetate	Flexible foams	
	Thermosetting emulsion surface coatings	35
Dibutyltin dimaleate	As above	35
Dibutyltin mercaptide	Hydrolysis-resistant catalyst used for 'storage stable' two-pot systems	27
Dioctyltin mercaptides	High stability delayed action catalysts for RIM systems	27
Tributyltin oxide	Cold curing non-cellular systems	36
	Urethane modified isocyanurate systems to improve combustion properties	37
Dibutyltin bis (isooctylmercapto) acetate	Rigid foam production	38

*Formulae − see text and Chapter 3.

9.4.2   Catalysts for silicone elastomers

Elastomeric polymers are an extremely important part of the full range of silicone polymer products.[39-41] Perhaps the most commonly encountered examples of these materials are the room temperature vulcanizing (RTV) elastomers. RTV silicones are used in applications ranging from surgical implants to bathroom caulking. The materials are supplied as either one- or two-component products usually incorporating an organometallic catalyst.[42]

One-component RTV elastomers are prepared by mixing the components and packaging them in the absence of moisture. Contact with moisture from the air initiates the curing process. The polymers used in nearly all commercial products are based on polydimethyl siloxanes.[43] Typical catalysts are dibutyltin(IV) dimaleate.

Two-component RTV silicone rubbers do not require atmospheric moisture to initiate the cure. There are several different approaches but in general each requires a catalyst (equations 9.1 and 9.2):

$$HO(Si(CH_3)_2O)_nH + CH_3SiOOCCH_3 \rightarrow CH_3SiO((CH_3)_2SiO)_nH$$
$$+ 3CH_3COOH \hspace{6cm} [9.1]$$

$$\begin{matrix} R^1 & & R \\ & \diagdown & \diagup \\ R^1\!-\!SiOH + HSi\!-\!R & \rightarrow & \\ & \diagup & \diagdown \\ R^1 & & R \end{matrix} \quad \begin{matrix} R^1 & & R \\ & \diagdown & \diagup \\ R^1\!-\!Si\!-\!O\!-\!Si\!-\!R + H_2 \\ & \diagup & \diagdown \\ R^1 & & R \end{matrix} \quad [9.2]$$

In both cases dibutyltin(IV) dilaurate is used as the catalyst.

9.4.3   Tin compounds as esterification catalysts

Both mono- and dibutyltin(IV) compounds are increasingly used as catalysts in the production of organic esters used as plasticizers and lubricants in the polymer industry. High temperatures (200–230 °C) are required but there are few side-reactions and hence the product is purer with better colour and odour. Typical catalyst levels are of the order of 0.05–0.3 per cent by weight.[44]

Dibutyl tin compounds are used in the transesterification and polycondensation of dimethylterephthalate to poly(ethyleneterephthalate) for packaging applications and in the manufacture of polyester-based resins.

9.4.4   Tin compounds as fluoroelastomer curatives

Tetraphenyltin and allyltriphenyltin ($CH_3CH = CH_2)(C_6H_5)_3Sn$ are curing agents for perfluororubbers containing nitrile cure sites. The polymer $CF_2CFOCF_2CF(CF_3)OCF_2CF_2CN\!-\!(CF_3OCF\!=\!CF_2)\!-\!C_2F_4$ can be successfully cured with 2.8 parts per hundred of the organotin compounds.[45]

9.5   Organotin stabilizers

Organotin compounds became generally available to the PVC processor in the immediate post Second World War years.[46] For the next few years these compounds were used to stabilize all forms of vinyl materials whether rigid or flexible. Mixed metal systems gradually reduced the use of organotin compounds in all but unplasticized PVC applications in which organotin mercaptides are supreme.[47]

These stabilizers are used with PVC, polyvinylidene fluoride (PVF), chlorosulphonated polyethylene, and polyvinyl acetate (PVA) materials with typical levels in the 0.5–5 per cent range. The stabilizers are used to reduce the decomposition of the polymer during processing, modify chain reactions to prevent the formation of HCl by unzipping and to stabilize the colour. They should have an affinity for the polymer and as far as possible be non-toxic.[48]

It is generally accepted that better heat stability is conferred by the use of dialkyltin compounds. Monoalkyl-, trialkyl- and aryltin compounds have been shown to be poor stabilizers for PVC. Furthermore, the lower alkyl homologues of trialkyltin compounds are toxic and are not used as stabilizers.[49] The most common practice is to use a blend of mono and dialkyl derivatives with common alkyl groups. Of the 30 000 tonnes of organotin compounds manufactured in 1980, an estimated 20 000 tonnes were used as PVC stabilizers.[50]

9.5.1   Preparation of intermediates

Until the early 1960s all the commercially available stabilizers were derivatives of the initial dibutyltin(IV) compounds. The lower dialkyl compounds were considered to be more toxic than the dibutyl while higher alkyl species provided no extra stabilization for increasing costs.

All the dibutyltin stabilizers are derived from the intermediate dibutyltin(IV) dichloride which has to be prepared by one of the two-step processes illustrated below (equations 9.3–9.6):

(a) $SnCl_4 + 4C_4H_9Cl + 4Mg \rightarrow (C_4H_9)_4Sn + 4MgCl_2$ [9.3]

$(C_4H_9)_4 Sn + SnCl_4 \rightarrow 2(C_4H_9)_2 SnCl_2$ [9.4]

NB   Sodium may be used in place of magnesium.

(b) $4(C_4H_9)_3 Al + 3SnCl_4 \rightarrow 3(C_4H_9)_4 Sn + 4AlCl_3$ [9.5]

$(C_4H_9)_4Sn + SnCl_4 \rightarrow 2(C_4H_9)_2 SnCl_2$ [9.6]

It is possible to produce methyltin stabilizer derivatives by direct dialkylation of tin with chloromethane. This process eliminates the costly two-stage synthesis shown above. Great care must be taken to control the reaction sequence which could readily lead to a mixture of methyltin chlorides, some of them being highly toxic (equation 9.7):[51]

$6CH_3Cl + 3Sn \rightarrow CH_3SnCl_3 + (CH_3)_2SnCl_2 + (CH_3)_3SnCl$ [9.7]

Methyltin derivatives are only used to any major extent in the USA; Europe and Japan prefer to continue using dibutyltin compounds. In the USA methyltin stabilizers are mainly used in PVC pipe applications where not only are they less expensive but more effective on a weight for weight basis than the dibutyltin derivatives.[52]

Certain di-n-octyltin derivatives show a low toxicity to man and two of these compounds have been sanctioned by various regulatory agencies throughout the world for food contact purposes.[53] The production process is similar to that for butyltin stabilizers and the volume consumption is correspondingly small.

9.5.2 Conversion to stabilizer

Once prepared, the alkyltin halides can be converted to oxides, alkoxides, phenoxides, mercaptides, carboxylates, etc. by any of the simple well-known preparative methods. The reason why so many different (see Table 9.7) organotin stabilizers have been used is because no single species confers all the required properties. In general, compounds that contain tin bonded to sulphur usually impart good heat stability but are poor with respect to photolytic degradation.[54] Non-sulphur-containing stabilizers are only poor as far as heat stabilization is concerned but they are unaffected photolytically. Mixed systems containing both tin−oxygen and tin−sulphur bonds work well.[54]

The environmental criterion concerning the use of organotin stabilizers is perhaps most stringent in connection with bottles used for squashes and other drinks. These considerations are more fully discussed in section 9.8.2.

Table 9.7   Some important organotin stabilizers*

Dibutyltin dilaurate
Dibutyltin maleate
Dibutyltin bis(n-alkyl maleate); alkyl is usually C_4 or C_8
Dibutyltin bis(lauryl mercaptide)
Thiabis (monobutyltin sulphide)
Dibutyltin sulphide
Dimethyltin bis(iso-octyl mercaptoacetate) and the dibutyl and
   di-n-octyl tin derivatives
Dimethyltin bis(β-alkanoyloxyethylmercaptide)
Dibutyltin β-mercaptopropionate and the di-n-octyl derivative
Thiabismonomethyltin bis(β-alkanoyloxyethylmercaptide)

---

*Formulae − see text and Chapter 3.

9.6   Organometallic polymers

Polymers containing metals as an integral part of the backbone structure are, in general, relatively new compounds. Transition metal polymers can be prepared from cyclopentadienyl and arene metal(II) complexes which undergo reactions characteristic of aromatic species. The presence of the metal atom often confers unusual properties, for example electrical conductivity and improved absorption of radiation. A wide range of metals has been investigated, but tin is by far the most successful species.[55-57]

Other metals include titanium, uranium, antimony, lead, zirconium, iron, bismuth, ruthenium, manganese, chromium, copper, platinum and palladium and, as might be expected, many of the properties of polymers containing these species are both unexpected and, in many cases, unexplained. Methyl methacrylate polymers copolymerized with the 4-methyl-4'-vinyl bipyridine complex of either europium or terbium were shown to be fluorescent and it is possible that there may be some applications in laser technology. Organometallic polymers, for example styryltriphenyllead and vinylosmocene may be incorporated into targets for use in fusion research.[57] Many of the other organometallic polymers will probably remain laboratory samples with no commercial use; others, the organotin compounds for example, will be used in quantity.

A full discussion of metal-containing polymers and their impact on the environment is not possible for two reasons. First, insufficient data are available and many of the compounds simply have not been allowed to have any impact on the environment, i.e. no use has been found for them. However, in the case of tin compounds in antifouling paints, these directly affect the environment, as also do the new chemotherapeutic organometallic species which are being developed. These two groups of materials with special applications will be discussed separately from the main section on organometallic polymers.

9.6.1   Vinylic polymers

In a recent review Pittman *et al.*[58] have summarized a considerable body of research concerning vinylic metal complex species. Table 9.8[55-63] gives examples of the monomer systems considered and indicates the more important polymeric systems which have been investigated.

Perhaps the more important vinylic polymers are prepared from monomers 1–8 (Table 9.8) by cationic or radical polymerization usually using 2,2-azobisisobutyronitrile as an initiator.[58,64-66] Copolymerization with styrene or methyl acrylate has been studied for all the monomers recorded in the table. Linear polymers with relatively uniform molecular weights have been obtained using $LiAlH_4$.[67] Block copolymers may be formed with both styrene and methyl methacrylate.[68]

Vinyl metallocenes, in general, form soluble polymers with molecular weights in the range 2000–50 000.[67] Solutions of the polymers tend to decompose in the presence of light and produce brittle films when evaporated. Polyvinyl ferrocene has been most extensively studied, the thermal properties have been reported and comprehensive degradation studies are discussed by Carraher.[69]

Only limited applications have been suggested for the vinylic polymers of

Table 9.8 Some transition metal vinyl monomers

Type	Name	Formula	HP*	ST†	MA†	Ref
1	Vinylferrocene	$C_5H_5FeC_5H_4CH = CH_2$	✓	✓	✓	58
2	Vinylruthenocene	$C_5H_5RuC_5H_5CH_2CH = CH_2$	✓	✓	✓	59
3	Ethynylferrocene	$C_5H_5FeC_5H_4C \equiv CH$	✓			58
4	Divinylferrocene	$Fe(C_5H_4CH = CH_2)_2$				58
5	Diisopropenylferrocene	$Fe(C_5H_4C(CH_3) = CH_2)_2$				58
6	Butadienylferrocene	$C_5H_5FeC_5H_4CH = CHCH = CH_2$				60
7	Methyleneallylferrocene	$C_5H_5FeC_5\,H_4C\underset{\parallel\ CH_2}{-}CH = CH_2$				60
8/9	Vinylcyclopentadienyl-manganesetricarbonyl	$(CO)_3\,MnC_5H_4CH = CH_2$	✓	✓	✓	61
	Vinylcyclopentadienyl-chromium- and tungstencarbonyls	$((CO)_3MC_5H_4CH = CH_2)_2$ $M = Cr,\ W$				61
10	Tricarbonyl (vinyl benzene) chromium	$(CO)_3CrC_6H_5CH = CH_2$				61

Table 9.8 (cont)

Type	Name	Formula	HP*	ST†	MA†	Ref
11	Vinylcyclopentadienyl-tungstenmethyl-tricarbonyl	$(CO)_3CH_3WC_5H_4CH = CH_2$	✓	✓	✓	62
12	Tricarbonylhexatriene iron stereoisomer	$Fe(CO)_3$		✓		63
13	Chloro(4-vinylphenyl)bis(tributyl phosphine) platinum	$Cl((C_4H_9)_3P)_2Pt$— —$CH=CH_2$				64
14	[(Acryloyl)methyl] ferrocene	$C_5H_5FeC_5H_4CH_2OOCCH=CH_2$	✓	✓	✓	65
15	[(Methacryloyl)methyl] ferrocene	$C_5H_5FeC_5H_4CH_2OOCC(CH_3)=CH_2$	✓	✓	✓	65
16	[(Acryloyl)ethyl] ferrocene	$C_5H_5FeC_5H_4CH_2CH_2OOCCH=CH_2$	✓			65
17	Tricarbonyl(benzene methacrylate) chromium	$(CO)_3CrC_6H_5CH_2OOCCH=CH_2$			✓	66

* HP – homopolymerizes.
† Copolymerizes with: ST – styrene; MA – methylacrylate.

compounds 1–14 (Table 9.8) because they have poor thermomechanical properties. The possible use as UV or radiation resistant coatings has been suggested. Compounds 9–12 may lose the CO ligands to form unsaturated species that can react with olefins, for example cyclopentadienyl dicarbonyl cobalt (η-$C_5H_5Co(CO)_2$) is used as a catalyst for the trimerization of olefins and also in stereospecific synthetic reactions.[70] Thermal decomposition of polymers containing iron, manganese and cobalt carbonyl units produces finely divided metal oxide in a polymer matrix,[71] and, since many of these residues are paramagnetic, they can be used in magnetic tape systems.

9.6.2 Metallocenemethylene polymers

Metallocenemethylene polymers based on iron or ruthenium are similar to the phenol–aldehyde resins and may have the structures shown in Fig. 9.1.

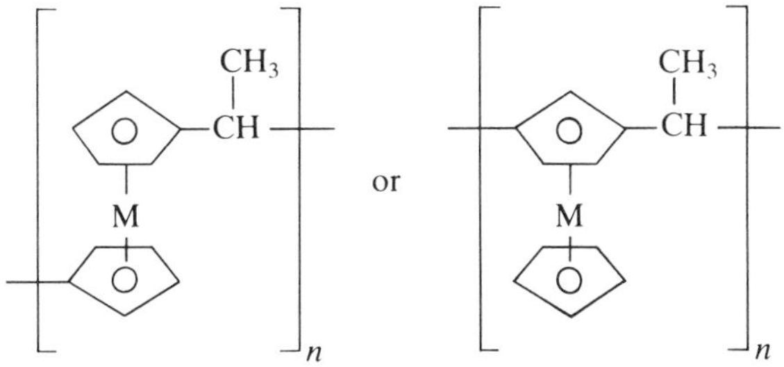

Fig. 9.1 Possible structures for metallocenemethylene polymers

A two-stage synthesis, starting with the generation of the metallocenyl carbonium ion followed by thermal polymerization, usually with an acid catalyst, is used to prepare these polymers. Polymers of this type are generally dark brown and partially soluble in aromatic solvents. The soluble fractions have molecular weights in the range 2000–20 000. The polymers soften at 220–300 °C and decompose above 300 °C. Thermogravimetric analysis shows 50 per cent weight retention at 550 °C.

Polymers of this type have been prepared as possible ablative materials for space capsule heat shields and ruthenocene derivatives were studied as radiation shields for communication satellites. However, cheaper ceramic based materials were shown to perform equally well and thus no large-scale application of these polymers has been found.[72]

9.6.3 Metallocenylene polymers

These polymers are based on any of the isomers shown in Fig. 9.2 and are of considerable interest because of their high rigidity and thermal stability. It is

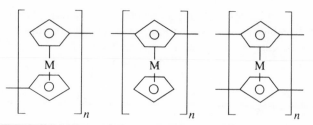

Fig. 9.2 Possible structures for metallocenylene polymers

also possible that partial oxidation could produce conjugated mixed valence systems which could act either as semiconductors or as metallic conductors.[73]

Poly(ferrocenylenes) are prepared by the direct polymerization of ferrocene in the presence of a stoichiometric amount of di-tert-butyl peroxide. This is a cheap, simple procedure but produces a highly branched and cross-linked material containing a mixture of structures. These species are usually light yellow in colour and have been prepared with molecular weights up to 10 000. Poly(ruthenocenylene) has been prepared, but to date the maximum molecular weight achieved is 2450. Efforts to prepare mixed polymers are under way.[74]

Linear poly(ferrocenylene) has the highest thermal stability and structural rigidity of any of the ferrocene-containing polymers. Possible applications as ablative materials and insulators may be envisaged. It may be possible, for example, to produce a polymer capable of withstanding extended periods at temperatures as high as 600 °C. Investigations of the conductivity of such polymers has shown that semiconductor properties can be produced as a result of partial oxidation.[74] Preparation of mixed metal polymers with varying degrees of oxidation could produce a wide range of semiconducting materials. The polymers are, however, somewhat intractable due to their low solubility in common solvents.

9.6.4 Polysilanyl and polysiloxanyl ferrocenes

Copolymers of ferrocenyl-siloxanylene monomers and ferrocenylene-silylene monomers should combine the advantages of high thermal stability and resistance to UV and γ radiation with the flexibility over a wide temperature range usually associated with organosilicon polymers. The best materials of this type have greater thermal stability than the vinylic polymers combined with much better flexibility. Structural details are shown in Fig. 9.3.

Precursors for these monomers can be prepared by direct reaction of either siloxanyl or silanyl halogen compounds with cyclopentadiene in the presence of butyllithium, followed by reaction with dimethylamine. Reaction with iron(II) chloride to form the ferrocenyl groups completes the synthesis. Polymerization is carried out using H_2PtCl_6 as the initiator.

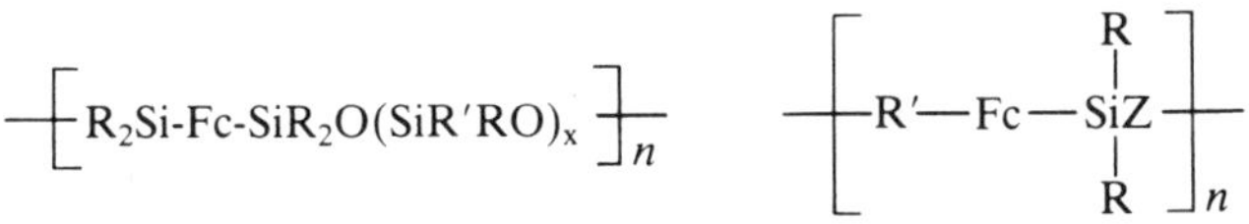

$$-\left[R_2Si\text{-}Fc\text{-}SiR_2O(SiR'RO)_x \right]_n- \qquad -\left[R'\!-\!Fc\!-\!\underset{\underset{R}{|}}{\overset{\overset{R}{|}}{Si}}Z \right]_n-$$

R=alkyl; aryl	R=alky, aryl
Fc=ferrocenyl group	R'=alkylene or arylene
	Z=alkylene or arylene

Fig. 9.3   Silicon–ferrocene combination system

The first ferrocenyl–siloxanylene polymers prepared were low molecular weight oils, possibly suitable for either hydraulic or high temperature lubricating oils. Materials with relatively high molecular weights (10^4–10^5), sufficient to cast as films, have also been produced.[75] No practical or commercial applications of these polymer systems have as yet been found.

9.6.5   Polymers containing trialkyltin esters

An extensive investigation of the polymers produced by a number of cross-linking agents used in conjunction with trialkyltin esters has been reported in the literature.[76] The most important reagents were the methacrylate and methylmethacrylate species in both groups but some idea of the extent of the materials investigated is shown by Table 9.9.

Table 9.9   Reagents used in the polymerization of trialkyltin esters

Tin ester	Cross-linking agent
$(C_4H_9)_3SnO_2CCH{=}CH_2$	$CH_2{-}CHCH_2O_2CCH{=}CH_2$ $\quad\backslash O /$
$(C_4H_9)_3SnO_2CC(CH_3){=}CH_2$	$CH_2{-}CHCH_2O_2CC(CH_3){=}CH_2$ $\quad\backslash O /$
$(C_4H_9)_3SnO_2CCH_2NH_2$	$1,3\text{-}(NH_2)_2C_6H_4$
$(C_4H_9)_3SnO_2C(CH_2)_3NH_2$	$(4\text{-}H_2NC_6H_4)_2CH_2$
$(C_3H_7)_3SnO_2CC(CH_3){=}CH_2$	$(4\text{-}H_2NC_6H_4)_2CH_2$

NB   Cross-combinations were also tried.

The majority of these polymers can be produced by direct radical initiated copolymerization of the monomers at temperatures between 60 and 150 °C. By careful control of the reaction conditions it is possible to produce random copolymers; these are blocks of random copolymers separated by blocks of cross-linking agents or blocks of tin polymer separated by random species. The copolymers of the methacrylate and/or the acrylate species all have molecular weights greater than 110 000. They are soluble in a wide range of solvents and produce very good films.

The trialkyltin ester based polymers are primarily used as antifouling compositions to prevent the growth of barnacles, etc. on ship bottoms and harbour installations. This application and some possible environmental consequences are discussed in section 9.8.4.

9.6.6   Condensation polymers

Condensation polymers may be prepared by the copolymerization of a difunctional metal halide with a difunctional Lewis base that contains a metallocene. This is possible because the halides of metals of Groups IVA, VA, IVB, have a high degree of covalent character.[77] Typical polymer structures are shown in Fig. 9.4.

$$R_2MCl_2 \begin{cases} +HO_2CZ\,CO_2H \longrightarrow [MR_2O_2CZ\,CO_2]_n \\ +HSZSH \longrightarrow [MR_2SZS]_n \\ +HOZOH \longrightarrow [MR_2OZO]_n \end{cases}$$

R=alkyl, aryl or C_6H_5 Z=alkylene or arylene

Fig. 9.4   Structures of typical condensation polymer systems

Condensation polymers cannot usually be prepared in aqueous solution since the formation of cyclic species would occur. Usually the metal halide is dissolved in a non-polar solvent (e.g. benzene) and the co-monomer is used either as the pure liquid or dissolved in a polar solvent (water or an aprotic solvent). Very rapid mixing of the two liquids leads to the formation of the polymer at the interface. The reaction is usually complete in a few seconds. Polymer yields depend on the initial concentration of the reagents, rate of mixing and the time allowed for the reaction to take place.

Polymers of Group IVB metallocene dihalides can be prepared by dissolving the dihalide in basic aqueous solutions and stirring with an organic phase containing an alcohol or amine. Polyoximes, polythiols and polyethers usually form at the interface. In all cases the resultant polymers are insoluble and can be collected by filtration, washed and dried.

The polymers of Groups IVA and VA are generally flaky, white powders which are not particularly soluble in organic solvents. Because of this limited solubility, they do not cast good films from solution. Thermal stability is relatively poor, decomposition occurring below 200 °C. The polymers are particularly susceptible to hydrolysis and may revert to the co-monomers.

Polymers of Group IVB metals, on the other hand, hydrolyse only slowly and polyethers of titanocene, for example, can form flexible films with molecular weights greater than 10^5. Thermal properties are similarly much better with decomposition not occurring below 350 °C.

Table 9.10 Possible uses for condensation polymers

Application	Comments
Bactericides	Tin compounds are good − arsenic and antimony compounds may be useful in this area
Semiconductors	Some possibility of limited application
Catalysts	Titanocene and Zirconocene are extremely powerful Ziegler−Natta catalysts and polymeric species may prove useful
Uranium recovery	The precipitation of uranyl ions from waste water using dicarboxylate ions makes this a possible technique for uranium recovery
Chemotherapeutic agents	Investigations of the pharmocological properties of these materials are underway (Ref. 78) and some results are discussed in Section 9.8.3
Polymeric dyes	Phenolphthalein and similar dyes may be polymerized with metal halides to form polymers of high stability. The absorbance and fluorescence of these materials are being investigated to see if they can be used in lasers

Condensation polymers are still very new and these materials represent some of the first attempts to incorporate uranium, bismuth, lead and antimony into the polymer backbone.[78] A wide range of potential applications has been suggested (see Table 9.10) but only a few of these have actually been tested.

9.7 Antifouling coatings

In recent years, underwater antifouling coatings have developed rapidly both in sophistication and complexity. New coating systems are tested in order to achieve maximum performance and availability of ocean-going vessels. Unlike the other proposed uses of organometallic polymers considered in this chapter, antifouling materials are designed to introduce a toxic substance into the natural environment in their immediate vicinity. Care must also be taken to ensure that the impact of this substance is restricted to the desired effect only.

9.7.1 Conventional coatings (binder-based)

The most common method for the prevention of marine fouling has been to paint the underwater structure of the vessel with cuprous oxide, triorganotin(IV) compounds, or a mixture of both held together in a polymeric binder. Tributyltin(IV) fluoride, triphenyltin(IV) hydroxide and triphenyltin(IV) fluoride are the preferred compounds for use in this application since they are most active against a wide range of fouling species. Other tin compounds including tributyltin carboxylates, bis (tributyltin) oxide $((n-C_4H_9)_3Sn)_2O$ and tributyltin(IV) acetate $(n-C_4H_9)_3SnOAc$ have also been used successfully. The tin compounds have two major advantages over cuprous oxide; since they are colourless they can be used in paints of any colour and, more importantly, they degrade to non-toxic inorganic tin compounds whereas copper is toxic in all its forms.[79]

An antifouling paint containing 3 parts of triphenyltin hydroxide dispersed in a polymer system comprising 50 parts styrene and 40 parts ethyl acrylate with 35 parts of copper oxide was used to protect steel plates for three years during immersion tests in the sea.[80]

Other binder systems used in conjunction with these antifouling compounds include vinyl, epoxy, chlorinated rubber, urethane, coal tar epoxy and low density polyethylene based systems. The release of tri-n-butyltin(IV) fluoride from one of these polymer based materials showed an initial peak after 10 days' immersion, a minimum release at 13 days and steady-state conditions after 24 days. The release was inhibited as water diffused into the polymer matrix where it reacted with the tributyltin(IV) fluoride to form tributyltin(IV) hydroxide. The immobile hydroxide species was subsequently dehydrated to bis (tributyltin) oxide which diffused into the aqueous layer to yield the steady state.[81] Eventually, coatings of this type become depleted and must be removed and replaced.

9.7.2   Eroding polymeric coatings

There is a great deal of interest in the use of eroding antifouling paints that are based on tributyltin acrylate or tributyltin methacrylate copolymers with various organic acrylate esters to produce a combined toxic material and binder.[82,83] These coatings erode as the triorganotin portion of the polymer is hydrolysed from the acrylic backbone to release the active species tributyltin(IV) chloride and bis (tributyltin) oxide.

The depleted surface layer of the polymer coating is swollen and easily eroded to expose a fresh surface of the triorganotin polymer. Coatings based on these polymers can be formulated to produce linear and controlled release of toxic materials with time, the surface is self-cleansing, there is no removal of depleted residue layers and there is 100 per cent use of the toxic constituents of the coatings. A typical formulation for these polmeric coatings is[84]:

Tributyltin methacrylate	350 parts
Methyl methacrylate	150 parts
5 per cent phenanthroline in xylene	450 parts

The use of organotin polymer systems may increase the effective life of the antifouling coating from 2–3 years to as much as 5 or even 6 years.

The potential toxicity of the organotin compounds used in antifouling paints is discussed in section 9.8 and other aspects are discussed in Chapter 3.

9.8   Interaction with the environment

As has already been seen, many of the organometallic species are still being actively developed and investigated. For these compounds it is too early to predict how they will interact with man and his surroundings.

It is, however, safe to say that any material which exhibits any commercial potential will automatically be subjected to a series of toxicological tests.[85] These tests will probably include *in vitro* examination of carcinogenicity and toxicity and it is usual to include some allergy tests. Such tests can measure only the acute effects of a short-term exposure – long-term effects cannot as yet be adequately tested. Furthermore, it is not always possible on the basis of such tests to assess the effects of impurities in a bulk material. The effects, for example, of the impurity β-naphthylamine are, however, well known to the rubber industry.

Most of the tests mentioned[85] are designed to protect the producer and the consumer, both of whom are considered as part of the environment! To date, with one or two exceptions, very little work has been carried out to assess the long-term effects of additives in plastics on the environment in general. Perhaps this is justified to the extent that in many cases the organometallic species is encapsulated by or chemically bound to a non-degradable material. However, such considerations do not apply to all forms of industrial waste and neither do they apply to the residue materials produced by the incineration of commercial refuse. In this latter case, the polymeric binder will be destroyed, leaving the metal atoms, usually as the oxide, but other compounds are possible which may be free to react in an uncontrolled way.

Incineration is one route by which the polymer bound species may interact more directly with the environment. Other routes by which the original organometallic additive, or a reaction product derived from this material, can be released into the environment, are summarized in Table 9.11.

Incineration will produce inorganic salts only from the organometallic species present and typically at the $\mu g g^{-1}$ level in the effluent.[86] The potential for environmental damage from this source does not, therefore, seem very large. A considerable amount of work has been carried out to investigate the materials leaching from plastics which are used in food contact applications. Senich[87] indicates that in general only very low levels of materials are extracted from polymer systems.

Burial in landfill sites produces a very complex situation in which both biological and/or thermal degradation can occur alongside leaching. In a long-term study of the materials leached from a model landfill site containing

Table 9.11   Access to the environment

Access route	Mechanism of access to the environment
Direct venting to the atmosphere	Work space ventilation to comply with health and safety legislation. Emission of process fumes
Spillage	Accidental emission of vapours, liquids and solids to the environment
Direct release into environment as the designated use	Antifouling materials, etc. deliberately released. Fungicides and preservative materials
Leaching	Possible migration of packaging into foodstuffs. Loss of organometallic from a polymer in a landfill site for example − access to soil and water courses
Degradation	Thermal processing, ageing, UV effects and biological action. Possible release to air, soil and water
Incineration	Direct release of unspecified combustion products usually as inorganic oxides

normal household refuse (2 per cent plastic) no materials originating from polymer additives were detected.[88]

Much of the information presented is, therefore, incomplete in nature but as detection equipment and analytical techniques improve, more data will become available. The Environmental Protection Agency in the United States, for example, has only recently organized a programme on the development of monitoring techniques for organotin compounds in the environment. It is convenient, therefore, to discuss the limited information concerning the impact of organometallic species in polymers, firstly, for materials with low levels of additives, residual catalysts or bond-promoting systems. Stabilizers, used at higher levels (0.5−8 parts per hundred parts of the polymer) will then be considered and, finally, materials with even higher levels of organometallic compounds − antifouling paints, potentially chemotherapeutic compounds and silicone elastomers.

9.8.1   Low level additives

There appears to be no evidence to suggest that any Ziegler−Natta catalyst which may remain, at very low levels, in olefinic polymers has had any environmental impact. The catalysts used in polyurethane foam systems appears to present no hazard since the fully cured foam presents no known hazard, is chemically inert and is insoluble in most solvents. The chemicals used in the production of polyurethanes, particularly the isocyanates, must be carefully handled to ensure minimal worker exposure. Exposure to other chemicals used in the production process will, therefore, be similarly low.

Various aspects of other fully cured urethane systems have been investigated and in general the materials appear to be safe for human use, one of the most recent applications being as a dental prosthesis.[89]

Dust from foams may produce respiratory problems[27] and there has been major concern over the flammability of the foams, which may well hide other problems. There are two general approaches to recycling used foam. The first is to macerate the scrap, compress it and rebond it to form carpet underlay. Some 40 000 tonnes annually are dealt with in this way in the USA.[27] The alternative approach is to hydrolyse the foam to the base polyol, amine, carbon dioxide and presumably the catalyst. Such a process is possible, but it has yet to be shown to be economically viable.[90]

Silane cross-linking agents in general exhibit moderate to low acute oral toxicity; an exposure limit of 20 mg m^{-3} in air is applied in some factory areas using these materials. The silanes may produce both skin and eye irritation, the effects being related to both exposure concentration and exposure period. A wise precaution, therefore, is to ensure adequate ventilation and that protective equipment is worn when required. A necessary consequence of factory ventilation is the direct injection of unspecified quantities of silanes into the airborne environment.

There is no evidence to suggest that long-term exposure to low levels of these silanes is harmful. Some toxicity data for vinyltrimethoxysilane to aquatic species is shown in Table 9.12.[91]

Table 9.12   Toxicity of vinyltrimethoxysilane ($(CH_2{=}CH)Si(OCH_3)_3$) to different species

Species	Exposure period (hours)	LC_{50}(mg dm^{-3})
Anabaena flos-aqua	336	1000
Selenastrum capricornutum	336	210
Blue gill sunfish	96	1000

See Ref. 91

9.8.2   Organotin stabilizers in PVC

Polyvinyl chloride (PVC) in terms of ecological and environmental considerations has much in its favour. Industrial waste generated during fabrication can be disposed of during recycling into other products or sanitary landfill. In landfill, PVC wastes are not biodegradable and, therefore, do not give off toxic products by this route.[92] This statement, however, ignores the PVC degradation products and the excess materials volatilized during processing, all of which are collected and vented to the atmosphere. Such careful collection is carried out to ensure that the workforce is not exposed to the vinyl chloride monomer (VCM).

Foakes[93] states that octyltin compounds have been subjected to careful study with regard to their toxicity and the exactness of the manufacturing techniques. This reaction followed the marketing of an organotin compound as a medicine in France in 1954. Impurities in the material affected some 1000 people and symptoms ranged from headaches to death from respiratory failure or cardiac collapse. It was found that some of the original material (dimethyltin dioxide) was contaminated with up to 10 per cent of the triethyltin monoiodide compound, which was at least ten times more toxic. Results obtained from toxicity tests on a range of tin stabilizers are summarized in Table 9.13.

Table 9.13   Toxicity studies for some organotin stabilizers*

Compound	LD_{50} rat $(mg\ kg^{-1})$
Diethyltin(IV) dilaurate	210
Dibutyltin(IV) dilaurate	175
Dioctyltin(IV) dilaurate	6000
Dibutyltin(IV) dioctylthioglycollate	510
Dioctyltin(IV) dioctylthioglycollate	2000
Monooctyltin(IV) tri-i-octylthioglycollate $\Big\}$ blend Dioctyltin(IV) di-i-octylthioglycollate	1200

* See Ref. 91

In Germany, the use of di-n-octyltin(IV) compounds in food-containing grades of PVC is allowed, provided that the levels of methyl-, ethyl- and aryltin compounds in any stabilizer are below currently available detection limits.[94] The American regulations are more stringent[95] and manufacturers are required to monitor any breakdown products during processing and to determine the extractability of any such breakdown products. Major breakdown products are usually octyltin chlorides.[92]

Allen *et al.*[96] have reported the investigation of the thermal decomposition of a range of stabilizers using Mössbauer spectroscopy. The thermal degradation studies of separate samples of PVC containing the following materials

- 1.2 per cent dibutyltin(IV) bis(iso-octylthioglycollate)
- 4 per cent dioctyltin(IV) bis(iso-octylthioglycollate)
- 2 per cent dibutyltin(IV) bis(iso-octylmaleate)

showed that in each case the stabilizer was converted into the dialkylmonochlorotin(IV) ester ($R_2SnCl(X)$—X = $SCH_2CO_2C_8H_{17}$ or $O_2CCH{=}CHCO_2C_8H_{17}$).

Comparison of samples prepared by hot milling and room temperature solvent casting processes revealed little degradation of the polymer during the initial milling process.

Photodegradation was studied by exposing PVC containing dibutyltin(IV) bis(iso-octylthioglycollate) or dibutyltin(IV) bis(iso-octylmaleate) to artificial sunlight produced by a Xenotest apparatus.[97] After 1000 hours' exposure, the maleate was chemically unchanged, whereas the thioglycollate was rapidly converted to the monochloroester (n-C_4H_9)$_2$SnCl(SCH$_2$CO$_2$C$_8$H$_{17}$). Prolonged exposure resulted in the formation of tin tetrachloride.

9.8.3   Biological activities of metal-containing condensation polymers

The materials so far discussed have exhibited an apparent biological inertness, but this is not always the case. Various metal-containing polymers exhibit a wide range of activity towards biological specimens, also biologically related organisms exhibit different responses to the same condensation polymers.[98]

The use of metal-containing compounds as medicines is widespread and the presence of metal-containing macromolecules in the body is well known. These metals include iron, vanadium, zinc and copper.

Objectives in the design of drugs are that they should have:

(a) high specificity;
(b) high activity;
(c) controlled duration of activity;
(d) wide concentration range of biological activity.

Metal-containing condensation polymers which may contain different metals in different chemical environments should exhibit a wide range of biological activity. Preliminary biological assay results for the organometallic polymer systems were obtained using bacterial and fungal inhibition data.

Nine condensation polymers were compared to show the variation of biological response with the state of the test material. The general nature of the structure of the material is shown in Fig. 9.5.

In the dry state only two of the antimony compounds and one of the tin compounds showed any inhibitory effects towards the test organisms. Similar differences in inhibition were observed for different members of the test samples when a range of solvents was utilized.

A range of tin dyes derived from xanthene dyes (cf. Fig. 9.6) were tested against a range of bacterial species. As expected, the greatest toxicity was observed for the dye derived from mercurochrome, probably as a result of the toxicity of the dye portion rather than the tin.

Polysaccharides modified with a range of organostannane halides were compared by monitoring the growth of fungi in solutions containing these materials. The results obtained, which are summarized in Table 9.14, indicate

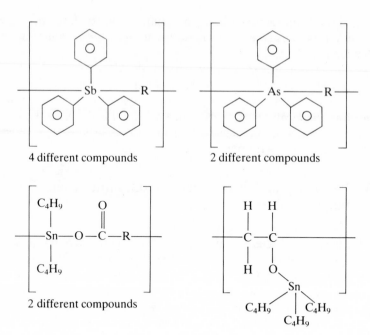

4 different compounds   2 different compounds

2 different compounds

Fig. 9.5   Structures of compounds for biological testing

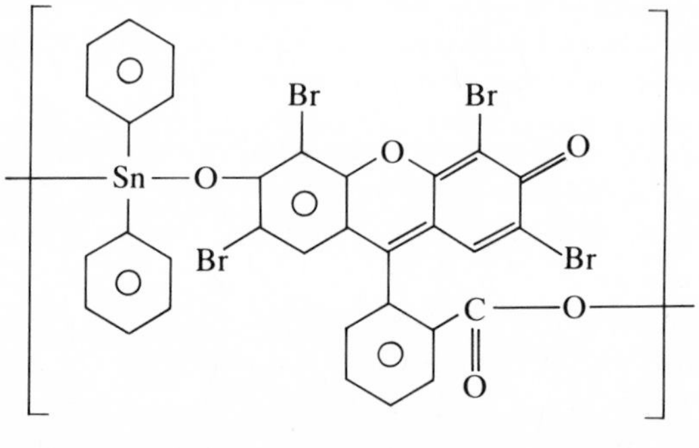

Fig. 9.6   An example of a tin-based xanthene dye

that the effectiveness of the material is inversely proportional to the length of the alkyl chain substitutents. These data indicate that some tin compounds could certainly be used to inhibit rot and mildew.

   These tests show that the compounds exhibited can exhibit quite dramatic biological action towards the test specimens. Care must be taken to ensure that the effects on other biological species, perhaps including man himself, do not outweigh the areas of usefulness so far identified.

Table 9.14   Some examples of fungal growth inhibition
for R_2SnCl_2-type compounds

R	Percentage inhibition : average
Dimethyl	100
Dipropyl	90
Dibutyl	70
Dioctyl	15
Dilauryl	0

9.8.4   Organotin antifouling compounds

Such compounds, if not used as directed, may produce a hazard both to man
− when applying the coatings − and to the environment in general. The bulk
of the information currently available relates to acute studies of the compounds: bis(tri-n-butyl)tin(IV) oxide, tributyltin(IV) fluoride and triphenyl-
tin(IV) fluoride.[9]

Lack of care during the coating operation may lead to the workforce being
exposed to high concentrations of organotin compounds. These compounds
are known eye and skin irritants and may cause irritation of the upper
respiratory tract. The current US Occupational, Safety and Health
Administration Threshold Limit Value (TLV) standard for exposure to all
organotin compounds is 0.1 mg of organotin compounds (as tin) m^{-3} air
averaged over an eight-hour work day. Providing the exposure levels are
maintained below this level, the materials can be handled safely. During
spraying operations the use of a full face, air supplied, respirator is recommended. Similar precautions are mandatory when removing residual material
from the hulls of ships.

Acute toxicity tests, LD_{50} and skin irritation studies have been carried out in
order to confirm the extent of the effects when related to compound type.
Both the organotin compounds alone and when compounded into the paint
formulation were tested and gave the results shown in Table 9.15.

It should be noted that the irritational effects are greater when the compounded polymer is considered versus the organotin compound on its own.
Acute toxicity tests may be misleading and a ninety-day subacute toxicity
study was carried out to check the dermal response data for tributyltin
fluoride.

The tin compound was applied to the skins of rabbits for a 7-hour period
each day, 5 days a week, for 13 weeks. Three dose levels were chosen, 14, 27.2
and 68.0 μg g^{-1}. Before the end of the test, 70 per cent of the rabbits exposed
to 68 μg g^{-1} died; the survivors eventually recovered with no apparent harmful effects. Rabbits exposed to 14 μg g^{-1} showed no effect whatsoever. All

these results confirm that care must be taken when handling these organotin compounds in order to prevent problems associated with acute exposure.

Chronic exposure considerations inevitably raise the question of carcinogenicity and, although such tests are now usually carried out *in vitro*,[85] mice were used to test solutions of tributyltin(IV) fluoride. These tests, designed to investigate skin lesions, showed that, under the particular test protocol, the tin compound was non-carcinogenic. All these tests confirm that, when normal personal hygiene is maintained, the organotin compounds can be safely handled at the application stage of their use.

The release of organotin compounds from antifouling paints corresponds to a very small quantity on a global scale, but is nevertheless environmentally significant in the immediate vicinity of the coated object. In order to discuss the effects of the release of organotin compounds, it is necessary to understand the reaction pathways which lead to degradation and which, therefore, control the persistence of toxic species in the environment.[100–104]

Halogenated tin compounds are believed to hydrolyse as they leach from the polymer and then react with dissolved carbon dioxide to produce carbonates (equations 9.8 and 9.9):

$$R_3SnF \rightarrow R_3SnOH \text{ (unstable)} \rightarrow R_3SnOSnR_3 + HF \qquad [9.8]$$

$$R_3Sn-O-SnR_3 + CO_2 \rightarrow R_3Sn-O-\overset{\displaystyle O}{\overset{\displaystyle \|}{C}}-O\,SnR_3 \qquad [9.9]$$

The bis (tri-n-butyltin) carbonate then undergoes a series of transformation reactions to produce tin oxide and the respective alkyl groups. Schematically this is shown as equation 9.10 illustrating sequential loss of alkyl groups at each stage:

$$R_3Sn-O-\overset{\displaystyle O}{\overset{\displaystyle \|}{C}}-O\,Sn\,R_3 \xrightarrow{UV} R_2Sn\,O \xrightarrow{UV} R\,Sn\overset{\displaystyle O}{\underset{\displaystyle OH}{<}} \xrightarrow{UV} Sn\,O_2 \qquad [9.10]$$

+ R_3CO_2	+ R	+ R
dibutyltin oxide	butyl stannoic acid	tin(IV) oxide

This reaction sequence has a half-life of 89 days at 20 °C[105] in sunlight but no apparent change in R_3—Sn moiety concentration was observed over a two-month period in the dark.

Photolytic degradation of triphenyltin hydroxide in water suggested a ready degradation to diphenyltin oxide accompanied by the formation of $(PhSnO_xH_y)_n$, a water soluble polymeric species.[106] Examination of the UV degradation of trimethyltin(IV) chloride revealed the production of a dimethyltin cation($(CH_3)_2Sn(H_2O)_4^{2+}$) which could be degraded further to hydrated stannic oxide.[107]

Biomethylation of trimethyltin salts to tetramethyltin has been reported[107,108] for experiments using estuarine sediments, but no similar observations have been reported in seawater, although methyltin species have been observed (see Ch. 3). There is insufficient evidence to date to confirm biomethylation of inorganic tin residues in the natural as opposed to laboratory model environment.[104] The possibility of bacteria being able to metabolize organotin compounds in some unexpected manner is being investigated by Porter.[108] For a discussion on biomethylation aspects see Chapter 3.

Methods which can be used to analyse material leaching from antifouling coats have been reviewed by Jewett *et al.*[109] Collection techniques include the use of high performance liquid chromatography packing material and solvent extraction with chloroform. The extract residues were then investigated by a range of techniques which included atomic absorption spectrophotometry (AA), thin layer chromatography (TLC) and infrared spectroscopy (IR).

Several studies of the impact of material leaching from antifouling paints have been reported. Concentration of tributyltin acetate at levels above $150 \mu g \, dm^{-3}$ killed all the fish present in under three hours. Invertebrate species exposed to the same levels survived for between five and eight days.[110] In a much larger toxicity study, Linden *et al.*[111] reported the initial concentration of material required to kill 50 per cent of the test species (bleak) in under 96 hours. The results obtained were:

- tributyltin(IV) fluoride $\qquad$ $6 \mu g \, dm^{-3}$
- tributyltin(IV oxide $\qquad$ $15 \mu g \, dm^{-3}$
- triphenyltin(IV) fluoride $\qquad$ $400 \mu g \, dm^{-3}$

Further laboratory tests have shown that tributyltin(IV) oxide levels of approximately $10 \mu g \, dm^{-3}$ and above affect the growth rate of mud crab larvae[112] while spat growth was affected by levels as low as $0.15 \mu g \, dm^{-3}$. In this latter case, the tissue concentrations of the organotin compounds were some 10^4 times that of the exposure level.[113] Comparison of these and other results[114] shows the differing toxicity of organotin compounds to different species.

Disposal of tin-containing coating, particularly as potential landfill material, has been examined.[115] In studies using clay, topsoil and sand, vertical migration was only observed for the sand. In clay soils 95–99 per cent of the triorganotin compound remained as dumped, suggesting that no environmental threat would result from this method of disposal.[116] Studies of the rates of degradation in soil showed that 50 per cent of both tributyltin oxide and triphenyltin(IV) acetate disappeared from silt and sand in about 20 weeks.[117] The degradation products appeared to be dibutyl derivatives and carbon dioxide.

For a discussion of other aspects of organotin antifouling compounds see Chapter 3.

9.8.5 Silicone elastomers

This chapter is concerned specifically with rigid or elastomeric organosilicon compounds; for information concerning oils, etc. the reader is referred to Chapter 6. A report[118] prepared by Syracuse University summarizes the general environmental impact of organosilicon compounds and provides an adequate literature review up to 1974. Further reports are currently being prepared by the Medical Research Council, the Swedish Water and Air Pollution Research Unit and by TNO in the Netherlands.

Silicone resins and elastomers are stable materials and do not usually become part of the mobile environment. Disposal by landfill or incineration is acceptable, although there is some evidence of polymer degradation when in contact with clay soils.[119] Since silicones are insoluble there is no major route by which leaching can occur. Incineration produces silica which is classed as a nuisance dust.

Silicones are biologically inert and do not react with body fluids or show any toxic effects. Medical grade Room Temperature Vulcanizing (RTV) silicones, for example, are used for surgical tubing, for the formation of prosthetic parts and the encapsulation of pacemakers.[120,121] Formulated silicones may contain ingredients that are irritating or toxic and some low molecular weight silicones show some biological activity which is, in general, similar to that shown by many solvents.[122]

The cross-linking catalyst dibutyltin(IV) dilaurate has been classified as non-toxic and is satisfactory for use in food contact applications. No such clearance has been given for dibutyltin(IV) diacetate, which has also been used as a curing catalyst. Silicone polymers used as seals may be post-cured for up to 24 hours in order to remove traces of the cross-linking agents, which may taint or affect the material in contact with the polymers. There has been some concern that the use of dichlorobenzoyl peroxide as a cross-linking agent may lead to the production of polychlorinated aromatic compounds.

9.9   Carcinogenicity tests applied to silicone compounds

Silicone compounds have been subjected to a wide range of many and varied toxicological tests, see Chapter 6. Within these tests, the current vogue is to examine chemical compounds for any evidence of carcinogenic, mutagenic or teratogenic propensities.[123,124] Such tests are carried out either by direct administration of the chemical to a population of test animals, usually rats or mice, or by *in vitro* testing, using different strains of *Salmonella* bacteria.[85]

It is impossible, in a survey of this type, to cite all the references to the volumes of work reporting the results of such tests performed on silicone compounds. The materials tested range from hair dyes[125] to solar heating fluids[126] and from herbicides[127] to intrauterine devices.[128] In the great majority of these investigations, no mutagenic, carcinogenic or teratogenic effect was observed for the dose regime studied. No increase in the number of tumours was detected in mice in these experiments after a 76-week exposure to a poly(dimethylsiloxane) antifoam agent at a 2−5 per cent level of the dietary intake.[129] Skin painting tests using mercapto functional silicone oils applied to mice skins showed no significant increase in tumours.[130] Similar negative results have been reported for hair dyes, a silane coupling agent,[131] herbicides,[127] silicone fluids in solar heating units and for silastic foam elastomers[132] used for encapsulation purposes.

Some work has been reported in which positive effects have been attributed to the presence of organosilicone compounds. For example, isomers of dimethyl-, methyl- and phenylcyclosiloxane exhibit some mutagenic characteristics,[133] while in a range of experiments on medical grade polydimethyl siloxane oils of different viscosities, an apparent dose-related response of *in utero* mortalities was observed at one dose level − but this could not be repeated for higher doses.[134] A study of the effects on rats of implanted silastic intrauterine devices showed that the exposed groups produced more malformed foetuses than the untreated controls. A second series of experiments confirmed this effect of the silastic elastomer. Sham operations performed on the control group produced no significant effects.[128] Similar positive observations have not however been reported elsewhere.

9.10   Summary

This work has shown that a large number of organometallic species are currently incorporated into polymeric systems. The development of new additives, stabilizers, catalysts and other system promoters continues as does the improvement and use of biocides and antifouling paints, etc. Furthermore, research into the development of new organometallic polymer systems, particularly for pharmaceutical applications, is actively being undertaken.

Many of the compounds being developed show significant potential for major interaction with the environment. This has been recognized by the major organotin producers who formed The Organotin Environmental Programme Association (ORTEPA) in order to monitor and exchange information pertinent to the effects of organotin compounds on the environment. Other organometallic species are, as yet, not used to the same extent as the tin compounds and possibly, therefore, do not justify the formation of similar organizations.

References

1. Kuntz I, Gerber A 1960 *J Polym Sci* **42**: 299−308
2. Morton A A 1950 *Ind Eng Chem* **42**: 1488−96
3. Morton A A, Lanpher E J 1960 *J Polym Sci* **44**: 233−9
4. Seyferth D 1972 *J Organometallic Chem* **41**: 155−275

5. Seyferth D 1973 *J Organometallic Chem* **62**: 25–174
6. Smith J D 1980 *J Organometallic Chem* **189**: 17
7. Oliver J P 1974 *J Organometallic Chem* **75**: 263–324
8. Lenz R W 1967 *Organic Chemistry Synthesis of High Polymers*. Wiley Interscience, New York
9. Wilke G 1963 *Angew Chem Intern Edn* **2**: 105
10. Porri L, Natta G, Gallazzi M C 1967 *J Polym Sci C* **16**: 2525–37
11. Dawans F, Teyssie P 1965 *J Polym Sci B* **3**: 1045–8
12. Dawans F, Teyssie P 1967 *CR Acad Sci Paris* **261C**: 4097–9
13. Natta G, Dall'Asta G, Manzanti G 1964 *Angew Chem* **76**: 765–72
14. Gunthe P 1970 *Angew Makromol Chem* **14**: 87–9
15. Fischer E O, Wagner W R 1976 *J Organometallic Chem* **116**: 21
16. Noll W 1968 *Chemistry and Technology of Silicones*. Academic Press, New York, pp 1–40
17. Mixer R, Bailey D 1955 *J Polym Sci* **18**: 573–7
18. Wagner G W 1953 *Ind Eng Chem* **45**: 367–8
19. Rosen M 1978 *J Coatings Technol* **50** (644): 70–1
20. Pluddemann E P 1982 *Silane Coupling Agents*. Plenum Press, New York, pp 1–48
21. Walker P 1980 *J Coating Technol* **52** (646): 49–51
22. Bjurksten J, Yaeger C L 1952 *Mod Plast* **29**: 124
23. Scott H G 1964 US Patent 3646155
24. Block G A 1981 *Organic Accelerators and Curing Systems for Elastomers*. (Trans Mosely R J) Rubber and Plastics Research Association of Great Britain, Shawbury, pp 232–96
25. Schill D 1974 *Proc Soil Science Soc Amer* **38** (4): 674–8
26. Synder G H, Ozaki H J, Hayslip N C 1974 *Proc Soil Science Soc Amer* **38** (4): 678–82
27. Woods G 1982 *Flexible Urethane Foams Chemistry and Technology*. Applied Science Publishers, London, pp 170–5
28. Annual Report 1978. International Tin Research Institute, Middlesex, England, p 20
29. Cox D F, Hostettler J 1959 ACS Meeting Proc Boston Mass, April
30. Frey H E, Maffly R L 1980. In *Economics Handbook*. Stanford Research Institute, Menlo Park, Calif, pp 580–1561A
31. Idemitsu Kusan Co Ltd 1983 JPN Koka Tokkyo Koho JP 58.77889
32. Warner Lambert Co 1983 Euro Pat Appl EP 89181
33. Bayer A G 1983 Ger Offn DE 3 2 12 735
34. Brecher L R 1977 *Plast Eng* March, 39
35. Dainippon Inland Chemicals Inc 1983 JPN Koka Tokkyo Koho JP 58 132051
36. VEB Synthewerke Schwerzheide 1983 Ger East DD 201473
37. Dabi S, Freirich S, Zilkha A 1982 *Proc IUPAC Macromol Symp 28th*, pp 262–4
38. Russo R V 1976 *J Cell Plast* **12**: 203–7

39. Olczyk W and Szpakawska H 1982 *Chem Stosow* **26** (3−4): 333−40
40. Van der Verk G J M 1966 *Ind Eng Chem* **58**: 29
41. Overmars H G J, Van der Want G M 1965 *Chimia* **19**: 126−9
42. Wacker Chemie GmbH 1963 US Patent 3082527
43. Beers M D 1977. In *Handbook of Adhesives* (ed Skeist I). Van Nostrand, New York
44. Gitlitz M M, Moran M K 1982. In *Kirk-Othmer Encyclopaedia of Chemical Technology*, Vol 23, pp 65−75
45. Aufdermarsh CA 1983 US Patent 4394489
46. Nass L I 1971. In *Encyclopaedia of Polymer Science and Technology*, Vol 12 (ed Bikales B M). Wiley Interscience, New York, p 725
47. Anon 1979. *Mod Plast* **56** (9): 73
48. Stepek J, Birkal C 1965 *Chem Listy* **59**: 1201−3
49. Barnes J M, Stoner H B 1958 *Br J Ind Med* **15**: 15−22
50. Cusack P A, Smith P J 1981. In *Speciality Inorganic Chemicals* (ed Thompson R). Royal Society of Chemistry, London, p 285
51. Anon 1976. *The Manufacture and Use of Selected Alkyl/Tin Compounds.* EPA 560/6-76-01. Office of Toxic Substances, Washington, DC
52. Dietz D R, Banzer J D, Miller E M 1979 *J Vinyl Technol* **1** (3): 761−3
53. Schulters G 1966 *Soc Plastics Eng J* **22** (2): 60−9
54. Stepek J, Daust H 1983 *Additives for Plastics*. Springer Verlag, New York, p 150
55. Sheats J E 1982. In *Kirk Othner Encyclopaedia of Chemical Technology* Vol 15, pp 184−219
56. Carraher Jr C E 1981 *J Chem Educ* **58**: 921−4
57. Sheats J E, Pittman Jr C U, Carraher Jr C E 1984 *Chem Brit* August: 709−14
58. Pittman Jr C U 1977. In *Organometallic Reactions*, Vol 6 (eds Becker E, Tsutsui M). Plenum Press, New York, pp 1−62
59. Sheats J E, Willis T C 1979 *Org Coat Plast Chem* **41**: 33−8
60. Pittman Jr C U, Suryanarayanan B 1974 *J Amer Chem Soc* **96**: 7916−19
61. Pittman Jr C U, Rounsefell T D, Lewis E A, Sheats J E, Edwards B H, Raush M D, Mintz A 1978 *Macromolecules* **11**: 560−5
62. Raush M D, Moser G A, Zaiko E J, Lipman A L 1970 *J Organometallic Chem* **23**: 185−92
63. Hart W P, Macomber D W, Raush M D 1980 *J Amer Chem Soc* **102**: 1196−8
64. Figita N, Sonogashira A 1974 *J Polym Sci Chem Ed* **12**: 2845−56
65. Pittman Jr C U 1971 *J Paint Technol* **43**: 29−35
66. Pittman Jr C U, Ayers O E, McManus S P 1973 *J Macromol Sci Chem* **A7**: 1563−79
67. Pittman Jr C U, Hirao A 1977 *J Polym Sci Polym Chem Ed* **15**: 1677−9
68. Pittman Jr C U, Hirao A 1978 *J Polym Sci Polym Chem Ed* **16**: 1197−201

69. Carraher Jr C E, Molloy H M, Taylor M L, Yelton R O, Schroeder J A, Bogdan M R 1979 *Org Coat Plast Chem* **41**: 197–201
70. Funk R L, Vollhardt K P C 1979 *J Amer Chem Soc* **101**: 215–17
71. Pittman Jr C U, Grube P I 1971 *J Polym Sci A1* **9**: 3175–86
72. Gal A, Cais M, Kohn D H 1971 *J Polym Sci A1* **9**: 1853–63
73. Pittman Jr C U, Voges R L, Elder J 1971 *Macromolecules* **4**: 302–9
74. Neuse E W, Bednarit L 1979 *Org Coat Plast Chem* **41**: 158–61
75. Gruber G and Hallersleben M L 1967 *Makromol Chem* **104**: 77–81
76. Yeager W L, Castelli V J 1978. In *Organometallic Polymers* (eds Sheats J E, Pittman Jr C U). Academic Press, New York, p 195
77. Carraher Jr C E 1977. In *Interfacial Syntheses* (eds Millich F, Carraher C E). Marcel Dekker, New York, pp 367–9
78. Carraher Jr C E 1980 *Org Coat Plast Chem* **42**: 427–31
79. Gitlitz M H 1981 *J Coat Technol* **53**: 678–9
80. Nippon Oils and Fats Co Ltd 1983 JPN Kokai Tokkyo Koho JP 5842668
81. Sherman L R 1983 *J Appl Polym Sci* **28**: 2823–6
82. Montemarana J A, Dyckmann E J 1975 *J Paint Technol* **47**: 59–62
83. Dyckmann E J, Montemarano J A 1973 *Amer Paint J* **20**: 6–10
84. van Londen A M, Johnsen S, Govers G J 1975 *J Paint Technol* **47**: 63–8
85. Ames B N, McCann J, Yamasaki E 1975 *Mutation Res* **31**: 347–63
86. Greenberg R R, Gordon G E, Zoller W H, Jacko R B, Nendorf D W, Yost K J 1978 *Environ Sci Technol* **12**: 1329–32
87. Senich G A 1982 *Polymer* **23**: 1385–7
88. Bromley J 1983. In *Industry and the Environment in Perspective* (ed Hester R E). Royal Society of Chemistry Special Publication 46, London, pp 153–89
89. Smith and Nephew Associated Co PLC 1983 Eur Pat Appl EP 91285
90. Ballisreri A, Foti S, Maravigna P, Montando G, Scamparrino G 1980 *J Polym Sci Polym Chem Ed* **18** (6): 1925
91. Firin R, Frye C L, Raum A L J 1983. In *Proc Martox Conf Environ Impact Organosilicon Compounds Aquatic Biosphere*, pp 1–36
92. Crider L B, Holbrook W C, Kent D C 1976 *Encyclopaedia of PVC*, Vol I. Marcel Dekker, New York
93. Foakes E H 1973. In *Developments in PVC Technology* (eds Hanson J H, Whelan A). Applied Science Publishers, London, pp 76–80
94. *Bundesgesund Heitsblatt* 1968 No 18: 271
95. Federal Register USA 1972 1B
96. Allen D W, Brooks J S, Clarkson R W, Mellar T M, Williamson A G 1980 *J Organometallic Chem* **199**: 299–310
97. Brooks J S, Clarkson R W, Allen D W, Mellar T M, Williamson A G 1982 *Polymer Degradation and Stability* **4**: 359–63
98. Carraher Jr C E, Giron D J, Cerutis D R, Burt W R, Verikatachalam R S, Gehrke T J, Tsusji S, Bloxall H S 1982 *Biological Activity of Polymers*. ACS Symp Series No. 186 pp 13–25
99. Sheldon A W 1975 *J Paint Technol* **47**: 54–8

100. Zuckermann J J, Reisdorf R P, Ellis H V, Wilkinson R R 1978. In *Organometals and Organometalloids: Occurrence and Fate in the Environment* (eds Brinckman F, Bellama JM). ACS Symp Ser No 82, pp 388–90
101. Brinckman F E 1981 *J Organometallic Chem Library* **12**: 343
102. Craig P J 1980 *Environ Tech Lett* **1**: 225–234
103. Plum H 1981 *Inf Chim* **220**: 135
104. Craig P J 1982. In *Comprehensive Organometallic Chemistry* Vol 2 (eds Abel E W, Stone F G A, Williams G). Pergamon Press, Oxford 979–1020
105. Maguire R J, Carey J H, Hale E J 1983 *J Agric Food Chem* **31**: 1060–2
106. Soderquist C J, Crosby D J 1980 *J Agric Food Chem* **28**: 111–14
107. Annual Report 1980. International Tin Research Institute, Middlesex, England, p 23
108. Porter G 1980 *Dim/NBS* March
109. Jewett K L, Blair W R, Brinckmann F E 1979 *Proc 6th Int Symp Contr Rel Bioact Mat* New Orleans, USA, pp 11–13
110. Good M L, Dundee D S, Swindler G S 1979. *Proc 6th Int Symp Contr Rel Bioact Mat* New Orleans, USA, pp 1–10
111. Linden E, Bengtsson B E, Svanberg O, Sundstrom G 1979 *Chemosphere* **8**: 843–4
112. Laughlin R, French W, Guard H F 1983 *Water Air Soil Pollut* **20**: 69–71
113. Waldock M J, Thain J E 1983 *Mar Pollut Bull* **14**: 411–12
114. Smith P J 1978 *Toxicological Data on Tin Compounds*. Intern Tin Res Inst Publn No 583
115. Harris L R, Andrews C, Burch D, Hampton D, Maegerlein S 1979 *US Dept Navy Ship Mat Eng Dept Res Dev Rep*, No DTNSRDC/SME-78/2A
116. Slesinger A E 1977 *Mar Coat Sem, Nat Paint Coat Assoc*, Biloxi, USA
117. Barug D, Vonk J W 1980 *Pestic Sci* **11**: 77–82
118. Anon. Assessment of Siloxanes 1974 Environmental Protection Agency Syracuse University Research Corporation USA
119. Buch R R, Ingrebrigton D N 1979 *Environ Sci Technol* **13**: 676–9
120. Braley S A 1970 *Rubber Chem Technol* **44**: 363–80
121. Rowe V K, Spencer H C, Bass L S 1948 *J Ind Hyg Toxicol* **30**: 332–5
122. Clark L C, Gollan F 1966 *Science* **152**: 1755–6
123. Tuchmann-Duplessis H 1983 *Am J Ind Med* **4**: 245–58
124. Wolff M S 1983 *Am J Ind Med* **4**: 259–81
125. Burnett C, Goldenthal E I, Harris S B, Wazeter F X, Strausburg J, Kapp R, Voelker R 1976 *J Toxicol Environ Health* **1**: 1027–40
126. Marshall T C, Clark C R, Brewster D W, Henderson T R 1981 *Toxicol Appl Pharmacol* **58**: 31–8
127. Ruchkovskii B S, Tsapenko V F 1982 *Visn Akad Nauk Ukr RSR* No 4: 43–5
128. Barlow S M, Knight A F 1983 *Fertil Steril* **39**: 224–30

129. Cutler M G, Collings A J, Kiss I S, Sharratt M 1974 *Food Cosmet Toxicol* **12**: 443–50
130. Parent R A 1980 *Drug Chem Toxicol* **2**: 369–74
131. Siddiqui W H, Hobbs E J 1984 *Toxicology* **31**: 1–8
132. Wang S Y, Smith D M 1980 *Energy Res Abstr* **5**(9) Abstr 14377
133. Batulin Yu M, Klyashchitskaya A L, Domilak M G, Chiskova E M 1977 *Biol Akt Soedin Elem* **IVB**: 68–75
134. Kennedy Jr G L, Keplinger M L, Calandra J C, Hobbs E J 1976 *J Toxicol Environ Health* **1**: 909–20

Chapter 10

Other organometallic compounds in the environment

10.1 Antimony

Several methylantimony(V) species have recently been detected in natural waters at the ng dm^{-3} level.[1-3] The analytical method used is the hydride generation (by sodium borohydride) of the methylstibine(III) derivatives from the aqueous sample. The hydrides are isolated in a cold trap and separated by temperature-programmed gas chromatography (GC). These compounds are sufficiently stable against hydrolysis and decomposition at room temperature for quantitation to be feasible. For antimony(V) a quantitative reduction step at pH 1−2 requires the presence of iodide ion. Antimony(III) only is reduced at pH 6 and the mono- and dimethylantimony(V) materials present in the sample are reduced at pH 1.5−2 without the presence of iodide. The only organoantimony species detected so far in the natural environment are monomethylstibonic acid ($CH_3SbO(OH)_2$) and dimethylstibinic acid (($CH_3)_2SbO(OH)$). The conditions allow differential reduction, and hence speciation, to take place. The cold trap and chromatographic column formed a single part of the apparatus and the packing used in the column was conventional (15 per cent OV-3 on chromosorb W/AWDMCS − 60−80 mesh). The detection system used for antimony was a quartz-cuvette AA detector (antimony 217.6 nm line) with a fuel-rich hydrogen−air flame. Detection limits were as follows (as nanograms) − antimony(V), 0.05; $CH_3SbO(OH)_2$ and ($CH_3)_2SbO(OH)$, 0.04. Using graphite furnace AA, limits in the range 0.12−0.15 ng were noted. Concentration limits (100 cm^3 sample) were from 0.3−0.6 ng dm^{-3}.

The range of antimony species present in seawater is from 1.1−1.7 nmol kg^{-1} with an overall world ocean average of 0.3 $\mu g\ dm^{-3}$.[4] Where lower values have been reported this may be due to incomplete reduction. In one river estuary[3] all four antimony species noted above were found to be present, at levels ranging from 1 to 12 ng dm⁻³ for the methyl derivatives. In contrast to arsenic (Ch. 5) the monomethyl form was about ten times as abundant as the dimethyl form. The dominant form is inorganic antimony(V). Samples were from the Ochlockonee River, USA. Other samples have also been reported from various rivers (Table 10.1). The occurrence of the thermodynamically unstable antimony(III) and the methyl forms was ascribed to biological (algal) activity. The methyl forms have also been postulated to occur in marine planktonic algae. The proportion of organic to inorganic forms of antimony is up to about 10 per cent.

More recent work[2] in areas of the Baltic Sea show lower antimony concentrations than in the open oceans. In oxic waters antimony(V) predominates but in anoxic basins the equilibrium shifts to the trivalent form, possibly

Table 10.1   Antimony species in natural waters*
*(Concentrations in ng(Sb)dm⁻³)

Location	Sb(III)	Sb(V)	MSA†	DMSA‡
Ochlockonee River	3.2	22.9	0.5	ND
	1.9	30.3	ND	ND
Trinity River	0.9	145.0	0.8	ND
Mississippi River	<0.3	148.0	2.3	ND
Escambia River	<0.3	12.8	0.8	ND
Apalachicola River	0.4	55.0	0.6	ND
Rhine River at Oppenheim	0.4	231.0	1.2	ND
Main River at Frankfurt	0.3	311.0	1.8	ND
Ochlockonee Bay estuary				
% salinity: 4.3	2.8	42.0	0.8	ND
12.3	6.6	81.5	1.4	ND
18.9	8.3	113.0	2.2	ND
23.8	5.4	122.0	5.1	0.6
30.2	8.8	126.0	10.9	1.1
33.4	11.1	136.0	12.6	1.5
Gulf of Mexico	4.4	149.0	5.3	3.2
Apalachee Bay	1.9	164.0	8.5	ND
Baltic Sea △	0.001– 0.1	0.01– 0.7	0– 0.05	<.01

* From Refs 1, 3.

† CH₃SbO(OH)₂

‡ (CH₃)₂SbOOH

△ Concentrations in nmol dm⁻³

Data reproduced with permission from Andreae M O et al 1981 Anal Chem 53: 1766–71. Copyright (1981) American Chemical Society.

complexed to sulphur species. Methyl species of antimony were detected throughout the water column and again formed up to 10 per cent of the total antimony present. In the Baltic methylantimony levels were fairly constant at about 0.03 nmol dm^{-3} to a depth of about 180 m with slight maxima near the surface and the anoxic zones. Antimony does not become taken up by phytoplankton here unlike arsenic and there is little evidence for methylation of antimony in the anoxic zone. Again the monomethyl form was about ten times more dominant than the dimethyl derivatives.

There has been almost no laboratory investigation to test for the feasibility of biological methylation for antimony as is known to exist for arsenic. However, some early work by Challenger and Barnard[5] provides strong circumstantial evidence that methylated forms of antimony may be produced by mould cultures. Phenylstibonic acid (C$_6$H$_5$SbO(OH)$_2$) as sodium salt was added to cultures of *Scopulariopsis brevicaulis* and *Penicillium notatum* on breadcrumbs and incubated in air. Volatile products were aspirated into nitric acid and analysed for antimony by the Marsh test. *P.notatum* gave a positive result. Reduced oxygen content of the atmosphere above the moulds prevented volatilization of any inorganic antimony. Further positive results were obtained with the soluble antimoniate salt, KSbO$_3$. Again a positive result was obtained with *P.notatum* only. Yields were so small, however, that with the analytical methods available at the time (1947), it was not possible to speciate or quantitate the antimony evolved. *S.brevicaulis* was able to produce positive results with dimethylstibonic acid salt ((CH$_3$)$_2$SbOONa) but this could have been due to redistribution of methyl groups only, if indeed the volatile product was a methylantimony derivative.

This latter work remains tantalizing and it is surprising that it has not been repeated with modern analytical methods for detecting and identifying the evolved products. There is evidence that methyl products were produced in this work but only on account of volatility and antimony content. Reduction to stibine itself could account for the observations. This is an area which appears to offer rich dividends for a reinvestigation.

The levels of methylantimony observed in seawater are much lower than the levels of methylarsenics reported in the same medium (Ch. 5). As for arsenic, methylation of antimony might be a postulated defence mechanism for organisms encountering the antimonate ion in phosphate-deficient environments. However, whereas arsenic and phosphorus are present as tetraoxo ions, antimony occurs as the hexahydroxo anion perhaps requiring a different mechanism. Organic compounds of arsenic and antimony are less toxic than the inorganic oxyanions. Methylation and storage in tissues prior to elimination can be postulated as the detoxification process for marine biota. To date only water samples have been surveyed in detail for methylantimony compounds but it seems likely that they are also present in algae, marine invertebrates and fish, based on the arsenic precedent. Investigations of marine, estuary and river sediments for methylation capacity in the case of antimony might also be useful, although negative results have been found for the arsenic case in marine muds. In view of the similar concentrations of arsenic and antimony in aquatic systems ($<$ ng g^{-1} in unpolluted waters) and the similarity of the arsenate and antimonate species, then a broadly similar environmental chemistry might be expected. So far there have been few investigations directed towards biota, with negative results reported.[6] Present evidence suggests that methylation does not occur by algae.

There is some use of antimony drugs as trypanocidal agents, and organo-antimonials have also been used against syphilis. Details are given in Fig. 10.1 but there are unlikely to be any general environmental consequences of such use.

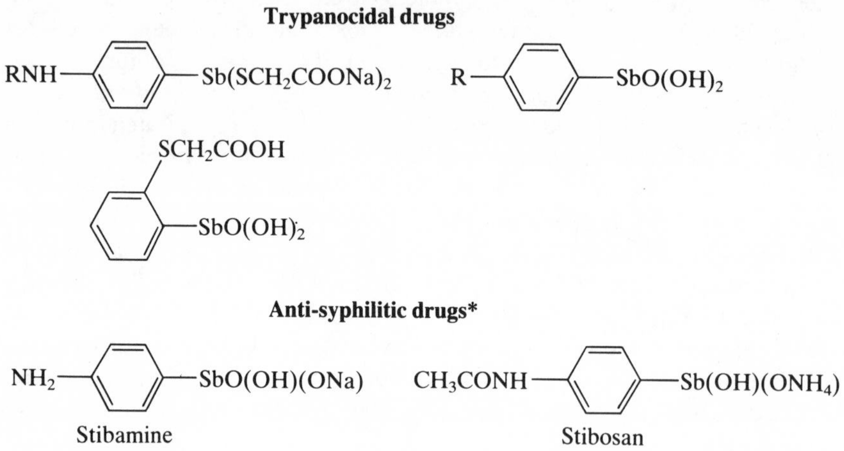

Trypanocidal drugs

RNH—⟨benzene ring⟩—Sb(SCH$_2$COONa)$_2$ R—⟨benzene ring⟩—SbO(OH)$_2$

⟨benzene ring⟩ with SCH$_2$COOH and —SbO(OH)$_2$

Anti-syphilitic drugs*

NH$_2$—⟨benzene ring⟩—SbO(OH)(ONa)   CH$_3$CONH—⟨benzene ring⟩—Sb(OH)(ONH$_4$)

Stibamine   Stibosan

*No longer used

Fig. 10.1   Organoantimony pharmaceuticals

10.2   Germanium

Inorganic germanium has been detected in natural waters by a combination of hydride generation, graphite furnace atomization and AA detection.[7] The germanium is detected as germane (GeH$_4$). In ocean waters levels of 0–10 ng dm^{-3} were found, with similar levels in estuarine and tap waters. More recently organic germanium species have been detected at the parts per trillion (pg g^{-1}) level by a development of this method.[8] The reduced methylgermanium hydrides ($(CH_3)_nGEH_{4-n}$) were cryogenically trapped on 15 per cent OV3 silicone oil on 60–80 mesh Chromosorb W/AWDMCS. The authors reported that passivation of the glass surfaces in the apparatus was necessary for the determination of monomethylgermane (CH$_3$GeH$_3$). This was done by regular cleaning in an acid bath.

It was found that mono- and dimethylgermanium species were vertically

Table 10.2 Germanium species in natural waters*
*(Concentrations in ng(Ge)dm^{-3})

Location	Ge$_{inorg}$†	CH$_3$Ge	(CH$_3$)$_2$Ge	(CH$_3$)$_3$Ge	Ge$_{total}$
Sargasso Sea	0.4–1.8	21.9–24.5	10.0–11.5	0	32.2–37.8
Bering Sea	2.2–12.3	18.2–20.3	8.0	0	28.4–40.6
Rainwater, Florida△	0.5–0.7	0	1.0–5.9	0.2–1.0	1.7–7.3
Peace River, Florida + estuary	0.7–7.1	0–18.6	0–10.0	0	7.1–29.3
Ochlockonee River, Florida + estuary	2.4–5.7	0–19.5	0–12.0	0	5.7–33.9
Tejo River + estuary, Portugal	0.6–7.4	0–19.6	0–6.6	0	7.4–26.8
Various ocean waters	~0–20	~30	~10	0	~50–60

* From Ref. 8. See also Lewis B L, Froelich P N, Andreae M O 1985 *Nature* **313**: 303–305

† Ge(OH)$_4$

△ These values may be artefactual. See Lewis *et al.* above

Data reproduced with permission from Hambrick G A (III) 1984 Anal Chem 56: 421–4. Copyright (1984) American Chemical Society.

homogeneous in concentration in the oceans. Trimethylgermanium was found only in rainwater samples, not in the oceans. Concentration ranges are shown in Table 10.2.

Monomethylgermanium had not been reported in river waters prior to 1984, although it is the main species in seawater where the trimethyl species is absent. It was also shown that diatoms do not take up organogermaniums although they do cycle silica and inorganic germanium. The organogermanium species are very stable and even persulphate oxidation was not successful in converting them to the acid ($Ge(OH)_4$).[9] More recently methylgermanium species have been reported from the Baltic Sea; again no trimethyl species were present.[2] It is believed that the methyl species are present as uncharged hydroxo forms rather than free ions. The similar concentrations of methyl-germaniums no matter the depth, seems to rule out the possibility that they are produced in sediments or anoxic basins in seawater.

To date there is no sound mechanistic explanation for the production of methylgermaniums. They are found in rainfall as well as in surface waters but no model experiments have been reported to simulate their production. There are no reports for reactions of CH_3CoB_{12} with suitable germanium species. CH_3CoB_{12} is a suitable methyl donor as germanium exists in the natural environment only in the tetravalent state requiring carbanion displacement to account for methyl metal production.

Natural leaching and inorganic germanium effluents account for the inorganic germanium content of rivers and estuaries. Inorganic (and some organic) germanium is also deposited from the atmosphere as a result of coal burning and in one marine location amounts to an order of magnitude more than is transported in rivers (Baltic Sea 3.6 v. 45×10^6 g yr^{-1}).[2] Relatively little, though, is known about the environmental chemistry of organogermanium species.

10.3   Thallium

Thallium(I) is isoelectronic with lead(II) ion and has been investigated for methylation reactions. CH_3CoB_{12} will demethylate on reaction with thallium(III) by direct substitution of methyl groups (as carbanions) for ligands.[10,11] The product of the reaction with a model CH_3CoB_{12} complex was the unstable monomethylthallium(III) cation (CH_3Tl^{2+}) and there was no evidence for dimethylthallium(III) formation,[12] although only the latter is stable in water. Trimethylthallium is unstable to both light and water, and hydrolyses to the very stable dimethylthallium cation (($(CH_3)_2Tl^+$) and methane. Monomethyl-thallium compounds (e.g. $CH_3Tl(OAc)_2$) decompose in solution to methanol, methyl acetate and thallium(I) acetate. Dialkylthallium species (R_2Tl^+) are among the most stable and unreactive organometallic species, being inert to water and atmospheric oxygen and it is these species that would be found if any environmental methylation beyond the unstable monomethyl species took place. Thallium(I) is not reported to react with CH_3CoB_{12},[10] and no experiments appear to have been reported describing its possible transformation in oxidizing moulds of the type which transform arsenic(III) to (V) by oxidative addition (Ch. 5).

Incubation experiments have been reported with a thallium(I) compound (T1OAc) under anaerobic conditions and in the dark with anaerobic bacteria from the sediment of a natural lake.[13,14] Between 7 and 21 days later water samples were found to contain dimethylthallium ions $((CH_3)_2Tl^+)$. The method of analysis was complexometric followed by extraction and spectrometric determination at 570 nm and the methylation yield was calculated at 3 per cent. It is possible in this experiment that the methyl source arose from the acetate grouping (see Ch. 2) as no control experiments with other ligands were reported. The mechanism proposed was an oxidative methylation rather than a cobalamin-promoted carbanion attack.

There are no reports to date of the occurrence of methylthallium species in the natural environment, despite the methyl species being less toxic than thallium(I). Other model experiments with thallium to give positive results have not so far been described. Dimethylthallium(III) has been reported to react with CH_3CoB_{12} to remove a methyl group but decomposition to starting material would occur so environmentally this process is tautological.[15]

10.4 Cobalt – methylcobalamin (CH_3CoB_{12}); 5′-deoxyadenosylcobalamin ($AdenCoB_{12}$)

Methyl transfer from CH_3CoB_{12} to metals and some mechanistic routes have been discussed in Chapter 1 and a structure has been presented (Fig. 1.1). The detailed reactions of CH_3CoB_{12} and model complexes for this molecule with various metals have been reviewed in Chapter 8. In that chapter the reactions of CH_3CoB_{12} or model complexes with mercury, tin, lead, platinum, gold, chromium, palladium, thallium, cadmium, copper and iron are discussed. The importance of these reactions in an environmental context is of course the potential of CH_3CoB_{12} in the generation of possible stable and persistent organometallics from inorganic precursors. CH_3CoB_{12} is the only known methyl carbanion-donating species in the natural environment and has attracted much interest on that account. Its significant reactions with important environmental metals are discussed in detail in the appropriate chapters (e.g. Chs 2 and 3).

In this chapter other roles of CH_3CoB_{12} in the environment and biochemistry are discussed, together with important pathways and naturally occurring levels. CH_3CoB_{12} is but one member of the biochemically essential series of B_{12} vitamins – the cobalamins. The other important derivatives are cyanocobalamin, hydroxocobalamin and 5′-deoxyadenosylcobalamin (Fig. 1.1 —CH_3 is replaced by —CN, —OH and the adenosyl (Aden) group). Vitamin B_{12} itself was first isolated as the cyano derivative ($CNCoB_{12}$) and together

with $HOCoB_{12}$ is nowadays considered as the vitamin – absence or insuffi-
ciency of which may manifest itself in human beings in the disease pernicious
anaemia or by other symptoms. In fact vitamin B_{12} was originally entitled the
'anti-pernicious anaemia factor'. Methyl- and 5'-deoxyadenosylcobalamin
are coenzymes of vitamin B_{12}. (Enzymes are protein catalysts for biochemical
reactions and many will only catalyse reactions of their substrates in the
presence of a specific non-protein organic molecule known as a coenzyme.
The whole system is then known as the holoenzyme and the protein part is the
apoenzyme. The coenzyme functions by binding to the apoenzyme and takes
an integral role in the chemical reaction, in that chemical changes in the
coenzyme complement those taking place in the substrate, e.g. in transphos-
phorylations in sugar metabolism, ATP coenzyme is dephosphorylated on a
1:1 molar basis as the sugar molecule is phosphorylated.)[16] Vitamin B_{12} itself
functions through its coenzymes in metabolism.

Hence the methyl and 5'-deoxyadenosyl (Aden) derivatives are naturally
occurring biological molecules of considerable importance and in no sense
should they be considered as environmental pollutants. Cobalamin derivatives
are in fact the most complex molecules synthesized in nature (apart from the
natural high polymers).[17] The formula of CH_3CoB_{12} is $C_{63}H_{91}O_{14}N_{13}PCo$. As
far as is known vitamin B_{12} and analogues are synthesized exclusively by
micro-organisms, unique among the vitamins in this respect. Details of some
organisms capable of vitamin B_{12} synthesis are given in Table 10.3.[18] Plants
contain no vitamin B_{12} and it is not synthesized in the tissues of animals
although it may be produced in animals via intestinal microbial flora.

All the cobalamins are crystalline hygroscopic materials which are soluble,
and heat and air stable. The methyl and 5'-deoxyadenosyl derivatives are
unstable to light in solution and from CH_3CoB_{12} this feature gives rise to
methyl radicals with metal methylation possibilities in the environment. They
are stable in acid and neutral solution but not in strong alkali.

Table 10.3   Some micro-organisms capable of B_{12} synthesis

Organism	Source
Streptomyces aureofaciens	Isolated from soils
Streptomyces griseus	Isolated from soils, river muds
Clostridium species	Isolated from soils
Propionibacterium shermanii	Produces highest yields; rumen organism
Propionibacterium variants	Found in dairy products, e.g. cheese
Butyribacterium rettgeri	Isolated from rat intestine
Pseudomonas denitrificans	Isolated from soil
Various	Isolated from sewage sludge

Data from References

Several important biochemical reactions are mediated by vitamin B_{12} coenzymes and some details are given in Fig. 10.2. It should be pointed out that cobalamin-dependent processes in man require input of vitamin B_{12} in the diet. Vitamin B_{12} derivatives produced in the human colon by bacterial flora are not useful as these are not absorbed. Hence external sources of vitamin B_{12} are required by man. In some animals cobalamin in the diet is supplemented by use of cobalamin produced by micro-organisms, e.g. in the gut or rumen.

A number of assay methods for vitamin B_{12} derivatives from natural products are known. These have included growth rate tests on birds and mammals and micro-organisms but they have now been mainly replaced by isotope dilution methods using a specific vitamin B_{12} binder. Many of the published data for cobalamin content of various matrices do not distinguish between the chemical forms being measured, i.e. they measure total $RCoB_{12}$ (R = CN, OH, CH_3, 5'-deoxyadenosyl and variants). Results for cobalamin content of various locations are presented in Table 10.4. In the context of this work it is levels in the natural environment rather than in organisms that are of most interest.

From the levels of cobalamin derivatives noted in Table 10.1[19-29] it may be deduced that CH_3CoB_{12} is present in environmentally important locations, i.e. sediments and natural waters. In addition, the methylation of mercury in fish becomes explicable in view of the cobalamin content of intestine and liver. What is questionable, however, is the low level at which pollutant metals (e.g. mercury) and cobalamins are present in mutual locations. Both exist at around or less than the part per million ($\mu g\ g^{-1}$) level in relevant locations, although metal levels in polluted zones may be some orders of magnitude greater than this. In the case of mercury, it seems likely that either CH_3CoB_{12} is not the sole source of methyl donated to mercury or that it is acting catalytically in the transfer of readily available single carbon groups to provide the methyl. Methylation in sediments where reactant mobility is low and the reactants are often at the sub-parts per million level suggests a biological rather than a purely chemical system, where the specificity of living, metabolizing cells towards substrates at low concentrations is utilized and locally high concentrations may be achieved. This argues for methylation by CH_3CoB_{12} being an enzymatic rather than an abiotic chemical process but little is known of the mechanisms actually occurring in the unmodified natural environment.

There have been a number of detailed surveys of the chemical and biochemical properties of vitamin B_{12} derivatives. The most recent and detailed is the two volume work edited by Dolphin.[18] The inorganic chemistry of vitamin B_{12} has been described by Pratt.[30] The biochemistry and pathophysiology has been reviewed in a work edited by Babior.[31] The role of vitamin B_{12} in biochemistry is covered at varying lengths in most undergraduate or graduate level texts in biochemistry or physiological chemistry.

1.

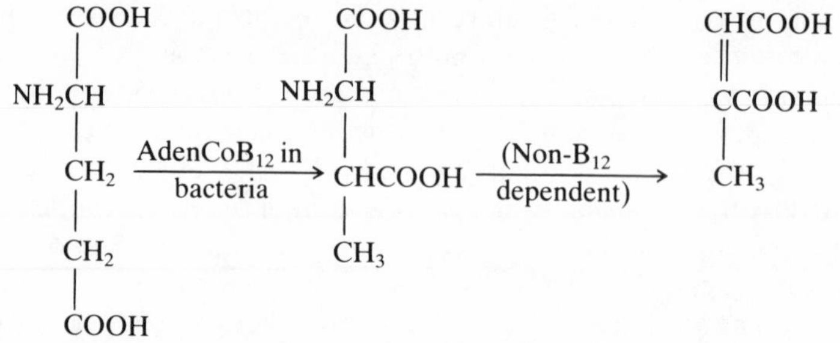

$$\begin{array}{ccc}
\text{COOH} & \text{COOH} & \text{CHCOOH} \\
| & | & \| \\
\text{NH}_2\text{CH} & \text{NH}_2\text{CH} & \text{CCOOH} \\
| & | & | \\
\text{CH}_2 & \text{CHCOOH} & \text{CH}_3 \\
| & | & \\
\text{CH}_2 & \text{CH}_3 & \\
| & & \\
\text{COOH} & &
\end{array}$$

$\text{CH}_2 \xrightarrow[\text{bacteria}]{\text{AdenCoB}_{12}\text{ in}} \text{CHCOOH} \xrightarrow[\text{dependent}]{\text{(Non-B}_{12}\text{)}}$

i.e. glutamate mutase; glutamic acid→β-methylaspartic acid (→mesaconic acid)

2. CH_3CHCOOH $\xrightarrow[\text{(animal tissue)}]{\text{AdenCoB}_{12}}$ CH_2COOH

$$\begin{array}{cc}
| & | \\
\text{C-SCoA} & \text{CH}_2 \\
\| & | \\
\text{O} & \text{C-SCoA} \\
& \| \\
& \text{O}
\end{array}$$

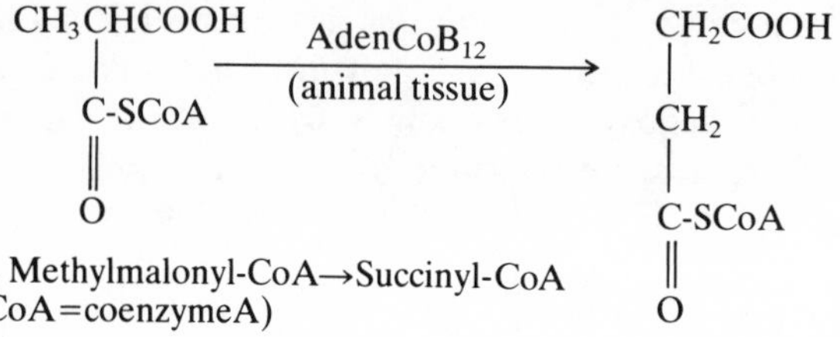

i.e. Methylmalonyl-CoA→Succinyl-CoA
(SCoA=coenzymeA)

3.
$\text{CH}_3 \text{ (from THF)} \xrightarrow{\text{AdenCoB}_{12}} \text{CH}_3\text{CoB}_{12} \xrightarrow[\text{HSCH}_2\text{CH}_2\text{CH(NH}_2)\text{COOH}]{\text{homocysteine,}} \begin{array}{c}\text{CH}_3\text{SCH}_2\text{CH}_2\text{-} \\ \text{(NH}_2)\text{COOH} \\ \text{(methionine)}\end{array}$

i.e. methionine synthetase
Methionine is the main CH_3 donor in biochemistry via *S*-adenosylmethionine. Methyl group source is THF (tetrahydrofolic acid) – demethylated THF is reused in other C_1 processes

4.

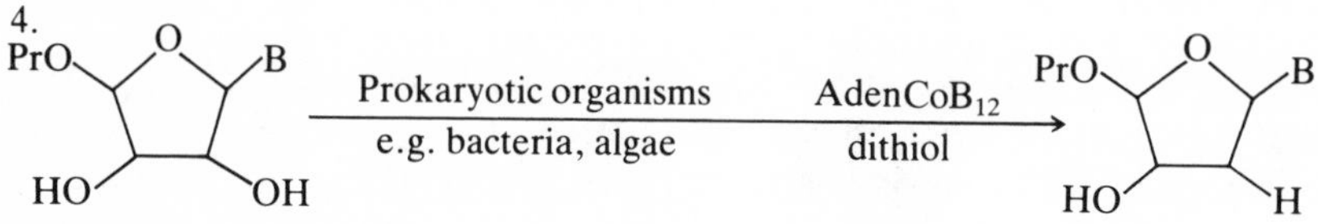

$\xrightarrow[\text{e.g. bacteria, algae}]{\text{Prokaryotic organisms}} \xrightarrow[\text{dithiol}]{\text{AdenCoB}_{12}}$

[PrO=di-, tri- phosphate. B=base. Process occurs during DNA formation]

5. $\begin{array}{c}\text{CO}_2\text{ ,} \\ \text{Glucose}\end{array} \longrightarrow \text{CH}_3\text{COCOOH} \xrightarrow{\text{CH}_3\text{CoB}_{12}} 2\text{CH}_3\text{COOH} + \text{CoB}_{12}^{\cdot}$

i.e. acetate biosynthesis

6. Aminomutases, i.e. $-\text{HCCH}_2- \underset{}{\overset{\text{AdenCoB}_{12}}{\rightleftharpoons}} -\text{H}_2\text{CCH}-$

$$\begin{array}{cc}
| & | \\
\text{NH}_2 & \text{NH}_2
\end{array}$$

7. Diol dehydrase, i.e.
$\text{RCH(OH(CH}_2\text{OH} \longrightarrow \text{RCH}_2\text{CHO} + \text{H}_2\text{O} \quad [\text{R=CH}_3,\text{H},\text{HOCH}_2]$

8. Ethanolamine–ammonia lyase, i.e.
$\text{NH}_2\text{CH(R)CH}_2\text{OH} \xrightarrow{\text{AdenCoB}_{12}} \text{RCH}_2\text{CHO} + \text{NH}_3 [\text{R=CH}_3,\text{H}]$

9. Methane biosynthesis, i.e. $\text{CH}_3\text{CoB}_{12} \longrightarrow \text{CH}_3\text{coenzme} \xrightarrow{\text{reduction}} \text{CH}_4$

10. α-Methyleneglutarate mutase, i.e.

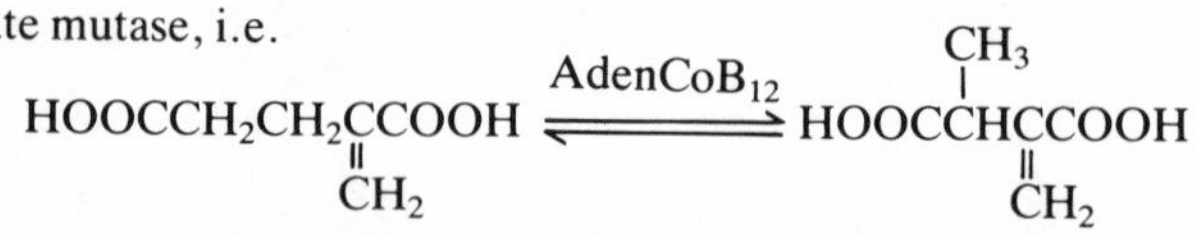

$$\text{HOOCCH}_2\text{CH}_2\text{CCOOH} \overset{\text{AdenCoB}_{12}}{\rightleftharpoons} \text{HOOCCHCCOOH}$$
$$\begin{array}{cc}
\| & \text{CH}_3 \quad \| \\
\text{CH}_2 & \text{CH}_2
\end{array}$$

Fig. 10.2 Biochemical processes catalysed by cobalamin coenzymes

Table 10.4   Some vitamin B_{12} concentrations

Location	Concentration	References, Comments
Pondwater	$0.1-1.0$ ng cm^{-3}	19
Marine fish-heart, liver	$0.26-4.25$ μg g^{-1} wet wt	20
Sewage sludge	$1-7$ μg g^{-1}	18
Legume root nodules	$20-80$ ng g^{-1}	21, 22 As adenosyl CoB$_{12}$
Seaweeds		23
Green alga (*Chlorella vulgaris*)	63 ng g^{-1} dry wt	24
Blue-green alga (*Anaberia cylindrica*)	6.3 ng g^{-1} dry wt	24
Animal serum		
horse	6300 pg cm^{-3}	25 From diet and absorption of B$_{12}$
sheep	5300 pg cm^{-3}	26 produced in gut
rabbit	156 000 pg cm^{-3}	27
Animal foods	$5-10$ pg g^{-1}	
Daily western diet	$5-30$ μg d^{-1}	28 $1-5$ μg of this is absorbed; $2-5$ μg is daily dietary requirement
Human body content	$2-5$ mg	29 Total body content (adult male)
Human body		
liver	1 mg (0.7 μg g^{-1} wet wt)	29
kidney	0.4 μg g^{-1} wet wt	
Fresh foodstuffs	up to 100 μg 100g^{-1}	18 Animal liver is highest. See Ref. 18 for details
Human plasma	$150-450$ pg cm^{-3}	
Other body fluids	$18-12\,960$ pg cm^{-3}	
Human solid tissue	$60-550$ ng g^{-1}	18 Mainly as CH$_3$ and adenosyl derivatives

Data from references

10.5 Phosphorus

The chemistry of compounds with carbon to phosphorus bonds is usually considered to be an area of organic rather than organometallic chemistry. The uses of organophosphorus compounds as plant growth control agents, flame retardents, insecticides, herbicides, nerve gases and antibiotics present specific problems in these areas and there have been numerous reviews of the direct toxicities and applications of these substances. These areas are, in the main, considered to be outside the scope of this work but some general consequences of the behaviour of the organophosphorus compounds in the natural environment will be noted. Aspects of phosphorus chemistry are also covered arising from its Periodic Table group relationship with arsenic and antimony.

Methylphosphorus compounds have been isolated from natural waters,[32,33] but were suggested to have arisen from the decomposition of stored nerve gases leaking to the water course. Hence biological methylation here was not the cause of the origin of the methylphosphorus bonds. Compared to arsenic and antimony, methylation of phosphorus by methyl carbonium ion addition to phosphorus(III) is less likely as the trivalent state for phosphorus is less easily achieved. In the arsenic and antimony biomethylation mechanisms there are successive reductions from the V to the III oxidation states but, although within the reach of biological reducing agents, the process is less favoured than for arsenic or antimony. Thayer[34] has recently compared the reduction potentials of these elements:

$$H_2As(v)O_4^-/HAs(III)O_2 \qquad E^{\ominus} = +0.662 \text{ volts;}$$
$$Sb(v)O_3^-/HSb(III)O_2 \qquad E^{\ominus} = +0.678 \text{ volts;}$$
$$H_2P(v)O_4^-/H_2P(III)O_3^- \qquad E^{\ominus} = -0.260 \text{ volts.}$$

This shows the comparative stability of the V to the III state for phosphorus compared to arsenic or antimony and there is to date no evidence for phosphorus(III) as a biochemical intermediate. The methylphosphorus species found in natural waters[32,33] were identified as dimethylmethylphosphonate by GC after derivatization. The species occurring in the natural environment was unknown but contained a methyl to phosphorus bond and existed at the $0.05-0.2$ μg dm^{-3} level.

A number of organophosphorus compounds have been separated from biota. They are all phosphoric acid or ester derivatives having only one carbon–phosphorus bond, viz. $RP(v)O(OH)_2$. The most prevalent compound is 2-aminoethylphosphonic acid ($NH_2CH_2CH_2PO(OH)_2$) which has been found mainly in marine species.[35,36] Its trivial name is ciliatine. Numerous pathways for the synthesis of this species in organisms have been proposed, the principal step being a conversion of a phosphoenol pyruvate (equation 10.1)

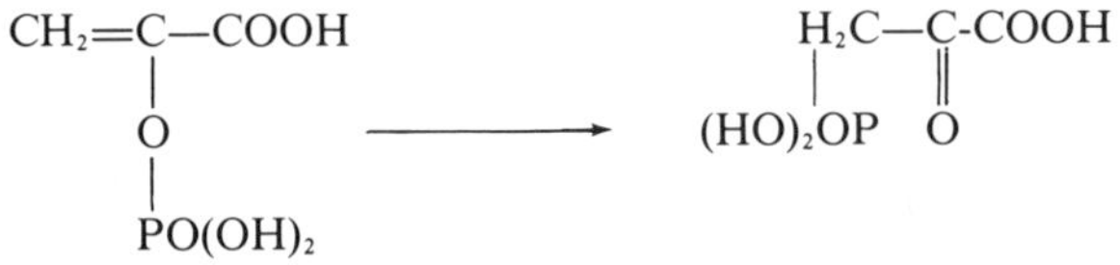

$$CH_2=\overset{\displaystyle |}{\underset{\displaystyle PO(OH)_2}{\underset{\displaystyle |}{\underset{\displaystyle O}{C}}}}—COOH \longrightarrow H_2C—\overset{\displaystyle \overset{\displaystyle O}{||}}{\underset{\displaystyle (HO)_2OP}{C}}-COOH \qquad [10.1]$$

A number of other species analogous to 2-aminoethylphosphonic acid have also been isolated, e.g. $NH_2CH(COOH)CH_2PO(OH)_2$ – phosphonoalanine.

Phosphonomycin is an organic phosphorus antibiotic isolated from the mould *Streptomyces fradiae*. This three-membered ring compound is thought to act by inhibiting bacterial cell wall synthesis. Its structure is given in Fig. 10.3 as are the structures of a number of other currently used organophosphorus drugs. The use of organophosphorus derivatives in medicine is reviewed in the work by Thayer.[34] In this recent work there is a detailed discussion of the environmental generation and impact of organophosphorus compounds and the reader is referred to that work for further details.

The overall use of organophosphorus derivatives are given in Table 10.5. These include organic derivatives as plant growth control agents, flame retardents, nylon stabilizers, herbicides, insecticides and nerve gases. Their use as flame retardents arises from formation of stable nitrogen–carbon bonds after reaction with amino groups on treated cellulose bases. It is fair to say that little is known about the general environmental consequences of disposal of these materials although direct toxicities have been investigated.

Structure	Names	Comments
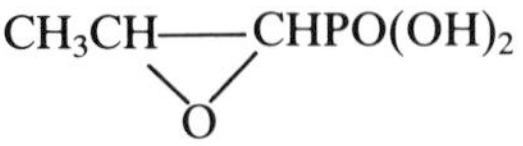 $CH_3CH\overset{\displaystyle \diagdown\;\diagup}{\underset{\displaystyle O}{\;}}CHPO(OH)_2$	Phosphonomycin	Propyl derivative has X2 antibiotic activity
 CH_3〈 〉$CH_2P(C_6H_5)_3Cl$		Trypanocide
$C_6H_5CH_2P(C_6H_5)_3X$		Trypanocide
$(HO)_2POCH_2COOH$	Phosphonoacetic acid	Antiviral
$[O_3PCO]Na_3$	Phosphonoformic acid	Antiviral
$HOOCCH_2CH(COOH)$ $NHCOCH_2PO(OH)_2$	*N*-Phosphonoacetyl-L-aspartic acid	Antitumour drug; undergoing clinical tests
$RSAuP(C_2H_5)_3$	Auranofin, Ridaura	Drugs for rheumatoid arthritis. Triorgano gold complexes

Fig. 10.3 Some organophosphorus drugs

Table 10.5   Some organophosphorus products

Compound	Use, comments
$(HOCH_2)_4P^+X^-$	Fire retardants for cotton; $X = Cl$, HO
$(CH_3O)_2POCH_2CH_2CONHCH_2OH$	Fire retardant for cotton
$R_2POCH{=}CH_2$	Fire retardant for cotton
$(C_2H_5O)_2POCH_2N(CH_2CH_2OH)_2$	Fire retardant for cotton
$C_6H_5PO(OH)H$	Nylon stabilizer
$CH_3PO(F)OR$	Nerve gas; $R = CH(CH_3)C(CH_3)_3$, $CH(CH_3)_2$
$C_6H_5PE(OC_2H_5)R$	Fungicides (e.g. $R = SCH_2C_6H_5$), insecticides (e.g. $R = OC_6H_5NO_2$) $E = O$ or S in $P{=}E$ linkage
$C_2H_5PS(OC_2H_5)R$	Insecticide; $R = $ e.g. SC_6H_5; $OC_6H_2Cl_3$
$Cl_3CCH(OH)PO(OCH_3)_2$	Insecticide; many $RPO(OCH_3)_2$ insecticides exist
$RPO(OH)_2$	Plant growth regulators; decay to release C_2H_4. $R = ClCH_2CH_2$; $CH_3CH_2CH_2$; $HOCOCH_2NHCH_2$; $NH_2CH_2NHCH_2$ for example. Numerous phosphonic acid derivatives are used as herbicides or growth regulators. Decay products are volatile or soluble, leaving low residues on plant tissues.
$[CH_2P(CH_2CH_2CH_2CH_3)_3]Cl$	Phosphon D, plant growth regulator
$(CH_3)_2CH\overset{+}{N}H_3$ $HOOCCH_2NHCH_2PO\bar{O}(OH)$	Roundup, glyphosphate-weed control agent; interrupts amino acid synthesis in plants
$ROCOCH_2NHCH_2PO(OH)_2$	R groups include $SCSR'$, COR', etc.
$n-C_4H_9NHC_6H_{10}PO(On-C_4H_9)_2$	Weed control agent, Buminafos; half-life in soil $= 8{-}11$ days
$C_2H_5OPOOCONH_2^- NH_4^+$	Herbicide, Fosamine NH; half-life in soil $= 10$ days

The use of organophosphorus compounds as nerve gases is outside the scope of this work. Their toxicity arises from inhibition of the enzyme acetylcholinesterase by formation of phosphorus–oxygen bonds (phosphorylation). This is also the cause of their effectiveness as insecticides. There are several works describing the uses of organic phosphorus compounds for the above purposes.[37,38]

A number of organophosphorus compounds are used as herbicides, fungicides or other plant growth regulators (Table 10.5). The properties of these are discussed in detail in the work by Thayer.[34] Numerous other phosphonic acid derivatives have been used as plant growth regulators as also have tetraorganophosphonium salts. Useful information on organophosphorus agrochemical usage is contained in a recent Royal Society of Chemistry (UK) publication.[39] An attractive feature of their use in this area is their rapid degradation to harmless products under conditions of use.

10.6   Manganese − methylcyclopentadienyl-manganesetricarbonyl (MMT)

Manganese is an essential trace element. It acts as a plant growth nutrient, activating enzyme systems which particularly involve carbohydrate and carboxylic acid degradation and nitrogen and phosphorus metabolism. It also helps maintain chloroplast structure in photosynthesis. Excess levels of manganese will damage plants but toxic levels vary widely, e.g. alfalfa 175 μg g^{-1}, carrots 7−10 000 μg g^{-1}. Essential levels for plants range up to 500 μg g^{-1}. Dietary manganese requirements for humans are thought to be in the range 3−7 mg d^{-1}, mainly for skeletal needs. In fact many animals can tolerate 100−1000 s of μg g^{-1} of manganese. So, unlike the probable situation for lead, manganese at proper levels is an necessary element for life.[40]

Poisoning by this element in man is most likely to occur by inhalation of manganese-containing dusts, and toxicity manifests itself by non-specific cerebral nervous system problems varying from apathy and debility to lack of muscular ccntrol. Respiratory conditions may also arise. Less acute poisoning may cause problems with haemoglobin formation.

There is considerable use of manganese as an inorganic material and this has led to TLV limits being established. These range from 0.3 mg m^{-3} to 10.0 mg m^{-3}. American proposals are for environmental standards for manganese in air at up to the 0.006 μg(Mn) m^{-3} value. The American wastewater limit is 1.0 μg g^{-1} and the allowable level in drinking water is 0.05 μg g^{-1}. Despite this, some test animals appear to be able to tolerate manganese levels of the order of 2000 μg g^{-1} for periods of up to six months, without obvious harm. Organic forms of manganese are more toxic than, e.g. manganese(II) chloride or potassium permanganate (which has an LD$_{50}$ to the rat by the oral route at 1090 mg mg^{-1}). $CH_3C_5H_4Mn(CO)_3$ has an LD$_{50}$ to the rat orally of 50 mg mg^{-1}, 23 mg kg^{-1} by injection and 350 mg kg^{-1} by inhalation. $C_5H_5Mn(CO)_3$ has an LD$_{50}$ of 150 mg kg orally to the mouse.[40−45] The chief area of environmental concern for both inorganic and organic forms of manganese is the role of aerosol-size particles in the atmosphere, where they catalyse the oxidation of sulphur dioxide to trioxide. In addition, if $CH_3C_5H_4Mn(CO)_3$ is used as a gasoline additive, there is concern over the damage caused to vehicle exhaust system catalysts (where they are used) by deposition of inorganic manganese after breakdown of the organic manganese additive.

Methylcyclopentadienylmanganesetricarbonyl (MMT, $CH_3C_5H_4(CO)_3$) has never been used as a gasoline additive in Europe and it is illegal in West Germany for both leaded and unleaded fuel. Any new gasoline additive introduced in Europe would require extensive screening before being used commercially and would have to satisfy EEC directives forbidding the introduction of new materials adding to air pollution. For this reason alone it is unlikely that organic manganese gasoline additives will now be introduced into EEC countries. A structure for MMT is given in Fig. 10.4.

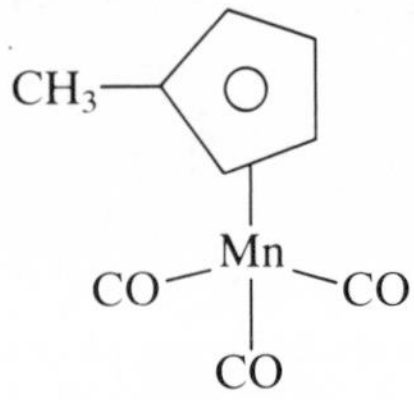

Fig. 10.4 Methylcyclopentadienylmanganesetricarbonyl (MMT)

The main use of MMT is as a boiler or gas turbine fuel additive to reduce deposits, smoke, particulates and sulphur trioxide formation. Concentrations used range from 0.025 g(Mn) gall^{-1} − 0.006 g dm^{-3} − (fuel oil) to 0.31 g(Mn) gall^{-1} − 0.08 g dm^{-3} − for gas turbine fuel. Use in fuel oil for boilers may reduce sulphur trioxide emission by around 50 per cent. At 20−100 μg g^{-1} in gas turbine fuel, smoke, particulates and carbon are reduced by about 50−90 per cent.[46,47]

Use of MMT in gasoline with lead alkyls has a synergistic effect in raising octane numbers more than predictable from either antiknock agent alone. This was achieved probably by manganese increasing the effective catalytic surface area of the lead particles.[46] Production level in the USA in 1974 was about 500 tonnes, but 1984 capacity is believed to be around 5000 tonnes. Present-day usage in Canada is around 1500 tonnes. MMT is used in gasoline at a maximum of 0.13 g(Mn) dm^{-3} in leaded fuel, about one-fiftieth of the lead concentraton. As an additive MMT is very effective, 0.15 g dm^{-3} raising research octane number by 1.5. For such low concentrations this is a very effective performance.

The primary manganese-containing product of the use of MMT is the oxide Mn_3O_4, and this has been claimed as the only non-hydrocarbon product.[48] Akin to its use in fuel oil, MMT is reported to reduce nitrogen oxide emission from gasoline.[49] In limited amounts it may be compatible with lead-sensitive car emission control devices,[49] but environmental concern over increasing manganese levels in urban atmospheres (to the 0.1−10.0 μg m^{-3} range) led to restrictions on its use. Environmental effects from inorganic manganese have been studied in some detail, either as a product of MMT use or from inorganic metal use. MMT is subject to rapid photochemical decomposition and it is unlikely to be persistent as such in the atmosphere, although its volatility makes it transportable.

Direct toxicity effects of MMT have been studied by several groups.[41-45] In rats, severe pulmonary haemorrhagic oedema occurred, particularly to the lungs. There were also pathological changes to lungs, liver and kidney. It was suggested that the toxicity was due to the unmetabolized material. MMT is subject to eventual metabolism by methyl group hydroxylation by cytochrome P450 enzymes but the direct toxic effects noted above occurred before metabolism. Unlike manganese salts, MMT inhibits oxidation of NAD^+-linked substances and associated energy transfer in mitochondria and this effect has been linked for other substances with haemorrhage in lung oedema. Such effects were not due to acute (total) manganese poisoning which occurs at much higher dosage rates. It has been estimated that at dose rates of about $15-150$ mg kg^{-1}, manganese concentrations in treated rats fell to normal level after about 14 days.

Where total manganese is not analysed as such, the normal method of analysis for MMT is gas chromatography coupled to a spectroscopic detection device. A hydrogen atmosphere flame ionization detector has been used in one study and was capable of detecting 6 mg $gall^{-1}$ (about 1.5 mg dm^{-3}) with a detection limit of 1.7×10^{-14} g s^{-1} for manganese.[50] In another study GC was interfaced to an argon plasma emission detector and gave a detection limit of 3 ng for manganese.[51]

It is unlikely that MMT in its organometallic form represents an important environmental hazard owing to its limited use and instability under natural conditions. However, use on a larger scale as a replacement for lead alkyls would then raise in a meaningful way many of the questions still not completely answered for lead. In order to pre-empt such questions it would be useful to determine in more detail for this material properties including its actual atmospheric lifetime, the amount (if any) not consumed in combustion and the normal breakdown pathway in the aqueous and atmospheric environments.

10.7 Cadmium

Health and toxicity problems arising from cadium have received much attention in recent years[52-54] but there has been little discussion of the possibility of organocadmium derivatives being present in the natural environment. This appears to be an unlikely proposition owing to the varied instability to water, air, and light of both mono- and dialkylcadmium species. However, the possibility should be considered as organocadmium derivatives have been stabilized in the laboratory by coordination with various oxo and sulphur ligands, e.g. tetrahydrothiophen. However, apart from these laboratory syntheses which establish the feasibility of organocadmium stabilization, there has been little attempt to investigate possible methylation processes for this element.

In one experiment[55] with CH_3CoB_{12}, a saturated solution of cadmium(II) chloride was reported to produce low concentrations of a volatile cadmium species at pH 9.6 in water at 37 °C. Up to 2×10^{-9} g of this volatile cadmium was collected on a cold trap in a side vessel under argon. The amount was too small for characterization, but at such small yields, even if they proved to be dimethylcadmium, they would have little direct environmental significance. Methylation at more reasonable yields in the presence of natural stabilizing ligands, followed by identification of the organocadmium product as a model for environmental alkylation, has yet to be achieved.

10.8   Acknowledgement

The author gratefully acknowledges the provision of material at an unpublished stage by Dr M O Andreae of Florida State University, USA.

References

1. Andreae M O, Asmode J-F, Foster P, Van't dack L 1981 *Anal Chem* **53**: 1766–71
2. Andreae, M O, Froelich P N Jr 1984 *Tellus* **36b**: 101–17
3. Andreae M O 1983. In *Trace Metals in Seawater* (eds Wong C S, Boule E, Bruland K W, Burton J L, Goldberg E D). Plenum Press, New York and London
4. Andreae M O 1985 (personal communication)
5. Barnard P 1947. PhD Thesis, University of Leeds, UK
6. Kantin R 1983 *Limnol Oceanog* **28**: 165–8
7. Andreae M, Froelich P N Jr 1981 *Anal Chem* **53**: 287–91
8. Hambrick G A III, Froelich P N Jr, Andreae M O, Lewis B L 1984 *Anal Chem* **56**: 421–4
9. Froelich P N Jr, Kaul L W, Byrd J T, Andreae M O, Roe K K 1985 *Estuarine Coastal Mar Sci* (in press)
10. Agnes G, Bendle S, Hill H A O, Williams F R, Williams R J P 1971 *J Chem Soc Chem Commun*: 850–1
11. Agnes G, Hill H A O, Pratt J M, Ridsdale S C, Kennedy F S, Williams R J P 1971 *Biochim Biophys Acta* **252**: 207–11
12. Abley P, Dockal E R, Halpern J 1973 *J Amer Chem Soc* **95**: 3166–70
13. Huber F, Kirchmann H, 1978 *Inorg Chim Acta* **29**: L249–50

14. Huber F, Schmidt U, Kirchmann H, 1978. In *Organometals and Organometalloids* (eds Brinkman F E, Bellama J M). ACS Symp Ser No 82, ACS, Washington DC, USA, pp 65–82
15. Thayer J S 1979 *Inorg Chem* **18**: 1171–2
16. Harper H A, Rodwell V W, Mayes P A 1977. In *Review of Physiological Chemistry*. Lange, Los Altos, CA, USA
17. Friedman H C 1975. In *Cobalamin: Biochemistry and Pathophysiology* (ed Babior B M). Wiley Interscience, New York, p 75
18. Dolphin D (ed) 1982 *B₁₂*, Vols 1 and 2. Wiley Interscience, New York
19. Robbins W J, Hervey A, Stebbins M E 1950 *Science* **112**: 455
20. Braekkan O R 1958 *Nature (London)* **182**: 1386
21. Kliewer M, Evans H J 1962 *Nature (London)* **194**: 108. ibid: **195**: 828
22. Kliewer M, Evans H J 1962 *Arch Biochem Biophys* **97**: 427–9
23. Southcott B A, Tarr H L A 1953 *Fish Res Board Can Progr Repts Pac Coast Sta* **95**: 45
24. Brown F, Cuthbertson W F J, Fogg G E 1956 *Nature (London)* **177**: 188
25. Alexander F, Davies M E 1969 *Brit Vet J* **125**: 169
26. Dawbarn M C, Hine D C, Smith J 1957 *Austr J Exp Biol Med Sci* **35**: 273–6, ibid: 321–6
27. Simnett K I, Spray G H 1961 *Brit J Nutr* **15**: 555
28. Heyssel R M, Bozian R C, Darby W J, Bill M C 1966 *Amer J Clin Nutr* **18**: 176
29. Beck W S 1982 Biological and Medical Aspects of Vitamin B₁₂. In *B₁₂*, Vol 2 (ed Dolphin D). Wiley Interscience, pp 12–15
30. Pratt J M 1972 *Inorganic Chemistry of Vitamin B₁₂*. Academic Press, London
31. Babior B M (ed) 1975 *Cobalamin: Biochemistry and Pathophysiology*. Wiley Interscience, New York
32. Verweij A, Boter J L, Denenhardt C E A M 1979 *Science* **204**: 616
33. Verweij A, Degenhardt C E A M, Boter H L 1979 *Chemosphere* **8**: 115–24
34. Thayer J S 1984 *Organometallic Compounds and Living Organisms*. Academic Press, New York
35. Kittredge J S, Roberts E 1969 *Science* **164**: 37
36. Ohki K, Kasai R, Nozawa Y 1979 *Biochem Biophys Acta* **558**: 273–81
37. Toy A D 1976 *Phosphorus Chemistry in Everyday Living*. Amer Chem Soc, Washington D C
38. Griffith E J, Beeton A, Spencer J M, Mitchell D T (eds) 1973 *Environmental Phosphorus Handbook*. Wiley Interscience
39. *The Agrochemicals Handbook*. 1983 Royal Society of Chemistry
40. *Kirk-Othner Encyclopaedia of Chemical Technology 1980*, 3rd edn, Vol 14. p 884
41. Moore W Jr, Hall L, Crocker W, Adams J, Stara J F 1974 *Environ Res* **8**: 171–7

42. Hysell D K, Moore W Jr, Stara J F, Miller R, Campbell K I 1974 *Environ Res* **7**: 158–68
43. Hakkinen P J, Morse C, Martin F M, Dalbey W E, Haschek W M, Witschi H R 1983 *Toxicol Appl Pharmacol* **67**: 55–69
44. Hanzlik, R P, Stitt R, Traiger G J 1980 *Toxicol Appl Pharmacol* **56**: 353–60
45. Hakkinen P J, Hascik W M 1982 *Toxicol Appl Pharmacol* **65**: 11–22
46. Ref 40, Vol 11, pp 665–6
47. Bailie J D, Zeitz A H Jr 1977. In *Energy Technol Handbook* (ed Considine D M). McGraw-Hill, pp 349–50
48. Otto K, Sulak R J 1978 *Environ Sci Technol* **12**: 181–4
49. Edwards H W, Harrison R M 1979 *Environ Sci Technol* **13**: 673–6
50. DuPuis M D, Hill H H Jr 1979 *Anal Chem* **51**: 292–5
51. Uden P C, Barnes R M, Disanzo F P 1978 *Anal Chem* **50**: 852
52. Nriagu J O 1981 *Cadmium in the Environment*. Wiley
53. Taylor D T 1983 *Ecotoxicol Environ Safety* **7**: 33–42
54. Cole J F, Volpe R 1983 *Ecotoxicol Environ Safety* **7**: 151–9
55. Robinson J W, Kiesel E L 1981 *J Environ Sci Health* **A16**: 341–52

Index

(References to *elements* are normally to their *organometallic* derivatives)